Periodic Table of the Elements with the Gmelin System Numbers

1 H 2																1 H 2	2 He 1
3 Li 20	4 Be 26											5 B 13	6 C 14	7 N 4	8 O 3	9 F 5	10 Ne 1
11 Na 21	12 Mg 27											13 Al 35	14 Si 15	15 P 16	16 S 9	17 Cl 6	18 Ar 1
19 * K 22	20 Ca 28	21 Sc 39	22 Ti 41	23 V 48	24 Cr 52	25 Mn 56	26 Fe 59	27 Co 58	28 Ni 57	29 Cu 60	30 Zn 32	31 Ga 36	32 Ge 45	33 As 17	34 Se 10	35 Br 7	36 Kr 1
37 Rb 24	38 Sr 29	39 Y 39	40 Zr 42	41 Nb 49	42 Mo 53	43 Tc 69	44 Ru 63	45 Rh 64	46 Pd 65	47 Ag 61	48 Cd 33	49 In 37	50 Sn 46	51 Sb 18	52 Te 11	53 I 8	54 Xe 1
55 Cs 25	56 Ba 30	57** La 39	72 Hf 43	73 Ta 50	74 W 54	75 Re 70	76 Os 66	77 Ir 67	78 Pt 68	79 Au 62	80 Hg 34	81 Tl 38	82 Pb 47	83 Bi 19	84 Po 12	85 At 8a	86 Rn 1
87 Fr 25a	88 Ra 31	89*** Ac 40	104 71	105 71													* NH_4 23

**Lanthanides 39	58 Ce	59 Pr	60 Nd	61 Pm	62 Sm	63 Eu	64 Gd	65 Tb	66 Dy	67 Ho	68 Er	69 Tm	70 Yb	71 Lu
***Actinides	90 Th 44	91 Pa 51	92 U 55	93 Np 71	94 Pu 71	95 Am 71	96 Cm 71	97 Bk 71	98 Cf 71	99 Es 71	100 Fm 71	101 Md 71	102 No 71	103 Lr 71

A Key to the Gmelin System is given on the Inside Back Cover

Gmelin Handbook of Inorganic Chemistry

8th Edition

Gmelin Handbook of Inorganic Chemistry

8th Edition

Gmelin Handbuch der Anorganischen Chemie

Achte, völlig neu bearbeitete Auflage

Prepared and issued by

Gmelin-Institut für Anorganische Chemie
der Max-Planck-Gesellschaft
zur Förderung der Wissenschaften

Director: Ekkehard Fluck

Founded by: Leopold Gmelin

8th Edition: 8th Edition begun under the auspices of the Deutsche Chemische Gesellschaft by R. J. Meyer

Continued by: E. H. E. Pietsch and A. Kotowski, and by Margot Becke-Goehring

Springer-Verlag Berlin Heidelberg GmbH 1987

STAFF GMELIN HANDBOOK

Dr. G. Bär, D. Barthel, Dr. N. Baumann, Dr. W. Behrendt, Dr. L. Berg, Dipl.-Chem. E. Best, K. D. Bonn, P. Born-Heck, Dipl.-Ing. V. A. Chavizon, E. Cloos, Dipl.-Phys. G. Czack, I. Deim, Dipl.-Chem. H. Demmer, R. Dowideit, Dipl.-Chem. M. Drößmar, U. Duggen, Dr. P. Eigen, M. Engels, Dr. H.-J. Fachmann, Dr. J. Faust, V. Frick, G. Funk, Dipl.-Ing. N. Gagel, Dr. U. W. Gerwarth, Dipl.-Phys. D. Gras, Dipl.-Bibl. W. Grieser, Dr. R. Haubold, B. Heibel, Dipl.-Min. H. Hein, G. Heinrich-Sterzel, A. Heise-Schuster, H.-P. Hente, H. W. Herold, U. Hettwer, Dr. W. Hoffmann, Dr. W. Huisl, Dr. R. Jotter, Dr. J. von Jouanne, Dr. B. Kalbskopf, Dipl.-Chem. W. Karl, H.-G. Karrenberg, Dipl.-Phys. H. Keller-Rudek, B. Kirchner, A. Klusch, Dipl.-Phys. E. Koch, Dipl.-Chem. H. Köttelwesch, R. Kolb, E. Kranz, Dipl.-Chem. I. Kreuzbichler, Dr. A. Kubny, Dr. M. Kunz, Dr. W. Kurtz, M. Langer, Dr. U. Lanzendörfer, Dr. B. Ledüc, Dr. A. Leonard, Dipl.-Chem. H. List, H. Mathis, E. Meinhard, C. Metz, K. Meyer, Dr. M. Mirbach, Dipl.-Chem. B. Mohsin, Dr. U. Neu-Becker, V. Neumann, K. Nöring, Dipl.-Chem. R. Nohl, Dipl.-Min. U. Nohl, Dr. U. Ohms-Bredemann, Dr. W. Petz, I. Rangnow, Dipl.-Phys. H.-J. Richter-Ditten, Dipl.-Chem. H. Rieger, B. Riegert, E. Rieth, A. Rosenberger, E. Rudolph, G. Rudolph, Dipl.-Chem. S. Ruprecht, Dr. B. Sarbas, V. Schlicht, Dipl.-Chem. D. Schneider, Dipl.-Min. P. Schubert, Dr. K. Schücke, Dr. B. Schwager, A. Schwärzel, Dipl.-Ing. H. M. Somer, Dr. C. Strametz, Dr. U. Taubald, M. Teichmann, Dr. D. Tille, Dipl.-Ing. H. Vanecek, Dipl.-Chem. P. Velić, Dipl.-Ing. U. Vetter, H.-M. Wagner, Dipl.-Phys. J. Wagner, R. Wagner, Dr. E. Warkentin, Dr. C. Weber, Dr. A. Wietelmann, Dr. M. Winter, Dr. B. Wöbke, K. Wolff

STAFF GMELIN ONLINE DATA SYSTEM

Dr. R. Froböse, Dr. P. Kuhn, Dr. G. Olbrich

Organometallic Compounds in the Gmelin Handbook

The following listing indicates in which volumes these compounds are discussed or are referred to:

Ag	Silber B 5 (1975)
Au	Organogold Compounds (1980)
Be	Organoberyllium Compounds 1 (1987) **present volume**
Bi	Bismut-Organische Verbindungen (1977)
Co	Kobalt-Organische Verbindungen 1 (1973), 2 (1973), Kobalt Erg.-Bd. A (1961), B 1 (1963), B 2 (1964)
Cr	Chrom-Organische Verbindungen (1971)
Cu	Organocopper Compounds 1 (1985), 2 (1983), 3 (1986), 4 (1987)
Fe	Eisen-Organische Verbindungen A 1 (1974), A 2 (1977), A 3 (1978), A 4 (1980), A 5 (1981), A 6 (1977), A 7 (1980), A 8 (1985), B 1 (partly in English; 1976), Organoiron Compounds B 2 (1978), Eisen-Organische Verbindungen B 3 (partly in English; 1979), B 4 (1978), B 5 (1978), Organoiron Compounds B 6 (1981), B 7 (1981), B 8 to B 10 (1985), B 11 (1983), B 12 (1984), Eisen-Organische Verbindungen C 1 (1979), C 2 (1979), Organoiron Compounds C 3 (1980), C 4 (1981), C 5 (1981), C 7 (1985), and Eisen B (1929–1932)
Ga	Organogallium Compounds 1 (1986)
Hf	Organohafnium Compounds (1973)
Nb	Niob B 4 (1973)
Ni	Nickel-Organische Verbindungen 1 (1975), 2 (1974), Register (1975), Nickel B 3 (1966) and C 1 (1968), C 2 (1969)
Np, Pu	Transurane C (partly in English; 1972)
Pt	Platin C (1939) and D (1957)
Ru	Ruthenium Erg.-Bd. (1970)
Sb	Organoantimony Compounds 1 (1981), 2 (1981), 3 (1982), 4 (1986)
Sc, Y, La to Lu	D 6 (1983)
Sn	Zinn-Organische Verbindungen 1 (1975), 2 (1975), 3 (1976), 4 (1976), 5 (1978), 6 (1979), Organotin Compounds 7 (1980), 8 (1981), 9 (1982), 10 (1983), 11 (1984), 12 (1985), 13 (1986), 14 (1987)
Ta	Tantal B 2 (1971)
Ti	Titan-Organische Verbindungen 1 (1977), 2 (1980), Organotitanium Compounds 3 (1984), 4 and Register (1984)
U	Uranium Suppl. Vol. E 2 (1980)
V	Vanadium-Organische Verbindungen (1971), Vanadium B (1967)
Zr	Organozirconium Compounds (1973)

Gmelin Handbook of Inorganic Chemistry

8th Edition

Be
Organoberyllium Compounds

Part 1

With 21 illustrations

AUTHOR

Hubert Schmidbaur, Anorganisch-Chemisches Institut der T.U. München

FORMULA INDEX

Driss Benzaid, Edgar Rudolph, Gmelin-Institut, Frankfurt am Main

EDITORS

Alfons Kubny, Marlis Mirbach, Gmelin-Institut, Frankfurt am Main

CHIEF EDITORS

Ulrich Krüerke, Adolf Slawisch, Gmelin-Institut, Frankfurt am Main

System Number 26

Springer-Verlag Berlin Heidelberg GmbH 1987

LITERATURE CLOSING DATE: 1986
IN SOME CASES MORE RECENT DATA HAVE BEEN CONSIDERED

Library of Congress Catalog Card Number: Agr 25-1383

ISBN 978-3-662-06026-1 ISBN 978-3-662-06024-7 (eBook)
DOI 10.1007/978-3-662-06024-7

Originally published by Springer-Verlag Berlin Heidelberg New York in 1987.
Softcover reprint of the hardcover 8th edition 1987

Preface

The present volume describes organoberyllium compounds containing at least one beryllium-carbon bond, except the beryllium carbides and cyanides. It covers the literature completely to the end of 1986 and includes most of the references up to mid-1987.

This Gmelin volume is different from all other volumes of the series on organometallic compounds in that it is dedicated to an area of research which has virtually come to a complete standstill. Organoberyllium chemistry has never been a very popular field, and only few workers have contributed to its slow growth, as is seen by the relatively small number of publications in the field. This very modest development became stagnant in the early 1970's and was followed by a rapid decline. This exceptional fate of a branch of organometallic chemistry is only partly due to the very limited number of potential applications of beryllium and its compounds. The compounds of this element are, in principle, at least as interesting and intriguing to scientists as those of other metals in the Periodic Table. No doubt the main reason for the apparent ban of all experimental organoberyllium chemistry is to be found in the established, and alleged, hazardous properties of beryllium compounds. Although similar hazards have been established for other organometallics where active research is still in process, e.g., mercury and lead, these observations were absolutely lethal for organoberyllium research. As early as 1975, theoretical papers on organoberyllium chemistry outnumbered the experimental reports; and since 1980, practical organoberyllium chemistry is seemingly no longer existent with the exception of a few recent papers.

The material has been arranged in the following way. Compounds with organic ligands bonded to Be by only one C atom (^{1}L ligands) are described in Chapter 1. Within the class of ^{1}L ligands, the compounds containing four and three ^{1}L ligands are treated first (Section 1.1), followed by the compounds with two ^{1}L ligands (Section 1.2) and one ^{1}L ligand (Section 1.3). These compounds represent the majority of the organoberyllium compounds. They belong mainly to the types BeR_2 and Be(R)X (R denotes a ^{1}L ligand and X denotes a group bonded by a non-carbon atom). The following chapters describe the compounds with organic ligands bonded by two or three C atoms (^{2}L and ^{3}L ligands, Chapter 2) and by five C atoms (^{5}L ligands, Chapter 3). Chapter 4 contains the compounds with an unknown structure that cannot be classified by this scheme. The large number of theoretical studies of hypothetical compounds is summarized in Chapter 5 at the end of the volume.

Due to free coordination sites at the Be atom, neutral compounds form many adducts with Lewis bases (symbol D). These adducts are described directly after the parent substance in the text. If the parent substances are arranged in tables, the adducts are found at the end of the table or in a separate table.

Monomeric formula units have been used to prepare the Empirical Formula Index on p. 216, which is followed by a Ligand Formula Index on p. 232.

Much of the data, particularly in tables, is given in abbreviated form without units; for explanations, see p. X. Additional remarks, if necessary, are given in the heading of the tables.

The author and editors dedicate this volume to Professor Geoffrey Coates, the pioneer of organoberyllium chemistry, to whom we owe the short and only period of expansion this field ever enjoyed.

Munich/Garching
August 1987

Hubert Schmidbaur

Explanations, Abbreviations, and Units

Abbreviations are used in the text and in the tables; units are omitted in some tables for the sake of conciseness. This necessitates the following clarification:

Temperatures are given in °C; otherwise, K stands for Kelvin. Abbreviations used with temperatures are m.p. for melting point, b.p. for boiling point, dec. for decomposition, and subl. for sublimation. **Densities** d are given in g/cm^3; d_c and d_m distinguish calculated and measured values, respectively.

NMR represents **nuclear magnetic resonance.** Chemical shifts are given as positive δ values in ppm when appearing low field of the following reference substances: $Si(CH_3)_4$ for ^1H and ^{13}C, $BF_3 \cdot O(C_2H_5)_2$ for ^{11}B, $[Be(H_2O)_4]^{2+}$ for ^9Be, and H_3PO_4 for ^{31}P. Multiplicities of the signals are abbreviated as s, d, t, q (singlet to quartet), quint, sext, sept (quintet to septet), and m (multiplet); terms like dd (double doublet)) and t's (triplets) are also used. Assignments referring to labelled structural formulas are given in the form C-4, H-3,5. Coupling constants J in Hz usually appear in parentheses behind the δ value, along with the multiplicity and the assignment, and refer to the respective nucleus. If a more precise designation is necessary, they are given as $^nJ(C, H)$ or J(1,3) and refer to labelled formulas.

Optical spectra are labelled as IR (infrared), R (Raman), and UV (electronic spectrum including the visible region). IR bands and Raman lines are given in cm^{-1}; the assigned bands are usually labelled with the symbols ν for stretching, δ for deformation, ϱ for rocking, ϰ for wagging, and τ for twisting vibrations. Intensities occur in parentheses either in the common qualitative terms (s, m, w, vs, etc.) or as numerical relative intensities. The UV absorption maxima, λ_{max}, are given in nm followed by the extinction coefficient ε ($L \cdot cm^{-1} \cdot mol^{-1}$) or log ε in parentheses; sh means shoulder.

Solvents or the **physical state** of the sample and the temperature (in °C or K) are given in parentheses immediately after the spectral symbol, e.g., R (solid), ^{13}C NMR (C_6H_6, 50°C), or at the end of the data if spectra for various media are reported. Common solvents are given by their formula ($c\text{-}C_6H_{12}$ = cyclohexane), except THF, which represents tetrahydrofuran.

Fragment ions of **mass spectra** (abbreviated MS) are given in brackets, followed by the relative intensities in parentheses.

References in tables are placed directly after the information in the same column.

Figures of molecular structures give only selected parameters. Barred bond lengths (in Å) or bond angles are mean values for parameters of the same type.

The following abbreviations are used for **theoretical calculation methods** and other expressions used in this context: CNDO = complete neglect of differential overlap; EHMO = extended Hückel molecular orbital; FSGO = floating spherical Gaussian orbital; MNDO = modified neglect of diatomic overlap; PNDO = partial neglect of differential overlap; PRDDO = partial retention of diatomic differential overlap; SINDO = symmetrically orthogonalized intermediate neglect of differential overlap; SCF = self-consistent field; STO = Slater type orbitals; and d-z = Gaussian type function with double-ξ set.

Table of Contents

Organoberyllium Compounds

General References:

The following list provides a survey of the general literature on organoberyllium compounds. Review articles are mentioned only when they include a specific discussion of organoberyllium compounds.

Bell, N. A, Beryllium, in: Wilkinson, G., Stone, F. G. A., Abel, E. W., Comprehensive Organometallic Chemistry, Vol. 1, Pergamon, New York 1982, pp. 121/54.

Armstrong, D. R., Perkins, P. G., Calculations of the Electronic Structures of Organometallic Compounds and Homogeneous Catalytic Processes, Part I, Main Group Organometallic Compounds, Coord. Chem. Rev. **38** [1981] 139/275.

Bertin, F., Thomas, G., Sur la chimie de coordination du béryllium. II. Composés organiques du béryllium et leurs complexes, Bull. Soc. Chim. France **1971** 3951/79.

Coates, G. E., Morgan, G. L., Organoberyllium Compounds, Advan. Organometal. Chem. **9** [1970] 195/257.

Fetter, N. R., Organoberyllium Compounds, Organometal. Chem. Rev. A **3** [1968] 1/37.

Coates, G. E., Some Advances in the Organic Chemistry of Beryllium, Magnesium, and Zinc, Record Chem. Progr. **28** [1967] 3/23.

Coates, G. E., Green, M. L. H., Wade, K., Organometallic Compounds, 3rd Ed., Vol. 1, Methuen, London 1967, pp. 103/21.

Everest, D. E., The Chemistry of Beryllium, Elsevier, Amsterdam 1964, pp. 31/101.

Balueva, G. A., Ioffe, S. T., Organic Compounds of Beryllium, Calcium, Strontium, and Barium, Usp. Khim. **31** [1962] 940/59; Russ. Chem. Rev. **31** [1962] 439/51.

Jones, R. G., Gilman, H., Methods of Preparation of Organometallic Compounds, Chem. Rev. **54** [1954] 835/90.

Also valuable compilations covering the literature from 1971 to 1981 are the annual surveys by Seyferth, D. (J. Organometal. Chem. **41** [1972] 1/7, **62** [1973] 19/24, **75** [1974] 1/4, **98** [1975] 117/23, **143** [1977] 129/39, **180** [1979] 1/9) and Oliver, J. P. (J. Organometal. Chem. **257** [1983] 1/15).

1 Compounds with Ligands Bonded by One Carbon Atom

1.1 Tetra- and Triorganylberyllates

General Remarks. This section summarizes the compounds of the types $M_2[BeR_4]$ and $M[BeR_3]$. Only two compounds of the composition $Li_2[BeR_4]$ have been prepared, the sample with $R=CH_3$ being fully characterized including a structure determination. $M[BeR_3]$ type compounds are not that well characterized, but there is sufficient evidence for their existence. Two additional compounds, formulated as $Li_3[Be(CH_3)_5]$ and $Li[Be_2(CH_3)_5]$ in the literature, are also included at the end of this chapter, although structural details are not available. $Li[Be_2(CH_3)_5]$ could also be a mixture of $Li[Be(CH_3)_3]$ and $Be(CH_3)_2$.

$Li_2[Be(CH_3)_4]$

$Be(CH_3)_2$ is added to a solution of $LiCH_3$ in ether in the mole ratio 1:2. The solution is frozen, and most of the ether is removed at $\sim 10^{-3}$ Torr during thawing. After addition of C_6H_{14} to double the remaining volume, the mixture is again frozen and evaporated while thawing. Colorless microcrystals are obtained which are dried at 25°C/10^{-3} Torr [7].

The solution obtained by mixing $LiCH_3$ with $Be(CH_3)_2$ in ether was investigated by ^1H NMR and IR spectroscopy. The ^1H NMR spectrum (Table 1, p. 5) consists of a single sharp resonance at room temperature which implies either a single compound or a rapid methyl exchange among different species in solution. At −65°C two resonances and at −96°C three resonances appear. From a tentative assignment of the three bands (compare $Li_3[Be(CH_3)_5]$, p. 6), the spectrum at −96°C (**Fig. 2**b, p. 4) can be interpreted in terms of an equilibrium mixture of 1:1, 2:1, and 3:1 complexes of $LiCH_3$ and $Be(CH_3)_2$. It follows from the measured areas that $Li[Be(CH_3)_3]$, $Li_2[Be(CH_3)_4]$, and $Li_3[Be(CH_3)_5]$ exist in a 1:1:1 mole ratio in solution at −96°C. $Li_2[Be(CH_3)_4]$ is also present in 3:1 and 1:1 mixtures of $LiCH_3$ and $Be(CH_3)_2$. The IR spectrum of the 1:2 mixture (**Fig. 1**d) differs only slightly from the IR spectrum of the 1:1 complex (Fig. 1c) [9].

X-ray powder diffraction data indicate a tetragonal body-centered cell with the dimensions a = 5.183(15) and c = 11.800(30) Å; Z = 2 gives $d_c = 0.869$ g/cm³. The space groups $I\bar{4}$-S_4^2 (No. 82) and $I\bar{4}2m$-D_{2d}^{11} (No. 121) were considered. From the calculated and observed intensities, and the resulting R values (0.12 for $I\bar{4}$ and 0.145 for $I\bar{4}2m$), the space group $I\bar{4}$ seems more probable.

The results of the full structure determinations for both space groups are shown in **Fig. 3**, p. 4. The structure resembles that of $Li_2[Zn(CH_3)_4]$. It follows from the listed BeC and LiC distances that the formulation as tetramethylberyllate is justified [7].

The complex is not pyrophoric but reacts with H_2O with ignition [7]. It reacts with $Li[AlH_4]$ in ether to give a precipitate of $Li_2[BeH_4]$ [9].

$Li_2[Be(C{\equiv}CC_6H_5)_4]$

From equimolar quantities of $Be(C{\equiv}CC_6H_5)_2 \cdot 2THF$ and $Li(C{\equiv}CC_6H_5)$ in THF, a yellow solution is obtained which, upon evaporation of the solvent, yields an oil. This product can be dissolved in C_6H_6, but subsequent addition of C_6H_{14} produces two liquid layers. Evaporation of the solvent gives a white solid, which dissolves only partially in $C_6H_5CH_3$. The insoluble part consists of impure $Li_2[Be(C{\equiv}CC_6H_5)_4]$ (by elemental analysis). The same material forms from the reactants in the mole ratio 1:2. The white precipitate is obtained from warm C_6H_6/C_6H_{14} and cannot be redissolved in C_6H_6. It does not melt below 300°C, and the IR spectrum has an absorption for $\nu(C{\equiv}C)$ at 2080 (vw, br) cm^{-1} [8].

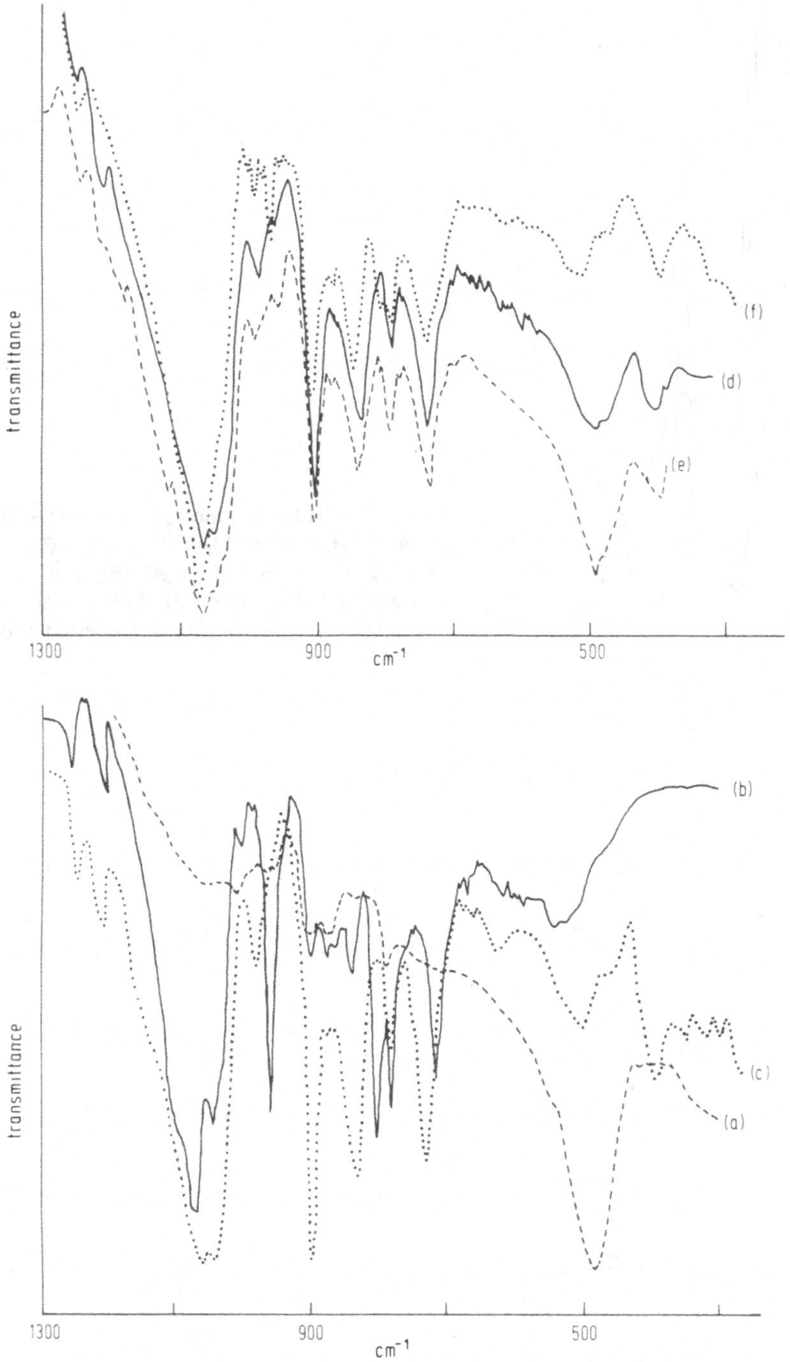

Fig. 1. Infrared spectra of $LiCH_3$, $Be(CH_3)_2$, and mixtures of $LiCH_3$ and $Be(CH_3)_2$ in ether: (a) $LiCH_3$; (b) $Be(CH_3)_2$; (c) 1:1 $LiCH_3 + Be(CH_3)_2$; (d) 2:1 $LiCH_3 + Be(CH_3)_2$; (e) 3:1 $LiCH_3 + Be(CH_3)_2$; (f) 1:2 $LiCH_3 + Be(CH_3)_2$ [9].

References on p. 7

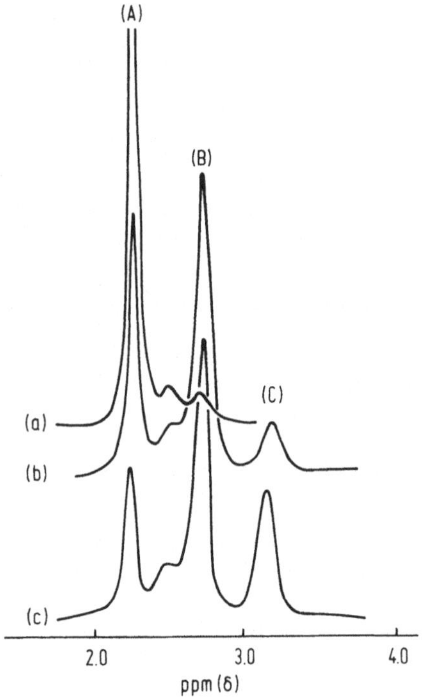

Fig. 2. ^1H NMR spectra in ether at −96°C: (a) $LiCH_3$ + $Be(CH_3)_2$; (b) $2LiCH_3$ + $Be(CH_3)_2$; (c) $3LiCH_3$ + $Be(CH_3)_2$; (A), (B), and (C) as in Formula I. The chemical shifts are shown upfield from the center of the ether triplet [9].

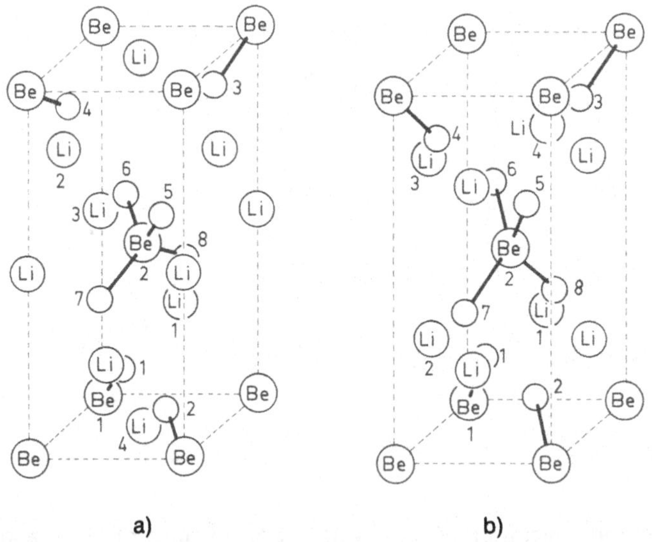

Fig. 3. Unit cell of $Li_2[Be(CH_3)_4]$; a) space group $I\bar{4}m$; b) space group $I\bar{4}2m$ [7].

Bond distances (Å) and bond angles (°):

Li–Li	3.93 ± 0.03	C(1)–C(2)	4.48 ± 0.05
Li(1)–C(1)	2.52 ± 0.06	C(1)–C(7)	4.5(5) ± 0.15
Li(3)–C(1)	2.52 ± 0.06	C(2)–C(7)	4.3(7) ± 0.15
Li(1)–Be(1)	3.93 ± 0.03	C(5)–Be(2)–C(6)	102.(2) ± 3
Li(4)–Be(1)	3.67 ± 0.03	C(5)–Be(2)–C(8)	113.(2) ± 3
Be–Be	5.18 ± 0.03	C(8)–Li(1)–C(1)	120.(5) ± 3
Be–C	1.8(4) ± 0.1	C(8)–Li(1)–C(7)	89.(2) ± 3
C(5)–C(6)	2.87 ± 0.05	C(8)–Li(4)–C(1)	102.(2) ± 3
C(5)–C(8)	3.0(8) ± 0.2	C(2)–Li(4)–C(1)	125.(4) ± 3

$Li[Be(CH_3)_3]$

The reaction of $Be(CH_3)_2$ with $LiCH_3$ in a 1:1 mole ratio in ether yields a clear solution the IR spectrum of which (Fig. 1c, p. 3) indicates the formation of a complex. The ^1H NMR spectrum consists of a single resonance at room temperature (Table 1). At −96°C a multiplet structure occurs (Fig. 2a). The peak at $\delta = 2.25$ ppm is assigned to the CH_3 group of the 1:1 complex and the peak at 2.67 ppm to CH_3 of $Li[Be(CH_3)_4]$. The peak appearing at 2.45 ppm is assigned to an impurity. It appears that at room temperature there is a rapid exchange between 1:1 and 1:2 complexes of $Be(CH_3)_2$ and $LiCH_3$, $Li[Be(CH_3)_3]$ being the predominant species. The 1:1 complex is also present in 2:1 mixtures of $LiCH_3$ and $Be(CH_3)_2$ (Fig. 1d, p. 3). Characteristic bands in the IR spectrum are at 500(ms), 450(w), and 400(ms) cm^{-1}.

Reaction with $Li[AlH_4]$ gives a solid which was shown to be $Li_2[BeH_4]$; whereas with AlH_3, $Li[BeH_3]$ is formed [9].

Table 1
^1H NMR Chemical Shifts of $LiCH_3$, $Be(CH_3)_2$, and Mixtures of $LiCH_3$ and $Be(CH_3)_2$ in Ether (in ppm from the center of the CH_3 triplet of $O(C_2H_5)_2$) [9].

sample	chemical shifts 20°C	−65°C (area)	−96°C (area)
$LiCH_3$	3.08	3.08	
$Be(CH_3)_2$	2.28	2.38	2.550
$LiCH_3 + Be(CH_3)_2$	2.21	2.22	2.25 (128) 2.67 (6)
$2LiCH_3 + Be(CH_3)_2$	2.40	2.27 (46) 3.07 (14)	2.23 (82) 2.69 (134) 3.14 (22)
$3LiCH_3 + Be(CH_3)_2$	2.55	2.30 (26) 3.08 (13)	2.23 (50) 2.70 (150) 3.15 (63)
$LiCH_3 + 2Be(CH_3)_2$			2.34

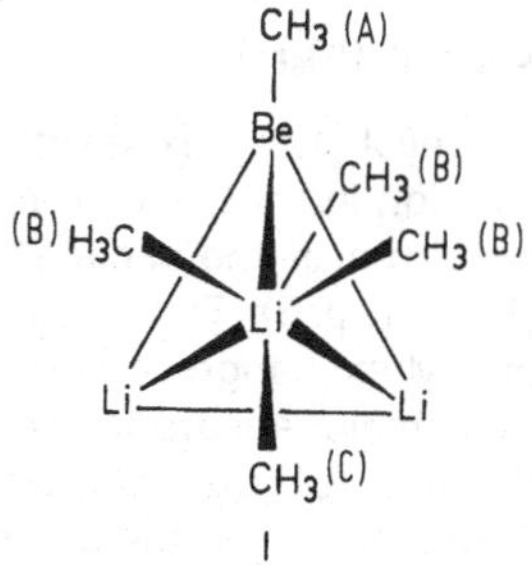

References on p. 7

$Na[Be(C_2H_5)_3]$

$Be(C_2H_5)_2$ is reacted with Na (1:1 mole ratio) at 90°C for 24 h. The black material obtained is treated with ether and filtered. The filtrate, when concentrated in a vacuum, yields a white crystalline material that is washed with C_7H_{16} and analyzes as $Na[Be(C_2H_5)_3]$. No other data are given [6].

$Li[Be(C_6H_5)_3]$

LiC_6H_5 and $Be(C_6H_5)_2$ form a 1:1 addition complex in ether [1 to 3]. After evaporation of the solvent the product can be crystallized from xylene [3] and is dried at 100°C in a vacuum. The rhombic prisms decompose at 198°C to give diphenyl. The complex is moderately soluble in c-C_6H_{12}, but more soluble in C_6H_6 and xylene. It dissolves in ether with warming. It crystallizes from hot dioxane with 4 molecules of the solvent. Ebullioscopic molecular mass determination in ether shows that the complex may be partly associated (found 320, calculated 247) [3]. The compound reacts slowly with $(C_6H_5)_2CO$ to give $(C_6H_5)_3COH$ after hydrolysis. Fluorene and Michler's ketone give no positive color test. Hydrolysis is slower than with $Be(C_6H_5)_2$ [2, 3].

$Li[Be(C_6H_5)_3]\cdot 4O_2C_4H_8$ ($O_2C_4H_8$=1,4-dioxane)

The adduct crystallizes from solutions of $Li[Be(C_6H_5)_3]$ in hot dioxane as colorless needles [3].

$Li[Be(C_6H_5)_2C_{13}H_9]\cdot 4O_2C_4H_8$ ($C_{13}H_9$=fluorenyl, $O_2C_4H_8$=1,4-dioxane)

The complex is formed from $Be(C_6H_5)_2$ and $LiC_{13}H_9$ ($C_{13}H_9$=fluorenyl) in ether. Addition of dioxane gives an orange-yellow powder.

The material decomposes at 160°C. It reacts with solid CO_2 to give $(C_6H_5)_2CHCOOH$ after hydrolysis [3].

$Na[Be(C_6H_5)_2C(C_6H_5)_3]\cdot 2O(C_2H_5)_2$

$Na[C(C_6H_5)_3]$ in ether is added to $Be(C_6H_5)_2$ in ether. Concentration gives red prisms that are dried in a vacuum at 30°C, m.p. 84 to 85°C [3].

Reactions of mixtures of $Be(C_6H_5)_2$ and $Na[C(C_6H_5)_3]$ in ether with butadiene, isoprene [4], and 2,3-dihydrobenzofuran [5] are listed with $Be(C_6H_5)_2$ on p. 71, since a participation by the title compound is not clear.

$Li_3[Be(CH_3)_5]$

$Be(CH_3)_2$ is mixed with $LiCH_3$ in ether in the mole ratio 1:3. ^{1}H NMR and IR spectra of the solution indicate the formation of an adduct of the net composition $Li_3[Be(CH_3)_5]$. The ^{1}H NMR spectrum of the same solution consists of a single sharp resonance at room temperature (Table 1, p. 5). This implies either the presence of a single compound with rapid positional methyl exchange or a rapid methyl exchange between different species in solution. At −96°C three major absorptions are observed (Fig. 2c, p. 4). The assignments of these resonances (A), (B), and (C) are made by comparison with $Be(CH_3)_2$ and $LiCH_3$ assuming a structure as in Formula I. From the areas of the absorptions (Table 1) it is evident that $Li_3[Be(CH_3)_5]$ is the major species in solution with ~10% dissociation to $LiCH_3$ and $Li_2[Be(CH_3)_4]$. When $LiCH_3$ and

$Be(CH_3)_2$ are admixed in a 2:1 ratio, it has been found from the measured areas that $Li[Be(CH_3)_3]$, $Li_2[Be(CH_3)_4]$, and $Li_3[Be(CH_3)_5]$ exist in a 1:1:1 mole ratio in solution at −96°C. The IR spectrum is recorded in Fig. 1e, p. 2. The strong band at 480 cm^{-1} is assigned to ν(LiC). Two medium strong bands at 405 and 335 cm^{-1} are tentatively assigned to ν(BeC). Reaction with $Li[AlH_4]$ produces $Li[BeH_4]$ [9].

$Li[Be_2(CH_3)_5]$

Stirring $LiCH_3$ and $Be(CH_3)_2$ in ether in a 1:2 mole ratio gives a mixture whose IR spectrum (Fig. 1f, p. 2) is identical with that of $Li[Be(CH_3)_3]$ (Fig. 1c, p. 2). The ^1H NMR spectrum consists of a single sharp resonance at room temperature as well as at −96°C indicating a single compound or very rapid exchange of CH_3 groups between $Li[Be(CH_3)_3]$ and $Be(CH_3)_2$ (Table 1, p. 5).

Reaction with $Li[AlH_4]$ or AlH_3 gives a mixture of hydrides. The expected $Li[Be_2H_5]$ was not found [9].

References:

[1] Wittig, G., Keicher, G. (Naturwissenschaften **34** [1947] 216).
[2] Wittig, G. (Angew. Chem. **62** [1950] 231/6).
[3] Wittig, G., Meyer, F., Lange, G. (Liebigs Ann. Chem. **571** [1951] 167/201).
[4] Wittig, G., Wittenberg, D. (Liebigs Ann. Chem. **606** [1957] 1/23).
[5] Wittig, G., Kolb, G. (Chem. Ber. **93** [1960] 1469/76).
[6] Strohmeier, W., Popp, G. (Z. Naturforsch. **23b** [1968] 38/41).
[7] Weiß, E., Wolfrum, R. (J. Organometal. Chem. **12** [1968] 257/62).
[8] Coates, G. E., Francis, B. R. (J. Chem. Soc. A **1971** 160/4).
[9] Ashby, E. C., Prasad, H. S. (Inorg. Chem. **14** [1975] 2869/74).

1.2 Diorganoberyllium Compounds and Their Adducts

1.2.1 Dialkylberyllium Compounds and Their Adducts

1.2.1.1 Dimethylberyllium, $Be(CH_3)_2$

Beryllium dialkyls were the first organoberyllium compounds to be reported in the literature as early as 1860/1861. After initial studies on the ethyl and propyl system [1, 2], $Be(CH_3)_2$ was synthesized by Lavrov in 1884 from beryllium metal and $Hg(CH_3)_2$ [3, 4]. This important reaction was repeated and improved almost 40 years later by Gilman and coworkers [7] and again by Burg and coworkers in 1940 [12]. Krause [5] and Gilman [7] introduced the Grignard reaction into organoberyllium chemistry and thus provided a broader preparative basis for studies in the chemistry of beryllium alkyls. Furthermore, the advent of organolithium reagents has furnished an even more convenient source for methylberyllium compounds [39, 62].

Along with $Be(C_2H_5)_2$ and $Be(C_5H_5)_2$, $Be(CH_3)_2$ is the most thoroughly investigated organoberyllium compound. Over one third of the literature on organoberyllium chemistry is dedicated to the preparation, properties, and reactions of this key compound.

Beryllium dialkyls were found to catalyze the polymerization of olefins by Ziegler et al. [79 to 81].

References on pp. 17/9

Preparation

$Be(CH_3)_2$ is formed in the reaction of Be metal with $Hg(CH_3)_2$ at 130°C in a sealed tube [3, 4]. From observations with the reactions of other mercury alkyls, it was assumed [7,70] that a mercuric chloride (or iodide) impurity is perhaps acting as a catalyst in this reaction. This assumption was confirmed in later experiments [35].

In contrast, however, investigations on a gram scale proved the synthesis to proceed as previously described [3, 4] without the influence of a catalyst at 120 to 125°C [12]. The product sublimes from the reaction mixture into colder parts of the tube and may be purified by a second sublimation in vacuum after opening of the reaction vessel and attachment to a vacuum system [12]. No yields are quoted [12], but the preparation appears to be almost quantitative after 24 h [20]. A reaction time of 90 h at 110°C is recommended as optimal conditions [39] with a crystal of $Be(CH_3)_2$ and a drop of mercury added as catalysts [40]. The reaction also proceeds in the absence of any catalyst if $Hg(CH_3)_2$ is heated under reflux over Be metal at slightly elevated pressure for 36 h [17, 18, 65]. Separation from Hg metal is facilitated if C_6H_6 is added to the reaction mixture, as the mercury then agglomerates [53]. The method was used for the preparation of pure material for gas phase structure determination [50] and spectroscopic investigations [48, 49]. Radioactive $^7Be(CH_3)_2$ [21] and $Be(CD_3)_2$ [35] have been prepared by this method [21].

Anhydrous beryllium chloride or bromide react with alkylmagnesium compounds to form beryllium alkyls [5,70]. For $Be(CH_3)_2$, this reaction is carried out in an inert gas atmosphere in ether with $BeCl_2$ and $Mg(CH_3)I$ [7,15] or $Mg(CH_3)Br$ [30] as the reagents. Distillation yields ethereal solutions of the product. Removal of the solvent by distillation, followed by vacuum treatment at 200°C [7,70] or 100 to 150°C [30] gives a pure crystalline material. Yields vary between 85 and 90% [7]. The compound for X-ray investigation was prepared via this method [16]. A special ether codistillation facilitates product separation particularly if the synthesis is carried out on a large scale (ca. 300 g) [10, 24]. $S(CH_3)_2$ has been used as the solvent for the more precious $Be(CD_3)_2$ because of the difficulty in removing trace amounts of ether [48]. Most reactions of $Be(CH_3)_2$ were carried out with ethereal solutions without the metal alkyl being isolated [28, 31 to 33, 54, 69].

$Be(CH_3)_2$ is also obtained from $BeCl_2$ and $LiCH_3$ (1:2 mole ratio) in ether. LiCl precipitates, and the product can be isolated from filtered solutions by evaporation to dryness and sublimation at 110°C/0.05 Torr [62].

Crystalline samples of $Be(CH_3)_2$ can be obtained from $Al_2(CH_3)_6$ and $(Be(N(CH_3)_2)_2)_2$. The reaction is of no synthetic value [47]. Traces of $Be(CH_3)_2$ appear to be formed from CH_3 radicals and Be metal heated to 150°C, but no sufficient details are available [11].

Be metal does not react with alkyl halides when exposed to ultrasound irradiation [13]. The reaction of various qualities of Be metal was carefully checked, but no positive result was obtained [6].

Association Equilibria

Solid $Be(CH_3)_2$ is a methyl-bridged polymer [14,16]. At 100 to 205°C, the vapor over the solid contains monomeric, dimeric, trimeric, and probably more highly associated species [17]. Under unsaturated conditions, the monomer is the major constituent of the vapor. This follows from electron diffraction studies [50], IR spectra [48], and mass spectroscopy [45, 48, 52].

The equilibrium constants for dimerization, $K_d = p_{dimer}/p^2_{monomer}$, and trimerization, $K_t = p_{trimer}/p^3_{monomer}$, were calculated as a function of total pressure from vapor densities determined from 190 to 200°C, the temperature at which all the solid disappeared. Typical values for 190°C are:

p in Torr	25	31	52	81	101	151
K_d in atm^{-1} ..	4.4	4.6	4.8	5.5	6.2	6.6
K_t in atm^{-2} ..	0.62	0.51	0.28	0.20	0.17	0.11

It follows that the equilibrium primarily involves monomeric and dimeric molecules. The true equilibrium constant of dimerization K_d^* for various temperatures was obtained by extrapolating K_d to $p \rightarrow 0$, assuming that the concentration of the trimer becomes negligible as the pressure approaches zero. A similar evaluation for a true K_t^* could not be performed due to experimental restrictions (too low vapor pressure below 170°C and thermal decomposition above 200°C). The true K_d^* is listed below along with K_t and the fractions f_d and f_t of dimer and trimer, respectively; K_t, f_d, and f_t are calculated for p = 0.1 atm [17]:

t in °C	170	175	180	185	190	195	200
K_d^* in atm^{-1} ..	12.3	9.1	6.7	4.9	3.7	2.8	2.1
K_t in atm^{-2} ..	52	29	21	11	6.1	3.6	2.4
f_d	0.48	0.454	0.408	0.376	0.338	0.298	0.256
f_t	0.16	0.132	0.126	0.090	0.063	0.045	0.036

(The values of f_t at the lower temperatures suggest that a small amount of tetramer is likely to be present.) The temperature dependence of K_d^* and K_t and the thermodynamic parameters are listed below, where mol refers to moles of dimer or moles of trimer, respectively [17]:

$$\log K_d^* = 10.92 - 532(5)/T$$
$$\Delta H_d^\circ = -24.4 \pm 1 \text{ kcal/mol}$$
$$\Delta G_d^\circ = -24400 + 50T \text{ cal}$$
$$\Delta S_d^\circ = -50 \pm 2 \text{ cal} \cdot mol^{-1} \cdot K^{-1}$$

$$\log K_t = 20.3 - 977(0)/T$$
$$\Delta H_t^\circ = -44.7 \pm 5 \text{ kcal/mol}$$
$$\Delta G_t^\circ = -44700 + 93T \text{ cal}$$
$$\Delta S_t^\circ = -93 \pm 10 \text{ cal} \cdot mol^{-1} \cdot K^{-1}$$

A molecular orbital calculation (within the approximation of PRDDO) gave dimerization energies in the range 8.6 to 12.5 kcal/mol from the total energies of $Be(CH_3)_2$ and $(Be(CH_3)_2)_2$ [71]. Another calculated value (within MNDO) of 25.1 kcal/mol is too high, suggesting that MNDO overestimates the Be–Be energy of interaction [66].

The heat of sublimation required to obtain one mol of monomeric vapor, $\Delta H_s = 23.5 \pm 1$ kcal/mol, is independent of the temperature within the narrow range of measurement [17].

Molecular Parameters and Physical Properties

$Be(CH_3)_2$ Monomer. From electron diffraction [50] and IR [48] studies of the unsaturated vapor at ~150°C, structures with a linear C–Be–C axis were derived, with point group symmetries D_{3h} (eclipsed), D_{3d} (staggered) [50], and D'_{3h} (free rotation of CH_3 groups) [48].

The following molecular parameters (D_{3h}) were determined by electron diffraction [50]: d(BeC) = 1.698 ± 0.005 Å, d(CH) = 1.127 ± 0.004 Å, α(BeCH) = 113.9° ± 1.5°, d(Be···H) = 2.386 ± 0.019 Å, d(C···C) = 3.357 ± 0.010 Å, d(C···H) = 3.978 ± 0.027 Å, d(H···H) = 1.785 ± 0.023 Å, d(H···H) = 4.304 ± 0.053 Å, d(H···H) = 4.660 ± 0.041 Å.

Geometrical parameters calculated by ab initio [61], FSGO [67], and MNDO [66, 77] methods are in fair agreement with the experimental values. Varying the angle of rotation in the ab inito calculations shows that essentially free rotation occurs as the rotational force constant is extremely small (1.3×10^{-4} mdyn/Å) [61].

The first ionization potential $E_i = 10.67 \pm 0.07$ eV was obtained from a mass spectroscopic study [45, 52]. E_i values calculated with Koopmans' theorem are (in eV): 10.67 (MNDO) [66],

References on pp. 17/9

10.77 (MNDO) [77], and 9.74 (ab initio) [61]. Values of one-electron energy levels (FSGO) are given in [67].

The following total energy values were calculated (in a.u.): −79.202 (FSGO) [67] and −93.7363 (PRDDO) [71]. The calculated enthalpy of formation is $\Delta H_f^\circ = 47.2$ kcal/mol (MNDO) [66].

Calculations show that the Be–C bond is mainly σ in character since the overlap population for π bonding is low (EHMO [51], ab initio [61]) and that the Be atom has a positive net atomic charge (ab initio [61], MNDO [66]). The calculated BeC bond energy is 440.5 kJ/mol [77].

The IR active fundamental vibrations of $Be(CH_3)_2$ are listed below along with the vibrations of $Be(CD_3)_2$ which were used to confirm the assignments based on D'_{3h} symmetry (wave numbers in cm^{-1}) [48]:

	$Be(CH_3)_2$	$Be(CD_3)_2$	D'_{3h}
$\nu_{as}(CH)$	2944	2193	ν_8
$\nu_s(CH)$	2813	2053	ν_5
$\delta_{as}(CH_3)$	not observed		ν_9
$\delta_s(CH_3)$	1222	994	ν_6
$\nu_{as}(BeC)$	1081	1150	ν_7
$\varrho(CH_3$ out-of-plane)	727	603	ν_{10}

The detection of the δ(CBeC), expected below 400 cm^{-1} (ν_{11}), was precluded by the KBr optics. Absence or weakness of the $\delta_{as}(CH_3)$ has also been reported for $Al_2(CH_3)_6$, and methylgallium compounds. P, Q, and R branches, although not resolved, were clearly observed for the $\delta_s(CH_3)$ band. The unexpected shift of ν_7 to higher frequency upon deuteration has been explained by an interaction with ν_6. Both vibrations are of the same symmetry class (A_{2u}), and the normally expected shifts (to 924 cm^{-1} for ν_6 and to 1050 cm^{-1} for ν_7) would be in violation of the noncrossing rule [48]. The spectrum is compared with that of $Be(CH_3)H$ in [55].

The mean vibrational amplitudes, $u = \langle u^2 \rangle^{1/2}$ in Å, were obtained from the electron diffraction study [50]:

Be–C	0.005 ± 0.010	Be···H	0.123 ± 0.018		0.120
C–H	0.063 ± 0.009	C···C	0.067 ± 0.014	H···H	0.200
		C···H	0.184 ± 0.022		0.200

The stretching force constant k(BeC) = 3.07 mdyn/Å results from ab initio calculations in a comparative study of $LiCH_3$, $Be(CH_3)_2$, and $B(CH_3)_3$ [61].

Calculations for the cation $[Be(CH_3)_2]^+$ by the MNDO method result in an optimized point group D_{3d}, $\Delta H_f^\circ = 795.6$ kJ/mol, $^2A_{2u}$ symmetry for the HOMO, distances of 1.687 (BeC) and 1.115 Å (CH), and a BeCH angle of 105.5° [73].

The BeC bond dissociation energy is calculated as 46 ± 3.2 kcal/mol on the basis of mass spectroscopic results [45, 52].

$(Be(CH_3)_2)_2$ Dimer. The MNDO method predicts, for a dimer of C_{2h} symmetry, terminal BeC bonds to be 1.698 Å and bridging BeC bonds to be 1.878 Å long. The corresponding CH bond lengths are 1.114 to 1.117 Å and 1.125 to 1.130 Å, respectively. The ionization energy $E_i = 10.67$ eV and the enthalpy of formation $\Delta H_f^\circ = -119.5$ kcal/mol have been calculated [66]. The total energy of −187.4589 a.u. was obtained by the PRDDO method using standard bond lengths [71].

The equilibrium $(Be(CH_3)_2)_2 \rightleftharpoons 2Be(CH_3)_2$ in the vapor phase is dealt with on pp. 8/9.

The IR bands appearing at 879, 798, and 585 cm^{-1} in the saturated vapor spectrum, but not present in the unsaturated vapor spectrum, are attributed to the dimer containing both bridging and terminal methyl groups. For $(Be(CD_3)_2)_2$, these IR bands appear at 760, 668, and 462 cm^{-1} [48], also [36].

An extended Hückel calculation results in a stabilization of the dimer relative to the monomer by 4.04 eV [78].

$(Be(CH_3)_2)_n$ Polymer. The density $d=0.88\pm0.1$ g/cm^3 of the solid (fibrous needles) at ambient temperatures was determined by the flotation method [16]. The compound sublimes easily. In vacuum, sublimation proceeds at t=110°C and p=10.05 Torr [62]. The following experimental values of the vapor pressure were determined by [17] and [30] (the values determined by [30] are slightly higher):

t in °C	100.2	115.0	125.4	130.2	135.1	140.6	145.3
p in Torr	0.62	1.72	3.39	4.98	6.81	9.82	13.1
t in °C	151.5	155	160	165	170	175	180
p in Torr	19.8	24.4	33.1	45.1	61.0	82.4	110.0

t in °C	115	120	125	135	140	145	150	155	160	165	170	175
p in Torr	1	3	5	8	11	16	23	27	35	49	64	88
t in °C	180	185	190	192	194	196	198	200	201	202	203	
p in Torr	123	135	151	195	227	273	317	372	380	395	398	

The vapor pressure values can be represented closely by the two linear equations $\log p=12.530-4771/T$ for 100 to 155°C and $\log p=13.292-5100/T$ for 155 to 180°C [17].

The sublimation temperature at 760 Torr, extrapolated from the experimental vapor pressure values, is 220 [30] or 217°C [17]. The virtual enthalpy of sublimation of $\Delta H_s=11.07$ kcal/mol at 493 K and the Trouton's constant of $\Delta H_s/T_s=22.5$ $cal\cdot mol^{-1}\cdot K^{-1}$ were calculated [30].

^1H NMR spectra of $(Be(CH_3)_2)_n$ are not available because of the insolubility of this species in noncoordinating solvents. ^1H NMR and ^9Be NMR spectra have been recorded in donor solvents such as $O(C_2H_5)_2$, $S(CH_3)_2$, $N(CH_3)_3$, and $P(CH_3)_3$, and these spectra are described under the respective $Be(CH_3)_2\cdot D$ adducts; see pp. 20/2 and 25.

If a local symmetry of D_{2h} for the repeating units of the polymeric structure is considered, twelve vibrations of the heavy atom skeleton are expected of which six are IR active and six are Raman active. Five bands are detected both in the IR [30] and Raman [30, 63] spectrum. The frequencies observed in the IR and Raman spectra and their assignments are compiled in Table 2. The IR spectrum at room temperature is compared with that of $(Be(CH_3)H)_2$ in [55].

Table 2
IR and Raman Vibrations (in cm^{-1}) of $(Be(CH_3)_2)_n$.

IR [30] at room temperature	Raman [30] at room temperature	Raman [63] at 20 K	assignment [63]
		2970(m)	} $\nu_{as}(CH_3)$
2912(vw)	2912(s)		} $\nu_{as}(CH_3)$
2885(w)		2900(vs)	$\nu_s(CH_3)$
		1440(br, m)	$\delta_{as}(CH_3)$

References on pp. 17/9

Table 2 (continued)

IR [30] at room temperature	Raman [30] at room temperature	Raman [63] at 20 K	assignment [63]
1255 (vs)	1255 (w)	1250 (m)	$\delta_s(CH_3)$
1243 (vs)			
835 (vs)	923 (w)	918 (m)	$\nu_{as}(BeC)$
		800 (w)	$\varrho(CH_3)$
		728 (w)	
		680 (w)	
567 (m, sh)			$\nu_s(BeC)$
535 (s)	510 (s)	505 (vs)	
	455 (m)	455 (s)	$\nu(BeBe)$
427 (s)	425 (m)		$\delta(CBeC_2Be)$
403 (s)	412 (m)	410 (s)	$\delta(BeC_2Be)$

The extrapolated force constants are k(BeC) = 0.84 mdyn/Å, indicating a low bond order for the Be–C interaction, and k(CH) = 4.55 mdyn/Å [30]. From symmetry arguments and based on the Be···Be distance observed, a significant metal-metal interaction is invoked in a general treatment of the situation in ligand-bridged polynuclear species, but no quantitative data are available [58].

From a single crystal X-ray diffraction study, it follows that solid dimethylberyllium consists of infinite chains of Be atoms with bridging methyl groups; see **Fig. 4.** The parameters are: d(BeC) = 1.93 ± 0.02 Å, d(Be...Be) = 2.09 ± 0.01 Å, α(CBeC) = 114° ± 1°, β(BeCBe) = 66°. The methyl-methyl distances between chains are 4.1 Å, which are normal van der Waals distances.

The polymer forms body-centered orthorhombic crystals with possible space groups Ibam-D_{2h}^{26} (No. 72) and Iba2-C_{2v}^{21} (No. 45): a = 6.13 ± 0.02, b = 11.53 ± 0.02, c = 4.18 ± 0.02 Å; Z = 4; R = 0.13. The hydrogen positions were not actually detected, but introduced at an optimized geometry and with a C–H distance of 0.9 Å. The crystal structure is shown in Fig. 4 [14, 16]; see also [23].

Chemical Reactions

The thermal decomposition of $Be(CH_3)_2$, carried out in a sealed tube or in an isoteniscope, yields 1 mol of CH_4 per mol $Be(CH_3)_2$ between 202 and 212°C. The nonvolatile primary decomposition product has the composition $BeCH_2$. Above 220°C, CH_4 is again evolved with formation of Be_2C (not Be metal [26]) as the final solid product [30]:

$$2\,Be(CH_3)_2 \xrightarrow{212°C} 2\,BeCH_2 + 2\,CH_4$$
$$2\,BeCH_2 \xrightarrow{220\ \text{to}\ 230°C} Be_2C + CH_4$$

Electron impact at (2 to 8) × 10^{-6} Torr produces essentially the parent ion $[BeC_2H_6]^+$ and the two fragment ions $[BeCH_3]^+$ and $[BeCH_2]^+$ as shown by the mass spectra in Table 3 [45, 48, 52]. The table gives the ion abundances of the Be-containing ions under various conditions. At low source temperatures, associated ions have been detected with relatively high abundance, e.g., $[Be_2C_3H_9]^+$. It seems likely that dimeric molecules vaporize from the solid surface. Be_8, Be_7, Be_6, and Be_5 ions, related to each other by successive loss of BeC_2H_6, have been identified. Measurements of the appearance potentials (AP in eV) gave AP = 10.67 ± 0.07 for $[BeC_2H_6]^+$, 12.67 ± 0.02 for $[BeCH_3]^+$, and 11.92 ± 0.05 for $[BeCH_2]^+$ [45, 52]. Calculated appearance potentials by the MNDO method are 9.98 eV for $[Be(CH_3)_2]^+$, 12.04 eV for $[BeCH_3]^+$, and 11.79 eV for $[BeCH_2]^+$. Electronic states for other fragment ions have also been calculated [77].

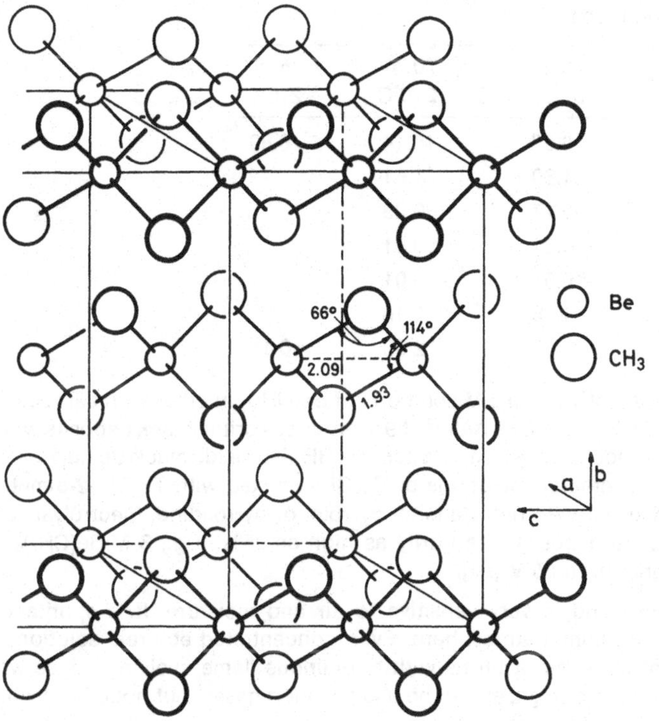

Fig. 4. Crystal structure of $(Be(CH_3)_2)_n$ polymer [14,16].

Table 3
Mass Spectra of $Be(CH_3)_2$. Relative Abundances of Be-Containing Ions at Various Impact Energies and Source Temperatures [45, 48, 52].

ion	70 eV 37 to 50°C	70 eV 210°C	15 eV 210°C
$[Be_8C_{12}H_{33}]^+$	0.48	0.09	2.50
$[Be_7C_{10}H_{27}]^+$	0.67	0.23	1.21
$[Be_6C_8H_{21}]^+$	0.15	0.06	
$[Be_5C_6H_{15}]^+$	0.48	0.03	
$[Be_2(CH_3)_3]^+$	0.82		
$[BeC_2H_6]^+$	22.19	17.91	78.37
$[BeC_2H_5]^+$	0.54	0.40	0.09
$[BeC_2H_4]^+$	3.43	2.65	2.69
$[BeC_2H_3]^+$	1.66	1.45	
$[BeC_2H_2]^+$	0.79	0.72	
$[BeC_2H]^+$	3.59	3.90	
$[BeC_2]^+$	0.18	0.29	
$[BeCH_3]^+$	30.13	31.93	2.52

References on pp. 17/9

Table 3 (continued)

ion	70 eV 37 to 50°C	70 eV 210°C	15 eV 210°C
$[BeCH_2]^+$	16.02	15.14	6.89
$[BeCH]^+$	1.30	1.61	
$[BeC]^+$	0.17	0.29	
$[BeH_3]^+$	<0.01	<0.01	
$[BeH_2]^+$	<0.01	<0.01	
$[BeH]^+$	1.68	2.18	
$[Be]^+$	1.60	2.63	0.11

Electrolysis of 1.2 and 3 M solutions of $Be(CH_3)_2$ in ether with cathode current densities of 0.05 A/dm^2 at 49 V and 0.09 A/dm^2 at 9 V produces brittle black deposits with a Be content of 77 and 63%, respectively. A 2 M solution in THF also yields black Be deposits. Similar results are obtained with ethereal solutions of $Be(CH_3)_2$ mixed with THF, 1,2-dimethoxyethane, acetal, bromoethane, xylene, triethylamine, pyrrole, or pyrrolidine. Electrolysis of $Be(CH_3)_2/BeCl_2$ in ether gives Be deposits of a purity as high as 95%, e.g., 3 M $Be(CH_3)_2$ and 2.3 M $BeCl_2$ at 0.1 A/dm^2 and 18 to 25 V [24].

The compound is very sensitive to air and moisture. It is spontaneously inflammable especially in a humid atmosphere. Even concentrated ethereal solutions will inflame in the presence of moist air. It burns with a luminous flame evolving dense white fumes of BeO. Nitrogen and the rare gases can be used as inert gases, but not CO_2 in which it inflames [7].

The controlled reaction with pure oxygen in ether solution at −78 or 25°C yields $Be(OCH_3)_2$ as a final product. Only small amounts of a peroxide (2.6 to 6.9%) are detected, depending on the reaction conditions. At 25°C, 1.1 mol equivalents of oxygen are absorbed, and 1.3 mol equivalents at −78°C [33]. For microanalysis of $Be(CH_3)_2$, a special apparatus has been designed for combustion in O_2 at 1050°C which gives very satisfactory results [22].

$Be(CH_3)_2$ also reacts vigorously with iodine. The products have not been rigorously identified, but the formation of BeI_2 and $Be(CH_3)I$ is reasonably assumed. The controlled reaction of $Be(CH_3)_2$ in ether with CO_2 yields CH_3COOH after hydrolysis [7, 70].

The reaction with H_2O is violent, CH_4 being evolved, and may inflame spontaneously [7, 70].

The addition of ethereal H_2O_2 to ethereal $Be(CH_3)_2$ produces a gelatinous precipitate of unknown composition. The precipitate gives no H_2O_2 with dilute acid [8]. Equimolar quantities of dry HCl gas can be condensed onto solid $Be(CH_3)_2$ at −183°C without any reaction. It is only above −5°C that CH_4 evolution starts. The reaction to give $Be(CH_3)Cl$ and CH_4 is completed after heating at 120 to 130°C for 1 h [19].

At −80°C, $Be(CH_3)_2$ forms a 1:1 addition compound with NH_3 in ether. No CH_4 is evolved at this temperature. The product precipitates upon addition of C_6H_{14}. Between −70 and 10°C, one mol of CH_4 is formed leaving a product of the composition $Be(CH_3)NH_2$. With excess NH_3, no $Be(NH_2)_2$ formation occurs, and a polymeric condensate $(BeNH)_n$ is generated instead [25]. In the reaction with excess HCN in ether, CH_4 is vigorously evolved, and $Be(CN)_2$ is produced quantitatively [28]. $Be(CN)_2$ is also produced in other inert solvents. From the reaction in ether with an equimolar amount of HCN, some $Be(CH_3)CN$ appears in the filtrate after removing $Be(CN)_2$ [29].

Ethereal solutions of $Be(CH_3)_2$ react with $LiAlH_4$ in the mole ratio 1:4.5 to form a white precipitate of BeH_2. Reaction with $Al(CH_3)_2H$ without a solvent or in i-C_5H_{12} gives $Al(CH_3)_3$ and possibly $Be(CH_3)H$ [15].

$Be(CH_3)_2$ and B_2H_6 react at 95°C in the absence of a solvent to form $Be(CH_3)BH_4$. In addition, $B(CH_3)_3$ [12, 55], $(B(CH_3)_2H)_2$, and/or $(B(CH_3)H_2)_2$ [12] are formed. Excess B_2H_6 converts the primary product into $Be(BH_4)_2$, along with a nonvolatile material of composition $BeBH_5$ [12]. The labelled compounds $Be(CH_3)^{10}BH_4$ and $Be(CH_3)BD_4$ are similarly obtained using $^{10}B_2H_6$ or B_2D_6, respectively [55]. The complex hydridoberyllates $Na_2[BeH(CH_3)_2]_2$ and $Na_2[BeD(CH_3)_2]_2$ are formed by the reaction of NaH and NaD with $Be(CH_3)_2$ in refluxing ether. If NaH, $Be(CH_3)_2$, and $BeCl_2$ are employed in the mole ratio of about 5:4:2, the product is "$Be_3(CH_3)_4H_2$" [34]. A suspension of LiH in ether treated with $BeBr_2$ and $Be(CH_3)_2$ gives an etherate of $Be(CH_3)H$ [72]. LiD instead of LiH gives $Be(CH_3)D$ [34].

$Be(CH_3)_2$ was found to rapidly redistribute with $BeCl_2$ and $BeBr_2$ in ether solution. The equilibrium $Be(CH_3)_2 + BeX_2 \rightleftharpoons 2Be(CH_3)X$ (X = Cl, Br) lies predominantly to the right, as determined by a combination of selective precipitation, molecular association, and low-temperature NMR techniques [38, 40]. The reaction is of synthetic value [57]. However, no reaction occurs when equimolar amounts of $Be(CH_3)_2$ and $BeCl_2$ are heated without solvent at 140°C for 1 h [65].

Redistribution is also observed with beryllium alkoxides, $Be(OR)_2$. Thus, $Be(CH_3)_2$ reacts with $Be(OCH_3)_2$ in boiling ether or $Be(OC_4H_9\text{-t})_2$ in $C_6H_5CH_3$ to produce $Be(CH_3)OCH_3$ and $Be(CH_3)OC_4H_9$-t, respectively [41, 74]. Ether solutions of $Be(CH_3)_2$ and $(Be(N(CH_3)_2)_2)_3$ yield $Be_3(CH_3)_2(N(CH_3)_2)_4$ [57], see p. 145.

$Be(CH_3)_2$ reacts readily with alcohols with evolution of CH_4 and formation of $Be(CH_3)OR$ species. Thus, a violent reaction occurs with CH_3OH even at the melting point of the alcohol (−98°C) [19, 41]. $(Be(CH_3)OCH_3)_2$ is isolated as the sole product if the reactant ratio is exactly 1:1. With excess CH_3OH the reaction proceeds to $Be(OCH_3)_2$ [19]. In ether solution, the reaction with CH_3OH starts at −100°C to give $(Be(CH_3)OCH_3)_4$ [41]. Ethanol, n- and i-C_3H_7OH, t-C_4H_9OH [41], and $C_6H_5CH_2OH$ [72] in ether produce the analogous tetramers; $Be_3(CH_3)_2(OC_4H_9\text{-t})_4$ can be obtained with t-C_4H_9OH in C_6H_6 if the stoichiometry is adjusted accordingly [57]. Excess t-C_4H_9OH in ether leads to $(Be(OC_4H_9\text{-t})_2)_3$ [41]. Treatment with $(C_6H_5)_2CHOH$ [72] and $(C_6H_5)_3COH$ [41] in ether results in formation of the $Be(CH_3)OCH(C_6H_5)_2 \cdot O(C_2H_5)_2$ and $Be(CH_3)OC(C_6H_5)_3 \cdot O(C_2H_5)_2$ adducts. In C_6H_6 the same reactants give the $(Be(CH_3)OCH(C_6H_5)_2)_2$ and $(Be(CH_3)OC(C_6H_5)_3)_2$ dimers [41].

Reactions of $Be(CH_3)_2$ with phenols R'OH in ether above −55°C have been used to prepare the bisphenoxides $Be(OR')_2$ in high yields, e.g., with 2-, 3-, and 4-$CH_3C_6H_4OH$, 2-, 3-, and 4-ClC_6H_4OH, 4-$NO_2C_6H_4OH$, and 2,6-$(NO_2)_2C_6H_3OH$ [31]. The $Be(CH_3)OC_6H_5 \cdot O(C_2H_5)_2$ adduct precipitates from the ether solution [41].

Other reactions of $Be(CH_3)_2$ in ether with organic hydroxy compounds (~1:1 mole ratio) involve $CH_3OCH_2CH_2OH$ to give $(Be(CH_3)OCH_2CH_2OCH_3)_4$; $N(CH_3)_2CH_2CH_2OH$ to give $(Be(CH_3)OCH_2CH_2N(CH_3)_2)_n$ (n~7); 8-hydroxyquinoline to give $(Be(CH_3)OC_9H_6N)_4$; and $(CH_3)_2C{=}NOH$ to give $(Be(CH_3)ON{=}C(CH_3)_2)_4$ or, with excess oxime, $(Be(ON{=}C(CH_3)_2)_2)_n$ [43].

Reactions with various ethers are described in the next section on $Be(CH_3)_2$ adducts on pp. 19/20. $Be(CH_3)OC_3H_7$ is formed in high yield from $Be(CH_3)_2$ and ethylene oxide in ether by an exothermic addition reaction. Dimethylberyllium is converted by t-$C_4H_9OOC_4H_9$-t in $C_6H_5CH_3$ into $Be(CH_3)OC_4H_9$-t and $CH_3OC_4H_9$-t [41].

Carbonyl compounds are added across the Be–C bond, e.g., yielding in ether solution, $Be(CH_3)OC_3H_7$-i with acetaldehyde or $Be(CH_3)OC_4H_9$-t with acetone [41]; $(t\text{-}C_4H_9)_2CO$ gives $Be(CH_3)OCCH_3(C_4H_5\text{-t})_2$ [60]. Benzophenone and benzoyl chloride give, after hydrolysis,

References on pp. 17/9

$CH_3C(C_6H_5)_2OH$. A positive color test is obtained with Michler's ketone [7]. Addition of $Be(CH)_3)_2$ to exo- and endo-2-methyl-7-norbornanone in ether at 25°C, followed by hydrolysis, results in the 7-methylated alcohols [64].

No well-defined products besides CH_4 could be obtained with CH_3SH [19] and i-C_3H_7SH, whereas t-C_4H_9SH afforded $(Be(CH_3)SC_4H_9\text{-t})_4$. $Be(CH_3)SC_6H_5 \cdot O(C_2H_5)_2$ is formed with C_6H_5SH in ether [42], while an excess of thiophenol gives $Be(SC_6H_5)_2$ [31]. $Be(CH_3)_2$ reacts with $N(CH_3)_2CH_2CH_2SH$ with formation of $(Be(CH_3)SCH_2CH_2N(CH_3)_2)_3$ [43]. There was no reaction between $Be(CH_3)_2$ in ether and $(C_6H_5)_2SO_2$ on standing at room temperature for one week [9].

NH_2CH_3 reacts with $Be(CH_3)_2$ already as low as > -183°C. At -90°C a solid appears, probably the 1:1 adduct, which immediately starts to decompose with CH_4 evolution. A solid, insoluble, apparently polymeric product is formed in this particular example. With $NH(CH_3)_2$, a 1:1 adduct is probably formed, which evolves CH_4 only above its melting point of 44°C to form the trimer $(Be(CH_3)N(CH_3)_2)_3$ [19]. The deuterated compounds $(Be(CD_3)N(CH_3)_2)_3$ and $(Be(CH_3)N(CD_3)_2)_3$ are similarly obtained from $Be(CD_3)_2$ and $NH(CH_3)_2$ and from $Be(CH_3)_2$ and $NH(CD_3)_2$, respectively [35].

The reactions of $Be(CH_3)_2$ with $NH(C_2H_5)_2$, $NH(C_3H_7)_2$, or $NH(C_6H_5)_2$ in ether give $(Be(CH_3)N(C_2H_5)_2)_3$ as a trimer, but $(Be(CH_3)N(C_3H_7)_2)_2$ and $(Be(CH_3)N(C_6H_5)_2)_2$ as dimers [37]. Piperidine and morpholine form only 1:1 adducts in ether at -20°C. Above 0°C, CH_4 is evolved and $Be(CH_3)NC_5H_{10}$ and $Be(CH_3)NC_4H_8O$ are generated. With excess amine and at temperatures above 40°C, the beryllium bisamides are formed [32]. For reactions with $N(CH_3)_3$, $N(CD_3)_3$, NC_5H_5, quinuclidine, and N-benzylidene methyl amine, see the adducts of $Be(CH_3)_2$ on pp. 21/5. The silylamines $CH_3HN(CH_3)_2Si\text{–}Si(CH_3)_2NHCH_3$ and $CH_3HN(CH_3)_2Si\text{–}Y\text{–}Si(CH_3)_2NHCH_3$, with Y = O, NCH_3, and CH_2, react with $Be(CH_3)_2$ in ether or ether/ligroin to give, with evolution of two mole equivalents of CH_4, the oligomeric beryllium heterocycles I to IV ($R = CH_3$) [68].

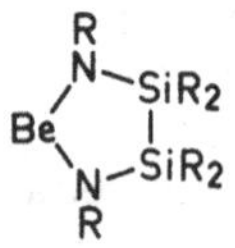
I

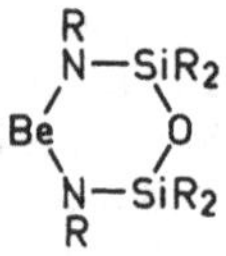
II

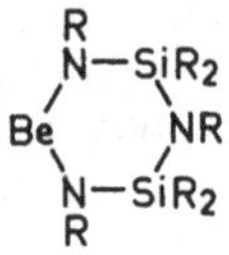
III

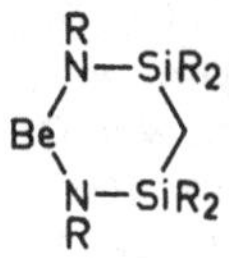
IV

An excess of $NH_2CH_2CH_2NH_2$ in a little ether at room temperature liberates 80% of the CH_3 groups from $Be(CH_3)_2$ to give a white polymeric material. With $NH_2CH_2CH_2N(CH_3)_2$ in excess, an exothermic reaction begins just below room temperature, yielding one equivalent of CH_4 and a dimeric compound of the structure type V. Compound V ($R = CH_3$) is similarly obtained with $NHCH_3CH_2CH_2NHCH_3$ in a little ether. $NH(CH_3)CH_2CH_2N(CH_3)_2$ reacts at 18°C to give the analogous compound of the type V [27]. For reactions with tetramethylethylenediamine, tetramethylorthophenylenediamine, bipyridine, and 2-picoline, see the adducts of $Be(CH_3)_2$ on p. 23/4. The pyridine derivative VI reacts with formation of $Be(CH_3)N(CH_3)CH_2CH_2C_5H_4N$ [69].

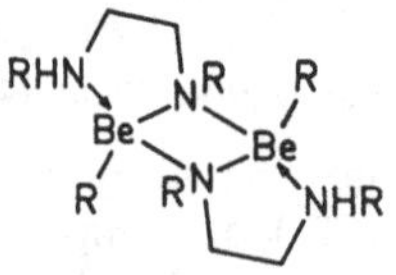
V

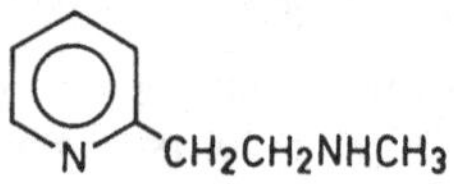
VI

The azomethine double bond, like the carbonyl group, can be added across the Be–C bond. N-benzylideneaniline and N-benzylidene-p-toluidine are rapidly taken up by $Be(CH_3)_2$ to form the amides $Be(CH_3)N(C_6H_5)CH(CH_3)C_6H_5$ and $(Be(CH_3)N(C_6H_4CH_3\text{-}4)CH(CH_3)C_6H_5)_2$, respectively, which yield the corresponding amines on subsequent hydrolysis. N-benzylidene-methylamine forms a 1:1 adduct (see p. 25), and N-benzylidene-t-butylamine does not react with $Be(CH_3)_2$ in C_6H_{14}/ether [60]. Reactions with $C_6H_5N{=}CH_2$ and 1-$C_{10}H_7N{=}CH_2$ in ether, followed by hydrolysis, give the expected amines $C_6H_5NHC_2H_5$ and 1-$C_{10}H_7NHC_2H_5$ [7]. The reaction with $C_6H_5N{=}NC_6H_5$ leads to C_2H_6 and a product of the formula $Be(N(C_6H_5)N(CH_3)C_6H_5)_2$, which was not fully characterized [9].

$P(CH_3)_2H$ reacts with $Be(CH_3)_2$ to give CH_4; other products could not be characterized [19, 69]. $P(CH_3)_3$ forms adducts; see p. 25. However, with $As(CH_3)_3$, no adducts could be identified [27].

An ethereal solution of $Be(CH_3)_2$ does not react with $CH_2{=}C(C_6H_5)_2$ on standing at room temperature for one week [9].

$Be(CH_3)_2$ cleaves the Si–O bond in $(-(CH_3)_2SiO-)_4$ to form $(Be(CH_3)OSi(CH_3)_3)_4$ on prolonged heating in $C_6H_5CH_3$ [69]. This compound is also generated from $Be(CH_3)_2$ and $(CH_3)_3SiOH$ [46, 69]. Reaction with $Sn(C_2H_5)_3H$ affords $Be(CH_3)H$ [59]. Reaction with closo-1,2-$C_2B_9H_{13}$ in C_6H_6/ether gives 3,1,2-$BeC_2B_9H_{11}\cdot O(C_2H_5)_2$ (see Chapter 2, pp. 152/3) [75, 76].

Ligand exchange occurs between $Be(CH_3)_2$ and other BeR_2 compounds: with $Be(C_4H_9\text{-}t)_2$ in C_6H_6 to give $(Be(CH_3)C_4H_9\text{-}t)_3$ [59], with $Be(CH_2C_4H_9\text{-}t)_2\cdot O(C_2H_5)_2$ in C_6H_6 to give $Be(CH_3)CH_2C_4H_9$-t (isolated as $N(CH_3)_2CH_2CH_2N(CH_3)_2$ adduct) [56], and with $Be(C_5H_5)_2$ in C_6H_6 [65] or without solvent [65, 82] at 50°C [65] to give $Be(CH_3)C_5H_5$ [65, 82].

$LiCH_3$ adds to $Be(CH_3)_2$ in ether to give solutions of $Li[Be(CH_3)_3]$, $Li_2[Be(CH_3)_4]$, $Li_3[Be(CH_3)_5]$, and $Li[Be(CH_3)_3]\cdot Be(CH_3)_2$, depending on the mole ratios (2:1 to 2:6) [62]. $Li_2[Be(CH_3)_4]$ could be isolated and its structure elucidated [44], see pp. 2/7.

References:

[1] Cahours, A. (Ann. Chim. [Paris] [3] **58** [1860] 5/82, 22).
[2] Cahours, A. (Compt. Rend. **76** [1873] 1383/7).
[3] Lavroff, D. (J. Russ. Phys. Chem. Soc. **16** [1884] 93/4).
[4] Lavroff, D. (Bull. Soc. Chim. France [2] **41** [1884] 548/58).
[5] Krause, E., Wendt, B. (Ber. Deut. Chem. Ges. **56** [1923] 466/72, 467).
[6] Gilman, H. (J. Am. Chem. Soc. **45** [1923] 2693/5).
[7] Gilman, H., Schulze, F. (J. Chem. Soc. **1927** 2663/9).
[8] Perkins, T. R. (J. Chem. Soc. **1929** 1687/91).
[9] Gilman, H., Schulze, F. (Rec. Trav. Chim. **48** [1929] 1129/32).
[10] Gilman, H., Brown, R. E. (J. Am. Chem. Soc. **52** [1930] 4480/3).

[11] Paneth, F., Loleit, H. (J. Chem. Soc. **1935** 366/71).
[12] Burg, A. B., Schlesinger, H. I. (J. Am. Chem. Soc. **62** [1940] 3425/9).
[13] Renaud, P. (Bull. Soc. Chim. France **1950** 1044/5).
[14] Rundle, R. E., Snow, A. I. (J. Chem. Phys. **18** [1950] 1125).
[15] Barbaras, G. D., Dillard, C., Finholt, A. E., Wartik, T., Wilzbach, K. E., Schlesinger, H. I. (J. Am. Chem. Soc. **73** [1951] 4585/90).
[16] Snow, A. I., Rundle, R. E. (Acta Cryst. **4** [1951] 348/52).
[17] Coates, G. E., Glockling, F., Huck, N. D. (J. Chem. Soc. **1952** 4496/501).
[18] Coates, G. E., Huck, N. D. (J. Chem. Soc. **1952** 4501/12).
[19] Coates, G. E., Glockling, F., Huck, N. D. (J. Chem. Soc. **1952** 4512/5).
[20] Rabideau, S. W., Alei Jr., M., Holley Jr., C. E. (LA-1687 [1954]; C.A. **1956** 3992).

[21] Muxart, R., Mellet, R., Jaworsky, R. (Bull. Soc. Chim. France **1956** 445/8).
[22] Head, E. L., Holley Jr., C. E. (Anal. Chem. **28** [1956] 1172/4).
[23] Rundle, R. E. (J. Phys. Chem. **61** [1957] 45/50).
[24] Wood, G. B., Brenner, A. (J. Electrochem. Soc. **104** [1957] 29/37).
[25] Masthoff, R., Vieroth, C. (Z. Chem. [Leipzig] **5** [1965] 142).
[26] Sanderson, R. T. (Chemical Periodicity, Reinhold, New York 1960, p. 292).
[27] Coates, G. E., Green, S. I. E. (J. Chem. Soc. **1962** 3340/8).
[28] Masthoff, R. (Z. Chem. [Leipzig] **3** [1963] 269/70).
[29] Coates, G. E., Mukherjee, R. N. (J. Chem. Soc. **1963** 229/33).
[30] Goubeau, J., Walter, K. (Z. Anorg. Allgem. Chem. **322** [1963] 58/70).

[31] Funk, H., Masthoff, R. (J. Prakt. Chem. [4] **22** [1963] 250/4).
[32] Funk, H., Masthoff, R. (J. Prakt. Chem. [4] **22** [1963] 255/8).
[33] Masthoff, R. (Z. Anorg. Allgem. Chem. **336** [1965] 252/8).
[34] Bell, N. A., Coates, G. E. (J. Chem. Soc. **1965** 692/9).
[35] Bell, N. A., Coates, G. E., Emsley, J. W. (J. Chem. Soc. A **1966** 49/52).
[36] Coates, G. E., Green, M. L. H., Wade, K. (Organometallic Compounds, Vol. I, Methuen, London 1967, p. 106).
[37] Coates, G. E., Fishwick, A. H. (J. Chem. Soc. A **1967** 1199/204).
[38] Ashby, E. C., Sanders, R., Carter, J. (Chem. Commun. **1967** 997/8).
[39] Ashby, E. C., Arnott, R. C. (J. Organometal. Chem. **14** [1968] 1/11).
[40] Sanders Jr., J. R., Ashby, E. C., Carter II, J. H. (J. Am. Chem. Soc. **90** [1968] 6385/90).

[41] Coates, G. E., Fishwick, A. H. (J. Chem. Soc. A **1968** 477/83).
[42] Coates, G. E., Fishwick, A. H. (J. Chem. Soc. A **1968** 635/40).
[43] Coates, G. E., Fishwick, A. H. (J. Chem. Soc. A **1968** 640/2).
[44] Weiß, E., Wolfrum, R. (J. Organometal. Chem. **12** [1968] 257/62).
[45] Chambers, D. B., Coates, G. E., Glockling, F. (Discussions Faraday Soc. No. 47 [1969] 157/64).
[46] Mootz, D., Zinnius, A., Böttcher, B. (Angew. Chem. **81** [1969] 398/9; Angew. Chem. Intern Ed. Engl. **8** [1969] 378/9).
[47] Atwood, J. L., Stucky, G. D. (J. Am. Chem. Soc. **91** [1969] 4426/30).
[48] Kovar, R. A., Morgan, G. L. (Inorg. Chem. **8** [1969] 1099/103).
[49] Kovar, R. A., Morgan, G. L. (J. Am. Chem. Soc. **91** [1969] 7269/74).
[50] Almenningen, A., Haaland, A., Morgan, G. L. (Acta Chem. Scand. **23** [1969] 2921/2).

[51] Cowley, A. H., White, W. D. (J. Am. Chem. Soc. **91** [1969] 34/8).
[52] Chambers, D. B., Coates, G. E., Glockling, F. (J. Chem. Soc. A **1970** 741/8).
[53] Coates, G. E., Francis, B. R., Murrell, L. L. (unpublished observations, from Coates, G. E., Morgan, G. L., Advan. Organometal. Chem. **9** [1970] 195/257).
[54] Kovar, R. A., Morgan, G. L. (J. Am. Chem. Soc. **92** [1970] 5067/72).
[55] Cook, T. H., Morgan, G. L. (J. Am. Chem. Soc. **92** [1970] 6487/92).
[56] Coates, G. E., Francis, B. R. (J. Chem. Soc. A **1971** 1305/8).
[57] Andersen, R. A., Bell, N. A., Coates, G. E. (J. Chem. Soc. Dalton Trans. **1972** 577/82).
[58] Mason, R., Mingos, D. M. P. (J. Organometal. Chem. **50** [1973] 53/61).
[59] Coates, G. E., Smith, D. L., Srivastava, R. C. (J. Chem. Soc. Dalton Trans. **1973** 618/22).
[60] Andersen, R. A., Coates, G. E. (J. Chem. Soc. Dalton Trans. **1974** 1171/80).

[61] Fitzpatrick, N. J. (Inorg. Nucl. Chem. Letters **10** [1974] 263/6).
[62] Ashby, E. C., Prasad, H. S. (Inorg. Chem. **14** [1975] 2869/74).
[63] Allamandola, L. J., Nibler, J. W. (J. Am. Chem. Soc. **98** [1976] 2096/100).

[64] Ashby, E. C., Noding, S. A. (J. Org. Chem. **42** [1977] 264/70).
[65] Drew, D. A., Morgan, G. L. (Inorg. Chem. **16** [1977] 1704/8).
[66] Dewar, M. J. S., Rzepa, H. S. (J. Am. Chem. Soc. **100** [1978] 777/84).
[67] Ray, N. K., Mehandru, S. P., Bhargava, S. (Intern. J. Quantum. Chem. **13** [1978] 529/36).
[68] Brauer, D. J., Bürger, H., Moretto, H. H., Wannagat, U., Wiegel, K. (J. Organometal. Chem. **170** [1979] 161/74).
[69] Bell, N. A., Coates, G. E., Fishwick, A. H. (J. Organometal. Chem. **198** [1980] 113/20).
[70] Schulze, F. (Iowa State Coll. J. Sci. **8** [1933] 225/8; C.A. **1934** 2325).

[71] Marynick, D. S. (J. Am. Chem. Soc. **103** [1981] 1328/33).
[72] Bell, N. A., Coates, G. E. (J. Chem. Soc. A **1966** 1069/73).
[73] Glidewell, C. (J. Organometal. Chem. **217** [1981] 273/80).
[74] Andersen, R. A., Coates, G. E. (J. Chem. Soc. Dalton Trans. **1974** 1729/36).
[75] Popp, G., Hawthorne, M. F. (J. Am. Chem. Soc. **90** [1968] 6553/4).
[76] Popp, G., Hawthorne, M. F. (Inorg. Chem. **10** [1971] 391/3).
[77] Bews, J. R., Glidewell, C. (J. Mol. Struct. **90** [1982] 151/63).
[78] Ohkubo, K., Shimada, H., Okada, M. (Bull. Chem. Soc. Japan **44** [1971] 2025/30).
[79] Ziegler, K. (Angew. Chem. **68** [1956] 721/60, 722).
[80] Ziegler, K. (Brennstoff-Chem. **33** [1952] 193/200).

[81] Ziegler, K., Gellert, H.-G. (Ger. 878560 [1950/53]; C. **1954** 662).
[82] Bartke, T. C. (Diss. Univ. Wyoming 1975; Diss. Abstr. Intern. B **36** [1976] 6141).

1.2.1.2 Adducts of $Be(CH_3)_2$

General Remarks. Adducts of $Be(CH_3)_2$ with O-, S-, N-, and P-donors were characterized and are described below. With t-C_4H_9CN and C_6H_5CN, adduct formation is assumed in toluene solutions, but no products could be isolated [17]. $Be(CH_3)_2$ complexes with amines containing reactive H are unstable, and further reactions with Be–C bond breaking occur [1, 2]. Only those adducts are listed here, which have been somehow characterized, but 1:1 adducts may also be intermediates in the reactions of $NHCH_3CH_2CH_2NHCH_3$ and $NHCH_3CH_2CH_2N(CH_3)_2$ with $Be(CH_3)_2$ [5]. No adduct formation could be observed with $As(CH_3)_3$ [2].

A systematic CNDO/2 study has been carried out on 1:1 adducts of $Be(CH_3)_2$ with amines. The adducts with NH_3, NH_2CH_3, $NH_2C_2H_5$, $NH(CH_3)_2$, and $N(CH_3)_3$ served as model compounds. It is shown that the calculated molecular properties of the adducts, e.g., the energy of formation, the Be–N distance, the amount of charge transferred, and the enhancement of the dipole moment are related to the ionization potential of the amine [18].

$Be(CH_3)_2 \cdot nO(CH_3)_2$ (n = 0.33, 0.5, 1, 1.5)

The reaction of $Be(CH_3)_2$ with $O(CH_3)_2$ was studied tensimetrically in a high vacuum apparatus, and different phases have been identified. $Be(CH_3)_2$ dissolves in an excess of $O(CH_3)_2$ at 5°C. On cooling, transparent crystals form which melt between 5 and 10°C/170 to 230 Torr. The composition of the solid varies in the experiments from 1:1.68 to 1:1.8. The composition of the liquids obtained on melting these solids was close to 1:1.5. On further heating the liquid dissociates until its composition is close to 1:1 at ~50°C/~275 Torr, and then decomposes into $Be(CH_3)_2$ and $O(CH_3)_2$. When the mole ratio of reactants is 1:2, a solid is formed, which is stable from −30 to −3°C/102 Torr. The solid is, within experimental error, a 1:1 adduct. It decomposes above −5°C and melts between 10 and 15°C with evolution of

References on pp. 25/6

$O(CH_3)_2$. Decomposition continues until the composition of the liquid is close to 2:1. At 43°C/164 Torr the liquid is suddenly transformed into solid $Be(CH_3)_2$ with evolution of all $O(CH_3)_2$. If the reactants are present in a 1:1 ratio, the adduct observed between −5 and 10°C has the composition 1:0.33. The adduct begins to decompose abruptly at 17°C/65 Torr [2].

$Be(CH_3)_2 \cdot nO(C_2H_5)_2$ (n = 1,2)

It is assumed that solutions of $Be(CH_3)_2$ in ether contain 1:1 and 1:2 adducts, if a large excess of ether is present. The ^1H NMR spectrum of such solutions shows a CH_3Be resonance at $\delta = -1.16$ ppm (J(C,H) = 107 Hz [13]) at 25°C [8, 13] or at $\delta = -1.17$ ppm at 35°C [10]. The signal shifts to higher field as the temperature is lowered: $\delta = -1.30$ at −75°C, −1.43 at −85°C [10], and −1.32 ppm at −87°C [13]. A broad shoulder appears at the high-field side below −54°C ($\delta = -1.43$ ppm at −87°C). The shoulder moves upfield with decreasing temperature accompanied by an increase in relative intensity [13]. The shoulder is attributed to a 1:2 adduct, and the sharp signal to a 1:1 adduct, which is assumed to prevail between 13 and 40°C [13].

The ^9Be NMR spectrum has only one resonance at $\delta = 20.8$ ppm (40°C), also ascribed to the 1:1 adduct [13,19].

A tensimetric study gave no indication of the formation of adducts of definite composition. Liquid ether has a somewhat lower vapor pressure in the presence of $Be(CH_3)_2$ compared to pure ether; this observation is attributed to the formation of $Be(CH_3)_2$–ether adducts in the liquid. At higher temperatures ether evolves into the gas phase, and a solid separates. At 40°C, the solid suddenly decomposes into $Be(CH_3)_2$ with an increase in pressure [2].

$Be(CH_3)_2 \cdot CH_3OCH_2CH_2OCH_3$

The adduct is prepared from the components in ether. It precipitates as colorless feathery needles, m.p. 100 to 101°C. Sublimation at 60 to 70°C/0.06 Torr gives prisms [5].

It is also formed by addition of the ligand, in excess, to ethereal $Be_3(CH_3)_4H_2$ followed by removal of the ether. At 48 to 50°C in a vacuum, the complex sublimes from the reaction mixture [22], mentioned in [21]; m.p. 101°C [22].

The adduct was found to be monomeric (by cryoscopy) in C_6H_6. It soon inflames when exposed to air. A chelate structure I is suggested [5].

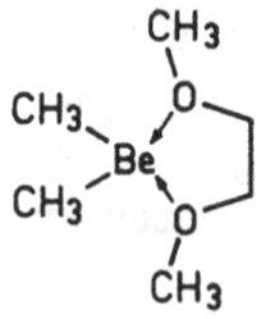

I

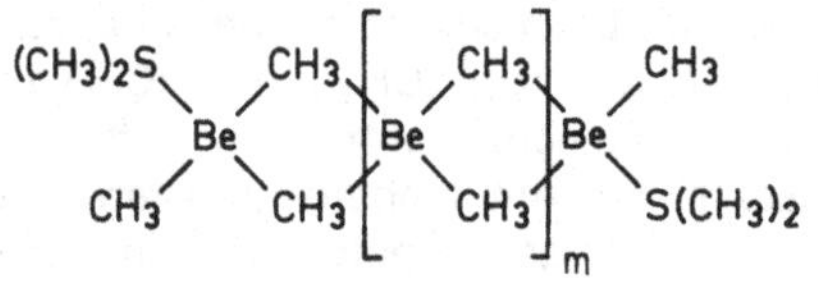

II

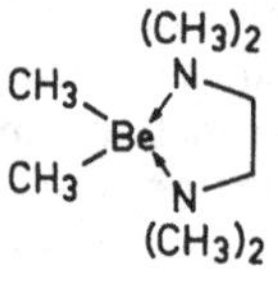

III

$Be(CH_3)_2 \cdot nS(CH_3)_2$ (n = 1,2) and **$(Be(CH_3)_2)_m \cdot 2S(CH_3)_2$** (m ≧ 2)

According to earlier tensimetric measurements, $Be(CH_3)_2$ and $S(CH_3)_2$ (mole ratio 1:2) show no reaction when kept in a closed system at 20 to 100°C [2]. $Be(CH_3)_2$, prepared by a Grignard reaction in $S(CH_3)_2$, always contained some sulfide, thus indicating complex formation [3, 11]. The solubility of $Be(CH_3)_2$ in $S(CH_3)_2$ also suggests the existence of adducts [12].

The existence of a 1:1 and a 1:2 complex was confirmed in a ^1H NMR spectroscopic study of solutions of $Be(CH_3)_2$ in $S(CH_3)_2$ in the temperature range −65 to 12°C, and the equilibrium constants for $Be(CH_3)_2 \cdot S(CH_3)_2 + S(CH_3)_2 \rightleftharpoons Be(CH_3)_2 \cdot 2\ S(CH_3)_2$ could be derived:

t in °C	−65	−54	−47	−27	12
K	0.710	0.470	0.344	0.205	0.093

The free enthalpy of formation of the 1:2 complex is $\Delta H = -3.23$ kcal/mol, the entropy change at −47°C is $\Delta S = -16$ e.u. The ^{1}H NMR signal at $\delta = -0.89$ to −0.81 ppm for $BeCH_3$ at −45 to 12°C is an average, due to rapid ligand exchange between the 1:1 and 1:2 complex. Below −45°C the exchange process is slowed down, and each distinct species is detected. At −54°C, $\delta = -0.77$ ppm for the 1:1 complex and $\delta = -1.30$ ppm for the 1:2 complex, whereby the ratio of the two signals changes in favor of the 1:2 complex as the temperature is lowered to −65°C [12]. For the 1:1 complex, J(C,H) is 105 Hz [13].

At −65°C new signals appear at $\delta = -0.31$, −0.41, and −0.80 ppm. An increase in intensity relative to the signal at $\delta = -1.30$ ppm is observed as the temperature is lowered. From the temperature dependence, the equilibrium $m\,Be(CH_3)_2{\cdot}2S(CH_3)_2 \rightleftharpoons (Be(CH_3)_2)_m{\cdot}2S(CH_3)_2 + (2m-2)S(CH_3)_2$ with $m \geqq 2$ is derived, for which the following equilibrium constants are given:

t in °C	−87	−85	−80	−74
K	1.00	0.96	0.76	0.56

The data give an enthalpy for the stepwise "polymerization" reaction of $\Delta H = -9$ kcal/mol. None of the equilibrium components has been isolated and characterized. The structure of the polymer shown in Formula II is tentative [12].

A ^{9}Be NMR spectrum of $Be(CH_3)_2$ in excess $S(CH_3)_2$ shows only one signal at $\delta = 11.6$ ppm at ambient temperature, assigned to the 1:1 complex [13, 19].

$Be(CH_3)_2{\cdot}NH_3$

$Be(CH_3)_2$ and NH_3 react in ether at −80°C to form a colorless solid 1:1 adduct without evolution of CH_4. The product is almost insoluble in ether and C_6H_{14}. Decomposition starts above −70°C, as followed tensimetrically. Between 0 and 10°C, $Be(CH_3)NH_2$ appears to be formed [7].

$Be(CH_3)_2{\cdot}NH_2CH_3$ and $Be(CH_3)_2{\cdot}NH(CH_3)_2$

The adducts are assumed to be formed by condensing the amine onto $Be(CH_3)_2$ in a high-vacuum apparatus at low temperature. But they are unstable, and $Be(CH_3)_2{\cdot}NH_2CH_3$ decomposes even at −90°C with evolution of CH_4 and formation of a polymeric material. $Be(CH_3)_2{\cdot}NH(CH_3)_2$ decomposes only above its melting point of 44°C with evolution of CH_4 and formation of the trimer $(Be(CH_3)N(CH_3)_2)_3$ [1].

$Be(CH_3)_2{\cdot}N(CH_3)_3$

The 1:1 adduct is prepared from $Be(CH_3)_2$ and a slight excess of the amine in a standard vacuum apparatus at room temperature. The excess of amine can be removed in a vacuum. Colorless rhombic crystals are obtained which melt at 36°C [2]. The adduct is also obtained beside $(Be(CH_3)H{\cdot}N(CH_3)_3)_2$ by reaction of an excess of $N(CH_3)_3$ with $Be_3(CH_3)_4H_2$. The products are separated by vacuum sublimation [21, 22].

M.p. 36°C [22]. Vapor pressures are measured in the range between 26 and 140°C. No simple equation could be derived, since the slope of the log p versus T^{-1} graph varies with temperature. The following values were measured (* solid):

References on pp. 25/6

t in °C	26	30	32	36	42	48	56	64
p in Torr	1.34*	1.83*	2.65*	3.92*	5.7	8.1	12.5	18.4
t in °C	72	80	90	100	110	120	130	140
p in Torr	26.2	36.6	53.7	76.6	101.2	137.4	178.5	222.0

At temperatures above 140°C, the sample volatilizes completely. The vapor pressure of the solid (26 to 36°C) corresponds to an enthalpy of sublimation of $\Delta H_s = 20 \pm 1$ kcal/mol. The enthalpy of evaporation of the liquid is $\Delta H \sim 12$ kcal/mol at 50°C and 8 kcal/mol at 130°C. No decomposition occurs in the temperature range mentioned above. As a rough estimate, the heat of coordination of $N(CH_3)_3$ to $Be(CH_3)_2$ is $\Delta H \sim -26$ kcal/mol $N(CH_3)_3$ [2].

Measurements of the vapor pressure indicate a monomer between 150 and 180°C [2], whereas cryoscopic measurements in C_6H_6 solution show degrees of association of 1.17 to 1.19 [14].

The ^1H NMR spectrum in c-C_6H_{12} shows a CH_3Be resonance at $\delta = -1.19$ ppm [8,13]. The ^1H NMR spectrum of neat molten $Be(CH_3)_2 \cdot N(CH_3)_3$ consists of two sharp signals at $\delta = -0.65$ (CH_3Be, J(C,H) = 105 Hz) and 2.87 ppm (CH_3N). The low chemical shift for CH_3Be may suggest the formation of a dimer [13], as it is also indicated by the increased magnitude of the heat of vaporization as the temperature is lowered [2]. The ^9Be NMR resonance appears at 19.9 ppm in c-C_6H_{12} [13,19].

In the IR spectrum in c-C_6H_{12}, bands are registered at 1192, 949, 824, 794, and 713 cm^{-1}, but the assignments are not straightforward [8]. In the vapor phase spectrum, ν(BeC) is assigned at 790 cm^{-1} [11].

The adduct reacts vigorously with air and moisture [2]. A mixture of the 1:1 adduct and a 10-fold excess of $N(CH_3)_3$ in $Si(CH_3)_4$ gives a ^1H NMR resonance at $\delta = -1.25$ ppm for CH_3Be at 25°C, but at −1.42 ppm (−43°C) and at −1.43 ppm (−60 to −70°C), indicating a temperature-dependent equilibrium between the 1:1 and 1:2 complex. Equilibrium constants for the reaction $Be(CH_3)_2 \cdot N(CH_3)_3 + N(CH_3)_3 \rightleftharpoons Be(CH_3)_2 \cdot 2N(CH_3)_3$ were calculated from the chemical shifts at 5 different temperatures:

t in °C	−21	−10	1	27	45
δ(CH_3Be) in ppm ..	−1.41	−1.40	−1.36	−1.26	−1.23
K	9.500	6.700	2.200	0.370	0.127

The enthalpy of reaction is $\Delta H = -10.4$ kcal/mol, and the entropy change is $\Delta S = -36.6$ e.u. at 40°C [13]. See also $Be(CH_3)_2 \cdot 2N(CH_3)_3$, below, for the reaction with $N(CH_3)_3$.

Reaction with HCN in C_6H_6 gives $Be(CH_3)CN \cdot N(CH_3)_3$ [23]. With equimolar amounts of $Be(CH_2C_4H_9\text{-}t)_2 \cdot N(CH_3)_3$ in c-C_5H_{10}, an equilibrium mixture with $Be(CH_3)CH_2C_4H_9\text{-}t \cdot N(CH_3)_3$ is formed, which was studied by ^1H NMR; see $Be(CH_2C_4H_9\text{-}t)_2 \cdot N(CH_3)_3$, pp. 58/9 [20]. $Be(CH_3)_2 \cdot N(CH_3)_3$ reacts with $(Be(C{\equiv}CCH_3)_2 \cdot N(CH_3)_3)_2$ in the mole ratio of 2:1 in toluene to give $(Be(CH_3)C{\equiv}CCH_3 \cdot N(CH_3)_3)_2$. Analogously, the reaction with $(Be(C{\equiv}CC_4H_9\text{-}t)_2 \cdot N(CH_3)_3)_2$ in C_6H_6 yields $(Be(CH_3)C{\equiv}CC_4H_9\text{-}t)_2$ [14].

$Be(CD_3)_2 \cdot N(CH_3)_3$ is prepared from $Be(CD_3)_2$ and $N(CH_3)_3$ as described for the nondeuterated analog above. It is purified by vacuum sublimation, m.p. 39 to 40°C. IR absorption bands are registered in c-C_6H_{12} at 1022, 949, 822, 732, and 633 cm^1. With NaH in boiling ether, $Na_2[Be(CD_3)_2H]_2$ is formed [8].

$Be(CH_3)_2 \cdot 2N(CH_3)_3$

The 1:2 adduct is formed from $Be(CH_3)_2 \cdot N(CH_3)_3$ and excess $N(CH_3)_3$ at 0°C in a very slow reaction. The vapor pressure is observed to fall until a 1:2 composition of the crystalline complex is reached [9]. A former tensimetric measurement on a closed system containing equimolar amounts of the 1:1 adduct and $N(CH_3)_3$ between −30 and 160°C gave no evidence for the existence of a 1:2 complex, though the stoichiometry of the experiment corresponds to this composition. Instead, the solid phase existing between −30 and 27°C appeared to have a 2:3 stoichiometry, as one half of the added $N(CH_3)_3$ remained in the gas phase [2]. This solid phase was reinterpreted to be a mixture of the 1:1 and 1:2 complex [9].

The 1:2 adduct melts sharply at 20°C under 1 bar $N(CH_3)_3$. Its vapor (decomposition) pressure is 9.7, 16.0, and 22.7 Torr at −9, −4.5, and 0°C, respectively [9].

^1H NMR spectra of solutions of $Be(CH_3)_2$ in $N(CH_3)_3$ have a single sharp line at $\delta = -1.43$ ppm at room temperature, which is only very slightly temperature-dependent ($\delta = -1.48$ ppm at −80°C). This line is ascribed to the CH_3Be resonance of the 1:2 complex [13]. The ^1H NMR spectrum of $Be(CH_3)_2 \cdot 2N(CH_3)_3$ in c-C_6H_{12} containing free $N(CH_3)_3$ is reported to show a single sharp resonance at $\delta = -1.28$ ppm (CH_3Be) [8].

The ^9Be NMR spectrum of the 1:2 complex in excess $N(CH_3)_3$ shows a resonance at $\delta = 12.0$ ppm [13, 19].

See p. 22 for a description of the equilibrium $Be(CH_3)_2 \cdot N(CH_3)_3 + N(CH_3)_3 \rightleftharpoons Be(CH_3)_2 \cdot 2N(CH_3)_3$.

$Be(CH_3)_2 \cdot NH_2CH_2CH_2N(CH_3)_2$

This adduct is reported to be observed as colorless crystals in the reaction of $Be(CH_3)_2$ and $NH_2CH_2CH_2(CH_3)_2$ at low temperatures prior to CH_4 evolution which starts just below room temperature to give the dimer $(Be(CH_3)NHCH_2CH_2N(CH_3)_2)_2$ [5].

$Be(CH_3)_2 \cdot N(CH_3)_2CH_2CH_2N(CH_3)_2$

The adduct is prepared from the components in ether in an exothermic reaction. The solvent is distilled until the adduct crystallizes, m.p. 81 to 82°C [5]. It is also formed by addition of the ligand to ethereal $Be_3(CH_3)_4H_2$. The ether is evaporated from the filtrate. Sublimation gives colorless prisms, m.p. 81°C [21, 22].

The ^1H NMR spectrum in c-C_6H_{12} has a signal at $\delta = -1.52$ ppm (CH_3Be) [8, 13]. The IR spectrum in c-C_6H_{12} has bands at 1186, 878, 771, and 698 cm^{-1} [8]. A chelate structure (Formula III) is suggested [5].

The adduct is partially associated in C_6H_6 (by cryoscopy). It fumes strongly in air, but does not inflame, and reacts vigorously with water with effervescence [5].

The deuterated analog $Be(CD_3)_2 \cdot N(CH_3)_2CH_2CH_2N(CH_3)_2$ is obtained by reaction of $Be(CD_3)_2$ with excess amine in ether at room temperature. It sublimes at 60°C in vacuum and melts at 79 to 81°C. The IR spectrum in c-C_6H_{12} shows bands at 969, 812, 712, and 630 cm^{-1}. A tentative assignment was made [8].

$Be(CH_3)_2 \cdot NC_5H_{11}$ (NC_5H_{11} = piperidine)

$Be(CH_3)_2$ and piperidine react in ether at −20°C. Colorless crystals are formed immediately.

The adduct is unstable at room temperature. On heating to 20°C, CH_4 is evolved, and $Be(CH_3)NC_5H_{10}$ can be isolated as a primary decomposition product [6].

$Be(CH_3)_2 \cdot NC_4H_9O$ (NC_4H_9O = morpholine)

$Be(CH_3)_2$ and morpholine react in ether at −20°C. Colorless crystals precipitate, which cannot be stored at room temperature. On heating to 20°C, CH_4 is evolved leaving $Be(CH_3)NC_4H_8O$ as a primary decomposition product [6].

$Be(CH_3)_2 \cdot 2NC_7H_{13}$ (NC_7H_{13} = 1-azabicyclo[2.2.2]octane)

$Be(CH_3)_2$ and quinuclidine (1-azabicyclo[2.2.2]octane) react in C_6H_6 at 50°C. Clear colorless crystals of the air-sensitive compound deposit on cooling.

The crystal and molecular structure was determined by X-ray diffraction. The crystals are monoclinic, with a = 11.82(2), b = 12.71(2), c = 12.00(2) Å, and β = 113.1(3)°; Z = 4, d_c = 1.05 g/cm³, space group $P2_1/c\text{-}C^5_{2h}$ (No. 14), R = 0.132 for 671 independent reflections. **Fig. 5** shows structural details of the molecule, which has a distorted tetrahedral coordination around Be. The most important feature is the large CBeC bond angle of 118.3(1.0)°. The NBeN angle is 110.8(1.2)° [15].

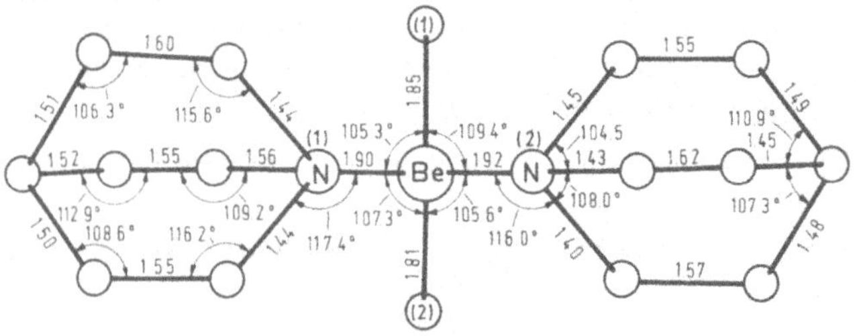

Fig. 5. Molecular structure of $Be(CH_3)_2 \cdot 2NC_7H_{13}$ (NC_7H_{13} = 1-azabicyclo[2.2.2]octane) [15].

Other selected nonbonded distances (in Å):

N(1)···N(2)	3.14(1)	N(2)···C(1)	3.08(2)
N(1)···C(1)	2.98(2)	N(2)···C(2)	2.97(2)
N(1)···C(2)	2.99(2)	C(1)···C(2)	3.14(2)

$Be(CH_3)_2 \cdot 1,2\text{-}(N(CH_3)_2)_2C_6H_4$

The exothermic reaction of the components in ether gives colorless crystals when the solution is concentrated. The adduct can be sublimed at 80 to 90°C/10^{-2} Torr; m.p. 103 to 104°C. It is a monomer in C_6H_6 (by cryoscopy) [9].

$Be(CH_3)_2 \cdot 2NC_5H_5$ (NC_5H_5 = pyridine)

$Be(CH_3)_2$ and pyridine give an exothermic reaction in ether. After filtration colorless needles gradually deposit, which melt to a yellow liquid at 91 to 92°C. The adduct fumes strongly in air and is readily hydrolyzed by H_2O [5].

$Be(CH_3)_2 \cdot NC_5H_4CH_3\text{-}2$ ($NC_5H_4CH_3$-2 = 2-picoline)

$Be(CH_3)_2$ is dissolved in c-C_5H_{10} containing 1 mol equivalent of 2-picoline. The adduct melts at 43°C after vacuum sublimation. It is a monomer in C_6H_6 (by cryoscopy) [14].

$Be(CH_3)_2 \cdot N_2C_{10}H_8$ ($N_2C_{10}H_8$ = 2,2'-bipyridine)

The adduct is prepared from the components in ether [5], or by addition of the ligand to $Be_3(CH_3)_4H_2$ [22] and is obtained as golden yellow needles from C_6H_6 [5].

It decomposes at 170°C and sublimes very slowly at 135 to 140°C/10^{-3} Torr [5]. The UV absorption maximum is at λ = 395 nm (ε = 2700 L·mol⁻¹·cm⁻¹) [4, 5].

$Be(CH_3)_2 \cdot N(CH_3){=}CHC_6H_5$

The components react in ether to yield the adduct. A pale yellow color forms which fades within seconds.

The white solid adduct sublimes at 45 to 50°C/10^{-2} Torr and melts at 60 to 62°C. The IR spectrum has an absorption at 1490 cm^{-1} in the 1650 to 1450 cm^{-1} region. The compound is slightly associated in C_6H_6 (by cryoscopy) [16].

$nBe(CH_3)_2 \cdot mP(CH_3)_3$ (n:m = 1:1, 1:2, 2:3, and 3:2)

The adducts are believed to appear as discrete solid phases in tensimetric measurements of the $Be(CH_3)_2$–$P(CH_3)_3$ system, each one being stable only in a certain range of temperature and pressure of $P(CH_3)_3$. $Be(CH_3)_2$ was dissolved in an excess of $P(CH_3)_3$ (ratio 1:6.3 and 1:4.8) and warmed from −183 to 150°C over 11 to 12 h. The observed succession of melting and freezing over the temperature range suggests formation of at least 4 adducts. With a ratio of 1:6.3 of the components, a rhombic crystalline solid forms at ca. −25°C which melts at 27 to 31°C/94 Torr with a subsequent rapid pressure increase. The composition of this solid is clearly 1:2, as determined from the residual pressure of the uncomplexed phosphine. On further heating, another solid appears at 42°C/108 Torr, which has a 2:3 composition. This solid melts again at about 70 to 75°C/~130 Torr. At this stage the composition is nearly 1:1. At 90 to 95°C/~148 Torr the liquid freezes again, the composition of the solid corresponding to 3:2. This phase melts rather gradually between 130 and 140°C/~100 to 200 Torr. The existence of the 1:1 and 3:2 adducts was confirmed by the isotherms at 25 and 50°C. The 25°C isotherm gives a dissociation pressure of 5.5 Torr for the 1:1 phase, and of 80 to 82 Torr for the 3:2 phase [2].

The ^1H NMR spectrum of $Be(CH_3)_2$ in excess $P(CH_3)_3$ shows a CH_3Be resonance at $\delta = -1.42$ ppm, and the ^9Be NMR spectrum of the same system has a resonance at $\delta = 3.6$ ppm. Both values are taken as evidence for tetracoordinated Be atoms in a 1:2 complex [13, 19]. The structures proposed in [2] for the adducts are only tentative.

References:

[1] Coates, G. E., Glockling, F., Huck, N. D. (J. Chem. Soc. **1952** 4512/5).
[2] Coates, G. E., Huck, N. D. (J. Chem. Soc. **1952** 4501/11).
[3] Bähr, G., Thiele, K. H. (Chem. Ber. **90** [1957] 1578/86).
[4] Coates, G. E., Green, S. I. E. (Proc. Chem. Soc. **1961** 376).
[5] Coates, G. E., Green, S. I. E. (J. Chem. Soc. **1962** 3340/8).
[6] Funk, H., Masthoff, R. (J. Prakt. Chem. [4] **22** [1963] 255/8).
[7] Masthoff, R., Vieroth, C. (Z. Chem. [Leipzig] **5** [1965] 142).
[8] Bell, N. A., Coates, G. E., Emsley, J. W. (J. Chem. Soc. A **1966** 49/52).
[9] Bell, N. A., Coates, G. E. (Can. J. Chem. **44** [1966] 744/5).
[10] Sanders, J. R., Ashby, E. C., Carter, J. H. (J. Am. Chem. Soc. **90** [1968] 6385/90).

[11] Kovar, R. A., Morgan, G. L. (Inorg. Chem. **8** [1969] 1099/103).
[12] Kovar, R. A., Morgan, G. L. (J. Am. Chem. Soc. **91** [1969] 7269/74).
[13] Kovar, R. A., Morgan, G. L. (J. Am. Chem. Soc. **92** [1970] 5067/72).
[14] Coates, G. E., Francis, B. R. (J. Chem. Soc. A **1971** 474/7).
[15] Whitt, C. D., Atwood, J. L. (J. Organometal. Chem. **32** [1971] 17/25).
[16] Andersen, R. A., Coates, G. E. (J. Chem. Soc. Dalton Trans. **1974** 1171/80).
[17] Coates, G. E., Smith, D. L. (J. Chem. Soc. Dalton Trans. **1974** 1737/40).
[18] Latajka, Z., Ratajczak, H., Romanovska, K., Tomczak, Z. (J. Organometal. Chem. **139** [1977] 129/33).

[19] Gaines, D. F., Coleson, K. M., Hillenbrand, D. F. (J. Magn. Resonance **44** [1981] 84/8).
[20] Coates, G. E., Francis, B. R. (J. Chem. Soc. A **1971** 1305/8).

[21] Bell, N. A., Coates, G. E. (Proc. Chem. Soc. **1964** 59).
[22] Bell, N. A., Coates, G. E. (J. Chem. Soc. **1965** 692/9).
[23] Coates, G. E., Mukherjee, R. N. (J. Chem. Soc. **1963** 229/33).

1.2.1.3 Diethylberyllium, $Be(C_2H_5)_2$

Preparation

Early attempts to synthesize $Be(C_2H_5)_2$ from Be metal and C_2H_5I met with only limited success [1], and it was later shown that this reaction is not a preparative route to ethylberyllium compounds [5 to 7].

The reaction between metallic Be and $Hg(C_2H_5)_2$ was performed for the first time by Cahours at 130 to 135°C. A colorless liquid with a boiling point of 185 to 188°C was distilled under CO_2 [2]. The reaction produces pure $Be(C_2H_5)_2$ when catalyzed with $HgCl_2$ [3, 39, 44]. At 125°C the reaction is complete after 36 [44] or 48 h [39]. The product is vacuum-distilled [39, 44]. A ^{7}Be labelled material was also obtained by this procedure [44].

A convenient laboratory procedure for the synthesis of $Be(C_2H_5)_2$ is the reaction of $BeCl_2$ [4, 6, 10, 68] or $BeBr_2$ [4] with $Mg(C_2H_5)I$ [6] or $Mg(C_2H_5)Br$ [10, 68] in ether [4, 6, 68], also mentioned in [10]. The product can be distilled from the reaction mixture in vacuum (~55°C/ 10^{-3} Torr [68]), and yields of up to 90% are achieved [10]. Removal of the last traces of ether by vacuum distillation is exceedingly difficult [10, 12, 14, 16, 20, 24, 29, 43, 54, 61]. Separation is achieved by boiling the mixture under reflux for 24 h at ~10^{-3} Torr (oil bath 50°C, condenser −15°C). Evolution of ether is insignificant after 18 h [43].

The compound is also prepared by the reaction of LiC_2H_5 with anhydrous $BeCl_2$ (mole ratio 2:1) in C_6H_6. After stirring at 35°C overnight, the solution is filtered and C_6H_6 removed under vacuum, leaving an oily liquid. Pumping is continued for several hours. Subsequently the compound is distilled at ~60°C/10^{-3} Torr [68].

C_2H_5 radicals (produced by decomposition of $Pb(C_2H_5)_4$ at 950°C) were introduced into a quartz tube containing a mirror of Be at 150°C. A white substance condensed in the liquid N_2-cooled tube placed after the quartz tube. The substance melts below room temperature to a colorless liquid and is assumed to be $Be(C_2H_5)_2$ [9].

Purification of $Be(C_2H_5)_2$ can be also achieved via isolation and pyrolysis of the complexes $(Be(C_2H_5)_2)_2 \cdot MX$ (MX = KF, RbF, $[N(CH_3)_4]F$, and KCN) at 120°C in a high vacuum [21, 35].

Association

The degree of association in the liquid state depends on the time of storage. Freshly distilled $Be(C_2H_5)_2$ is dimeric, as shown by its IR and Raman spectra [54], from cryoscopic measurements in the noncomplexing solvents C_6H_6 [15, 43] and c-C_6H_{12} [15], and from dipole moment measurements in C_7H_{16} and C_6H_6 [12, 15]. If the liquid is stored for eight months, the degree of association rises to >3, as shown by cryoscopic measurements in c-C_6H_{12} and C_6H_6 [15].

A Raman study and the low vapor pressure [10] indicate that higher associated species are present in the liquid [10, 54].

A mass spectroscopic study shows that the unsaturated vapor contains trimeric, dimeric, and monomeric species. The abundance of monomeric ions increases with the source temperature, whereas trimeric ions, and the much more abundant dimeric ions, decrease with temperature [46, 49].

In the strong complexing solvent dioxane, coordinated monomers form from liquid $Be(C_2H_5)_2$ stored for 2 days (degree of association 2) and for 8 months (degree of association >3). This was established from cryoscopic [15] and dipole moment [12, 15] measurements.

Physical Properties

$Be(C_2H_5)_2$ is a colorless mobile liquid [6]. The vapor pressure could only be followed in the temperature range from 20 to 85°C, when decomposition becomes noticeable. For the mentioned range, the vapor pressure can be expressed by $\log p = -2200/T + 7.59$ (p in Torr, T in K) [10]. The vapor pressure deduced from the boiling point data in the literature is given as $\log p = 14.496 - 5102/T$ (p in Torr, T in K) [66].

The boiling point of $Be(C_2H_5)_2$ without traces of ether was measured at 65°C/10^{-3} Torr [43]. For samples containing traces of ether the following values are given: 56 to 57°C/10^{-3} Torr [29], 60°C/0.15 Torr [44], 61°C/0.15 Torr [39], 63.0°C/0.3 Torr (2% ether), 62.5 to 63°C/0.3 Torr (3.9% ether) [10], 93 to 95°C/4 Torr [6, 13], and 110°C/15 Torr [6]. A value of 194°C/760 Torr is extrapolated from the vapor pressure data [10], but distillation under ordinary pressure at 180 to 240°C leads to considerable decomposition. The melting region is −13 to −11°C [6].

Dipole moment measurements gave values of $\mu = 1.0$, 1.7, and 4.3 D for C_7H_{16}, C_6H_6, and dioxane solutions, respectively, at 20°C [12]. These data are assigned to dimers for the first two noncoordinating solvents, but to the monomer in dioxane [15].

The neat liquid has a very low specific conductivity of $\varkappa = 8.4 \times 10^{-10}\ \Omega^{-1} \cdot cm^{-1}$ at 80°C [35] and $1 \times 10^{-10}\ \Omega^{-1} \cdot cm^{-1}$ at 20°C [28]. $Be(C_2H_5)_2$ containing 3% ether shows $\varkappa = 1 \times 10^{-7}\ \Omega^{-1} \cdot cm^{-1}$ at 20°C. The specific conductivities of solutions in C_6H_6, dioxane, $N(C_2H_5)_3$, ether, and THF were also determined. A 0.1 M solution in THF has $\varkappa = 1 \times 10^{-6}\ \Omega^{-1} \cdot cm^{-1}$ at 20°C [25]; see also [28].

The ^1H NMR spectra in $C_6D_5CD_3$ and in c-$C_6D_{11}CD_3$ show significant temperature changes and second-order splittings. In $C_6D_5CD_3$ (40%) at ambient temperature, $\delta = 0.26$ (q, CH_2; $^1J(H,H) = 8$ Hz) and 1.06 ppm (t, CH_3; $^1J(H,H) = 8$ Hz). In c-$C_6D_{11}CD_3$ (40%) at ambient temperature, $\delta = 0.57$ (q, CH_2; $^1J(H,H) = 7$ Hz) and 1.55 ppm (t, CH_3; $^1J(H,H) = 7$ Hz). In both solvents the signals broaden as the temperature falls. In c-$C_6D_{11}CD_3$, all detail of the CH_2 signal is lost at −10°C, but two broad components can be distinguished at −40°C, and three at −60°C. In contrast, in $C_6D_5CD_3$, the CH_2 signal consists of three or more peaks at 80 to −55°C. At −55°C five components can be distinguished. The area under the CH_2 signal is at all temperatures about ⅔ of the area under the CH_3 signal in c-$C_6D_{11}CD_3$, but in $C_6D_5CD_3$ the area is much less than ⅔ at $\leqq -40$°C. The CH_3 signal in c-$C_6D_{11}CD_3$ broadens without loss of the three-component structure, and the area increases relative to the area of CH_2 as the temperature is lowered. In $C_6D_5CD_3$, however, the triplet persists down to 0°C and collapses to a single broad peak at −20°C. At −40°C it splits into two quite distinct triplets at $\delta = 0.87$ ($^1J(H,H) = 7$ Hz) and 1.29 ppm ($^1J(H,H) = 7$ Hz) with an area ratio of 1:2. To explain these observations, the presence of several oligomers with nonequivalent C_2H_5 groups is suggested [43]. A ^1H NMR spectrum in c-C_5H_{10} shows signals at $\delta = 5.63$ (t, CH_3; J(H,H) = 7.7 Hz) and 6.35 ppm (q, CH_2; J(H,H) = 7.6 Hz) relative to external C_6H_6 [33]. Values of $\delta = 0.90$ and 1.56 ppm, without further specification, are also given [45].

IR and Raman spectra were taken from freshly distilled $Be(C_2H_5)_2$, and an assigment based upon a dimeric species of D'_{2h} pseudosymmetry was attempted. The vibrations and their

References on pp. 33/4

assignments are summarized in Table 4 [54]. A former Raman study between 4000 and 200 cm^{-1} was performed with liquid $Be(C_2H_5)_2$ containing 2 and 20% ether. No significant difference was found for the two samples. The large number of lines rules out the sole existence of monomers, but supports the idea that oligomers of a complicated structure are present. The bands could not be assigned [10].

Table 4
IR and Raman Vibrations (in cm^{-1}) of Liquid $Be(C_2H_5)_2$ [54].

IR	R	assignment	IR	R	assignment
2946(s)	2940(m)		969(m)		ν(CC), B_{1u}
2900(sh)			925(m)		ν(BeC), B_{1u}
2880(s)	2875(s)	$\nu(CH_2, CH_3)$		915(w)	ν(CC), A_g
2865(s)	2864(vs)			875(m)	ν(BeC), A_g
2795(sh)	2785(sh)		836(w)		
	2740(w)		724(vs)	720(vw)	$\varrho(CH_3)$
1469(m)	1470(sh)	$\delta_{as}(CH_3)$	670(s)		$\varrho(CH_2)$
1455(m)	1460(s)		630(w)		
1402(s)	1405(m)	$\delta(CH_2)$		615(vw)	
1385(m)	1385(sh)	$\delta_s(CH_3)$	538(m)		ν(ring), B_{1u}, B_{3u}
1326(w)		$\delta(CH_2$ bridged)	508(w)		
1208(s)	1204(m)	$\gamma,\tau(CH_2)$		499(m to s)	ν(ring), A_g
997(s)		ν(CC), B_{3u}		470(m)	δ,γ(CC), B_{1g}
	972(m to s, br)	ν(CC), A_g		234(w to m)	δ(BeCC), B_{3g}

It is suggested in the literature that the dimer and the oligomers contain $-CH_2(CH_3)-$ bridges in analogy to $Be(CH_3)_2$, e.g. [10].

The first ionization potential of 9.46 ± 0.05 eV was calculated from mass spectroscopic data [46, 49].

Chemical Reactions

Thermal decomposition of $Be(C_2H_5)_2$ containing 14% ether starts above 80°C. At 190 to 220°C, no H_2 or CH_4 is yielded. The gaseous mixture of the volatile products consists of C_2H_6, C_2H_4, and C_4H_8 in the ratio 2:1:1.5, and traces of hex-3-ene, cyclohexa-1,3-diene, and C_6H_6. The residue is a yellow viscous oil and some yellow crystalline material. The oil is volatile in vacuum; the condensate reacts vigorously with H_2O to form H_2, C_2H_6, C_2H_4, and $Be(OH)_2$. Obviously a complicated mixture of products is present with a number of structural units like BeH, $BeCH_2CH_2Be$, BeC_2H_5, etc. [10].

The ions produced by electron impact at (2 to 8) × 10^{-6} Torr of gaseous $Be(C_2H_5)_2$ depend on the source temperature. At low source temperature (37 to 50°C or 45 to 55°C at 70 eV), ions derived from electron deficient associated species are produced, the trimer being the upper limit. At higher source temperatures, only ions derived from the dimer and monomer are observed (196 or 210°C at 70 eV). The same effect is observed at 169°C/12 eV and 210°C/15 eV. Whereas at 169°C the fragment with the highest mass is $[Be_2(C_2H_5)_3]^+$, only mononuclear species are observed at 210°C. A selection of the observed fragments is shown in Table 5. Measurements of the appearance potentials (AP in eV) gave AP = 9.46 ± 0.05 for $[Be(C_2H_5)_2]^+$,

10.35 ± 0.03 for $[BeC_2H_4]^+$, and 11.51 ± 0.05 for $[BeC_2H_5]^+$. The Be–C bond dissociation energy is calculated to be 47.3 ± 3.2 kcal/mol [46, 49].

Table 5
Mass Spectra of $Be(C_2H_5)_2$. Relative Abundances of Be-Containing Ions at Various Impact Energies and Source Temperatures [46, 49].

ion	70 eV 45 to 55°C	70 eV 196°C	12 eV 169°C	15 eV 210°C
$[Be_3(C_2H_5)_5]^+$	0.14			
$[Be_3(C_2H_5)_4H]^+$	0.06			
$[Be_2(C_2H_5)_4]^+$	0.38			
$[Be_3(C_2H_5)_3H_2]^+$	0.06			
$[Be_2(C_2H_5)_3]^+$	5.46	0.18	0.26	
$[Be_2C_6H_{14}]^+$	0.30			
$[Be_3(C_2H_5)_2H_3]^+$	0.06			
$[Be_2(C_2H_5)_2H]^+$	11.11	0.42		
$[Be_2C_4H_9]^+$	0.22		0.16	
$[Be_2C_4H_7]^+$	0.29			
$[BeC_4H_{10}]^+$	3.72	8.02	21.28	
$[BeC_4H_9]^+$	0.48	0.34	0.41	
$[BeC_4H_8]^+$	0.31	0.19	0.28	
$[BeC_4H_7]^+$	0.24	0.10	0.04	
$[Be_2C_3H_9]^+$	0.12			
$[BeC_4H_6]^+$	0.31	0.28	0.16	
$[BeC_4H_5]^+$	0.19	0.18		
$[BeC_4H_4]^+$	0.12	0.03		
$[Be_3(C_2H_5)H_4]^+$	0.13			
$[BeC_4H_3]^+$	0.13	0.11		
$[BeC_4H]^+$	0.07			
$[BeC_3H_8]^+$	0.05	0.07		
$[BeC_3H_7]^+$	1.19	2.02	4.03	
$[BeC_3H_6]^+$	0.31	0.48	0.95	
$[BeC_3H_5]^+$	0.46	0.48	1.57	
$[Be_2C_2H_7]^+$	6.02	0.19		
$[BeC_3H_4]^+$	0.24	0.14	0.40	
$[Be_2C_2H_6]^+$	0.27	0.06		
$[BeC_3H_3]^+$	0.50	0.41		
$[Be_2C_2H_5]^+$	0.72	0.02		
$[BeC_3H_2]^+$	0.20	0.15		
$[BeC_2H_7]^+$				0.10
$[BeC_2H_6]^+$				0.43
$[BeC_2H_5]^+$				9.33
$[BeC_2H_4]^+$				30.67

References on pp. 33/4

Table 5 (continued)

ion	70 eV 45 to 55°C	70 eV 196°C	12 eV 169°C	15 eV 210°C
$[BeC_2H_3]^+$				0.37
$[BeC_2H_2]^+$				0.19
$[BeCH_3]^+$	1.07	1.09	0.09	
$[BeCH_2]^+$	0.54	0.56	0.04	
$[BeCH]^+$	0.07	0.08		
$[Be_2H_3]^+$	0.11	<0.01		
$[BeC]^+$	0.01	<0.01		
$[BeH_2]^+$	<0.01	<0.01		
$[BeH]^+$	1.64	1.05		
$[Be]^+$	1.82	1.30		

$Be(C_2H_5)_2$ is spontaneously inflammable in air [2, 6], especially in a humid atmosphere, where the compound burns with luminous flames evolving white fumes of BeO [6]. The reaction with I_2 is also vigorous [2, 6], and $Be(C_2H_5)I$ is assumed to be the product [6]. Reaction with Na in ether at 90°C gives Be and BeC_2. From the filtrate, a compound of the composition $Na[Be(C_2H_5)_3]$ was isolated [62]. Rare gases or N_2 can be used as inert gases in experiments with the solid compound or with solutions [6, 10]; but with CO_2, a reaction occurs which gives, after hydrolysis, $(C_2H_5)_3COH$ [6]. The violent reaction with H_2O [2, 6, 10] produces C_2H_6 gas [6,10]. No peroxides were detected in the solid products of controlled reactions of $Be(C_2H_5)_2$ with H_2O_2 in ether [8]. When HCl gas is condensed onto pure $Be(C_2H_5)_2$ at −185°C, a violent reaction and thick fumes of $BeCl_2$ are observed [10]. NH_3 forms a 1:1 adduct at −80°C; no gas is evolved. Above −60 to −50°C, decomposition ensues leading to C_2H_6 and $(Be(C_2H_5)NH_2)_n$ polymer. Above 10°C and with or without excess NH_3, a polymeric Be imide is formed [27].

$Be(C_2H_5)_2$ reacts with an ethereal solution of $BeCl_2$ [6, 38, 50], $BeBr_2$, or BeI_2 [6, 44, 50] in 1:1 ratios to give solutions of $Be(C_2H_5)Cl$ [6, 38, 50], $Be(C_2H_5)Br$ [6, 44, 50], and $Be(C_2H_5)I$ [6]. The latter is deposited as the 1:1 adduct with 1,4-dioxane [44], and the former as the 2,2′-bipyridine [38] or $N(CH_3)_2CH_2CH_2N(CH_3)_2$ [50] adduct. The equilibrium $Be(C_2H_5)_2 + BeCl_2 \rightleftharpoons 2\,Be(C_2H_5)Cl$ lies well to the right [38]. Reaction of $Be(C_2H_5)_2$ with $BeCl_2$ in the mole ratio 4:1 with stirring for 12 h, subsequent addition of c-C_6H_{12}, and stirring for another 24 h yields pure polymeric $(Be(C_2H_5)Cl)_n$ [61].

$Be(C_2H_5)_2$ reacts with $Be(OC_4H_9\text{-}t)_2$ (mole ratio 1:1) to yield colorless prisms, m.p. 76°C (from C_6H_{14}), which contain Be:C_2H_5 in the ratio 2.7:2.0. The product was not further characterized [59].

$Be(C_2H_5)_2$ reacts with excess $LiAlH_4$ in ether to form a suspension of solvated BeH_2 [53], in analogy to the reaction of $Be(CH_3)_2$ [11]. $Al(C_2H_5)_2H$ gives with ether-free $Be(C_2H_5)_2$ an addition product $BeAl_2(C_2H_5)_6H_2$ of unknown structure (formulated as $Be(C_2H_5)_2 \cdot 2\,Al(C_2H_5)_2H$) [47]. $Be(C_2H_5)_2H$ (as etherate) is obtained from $Be(C_2H_5)_2$, $BeBr_2$, and LiH in ether [34, 53]. The reaction with $Sn(CH_3)_3H$ in ether at 40 to 75°C [31], or in C_6H_{14} at 60 to 70°C [36], yields only impure $Be(C_2H_5)H$ and $Sn(C_2H_5)_4$ [31, 36]. Impure $M_2[Be(C_2H_5)H_3]$ is generated in very low yield with $M[Al(C_2H_5)_3H]$ in ether, where M = Li, Na [42]. NaH adds to $Be(C_2H_5)_2$ in boiling ether with formation of the salt $Na_2[Be_2(C_2H_5)_4H_2]$ [22], while NaD gives the analogous deuterated complex [31]. Reactions of $Be(C_2H_5)_2$ with $Na[Be(C_2H_5)_2H]$ in the mole ratios 1:1, 2:1, 3:1, and

4:1 in C_6H_{14} lead to solid products with the empirical formula $Na[Be(C_2H_5)_2H]\cdot n\,Be(C_2H_5)_2$ (n = 1, 2) and products with ratios of $Na:Be:C_2H_5:H = 2.92:7.85:4.00:2.95$ and 1.00:3.00:5.92:1.00 [48]. Excess LiH also absorbs $Be(C_2H_5)_2$ from an ethereal solution at 110°C to give a monoetherate of $Li_2[Be_2(C_2H_5)_4H_2]$ [42]. No reaction was detected between MgH_2 and $Be(C_2H_5)_2$ in ether [22]. The following alkali and quaternary ammonium salts MX were also found to form complex salts $M[Be_2(C_2H_5)_4X]$: KF [17], RbF [18, 21], CsF [18, 21], $[N(CH_3)_4]F$ [35], $[N(CH_3)_3CH_2C_6H_5]F$ [35], $[N(CH_3)_4]Cl$ [35], $[N(C_2H_5)_4]Cl$ [21], $[N(CH_3)_3CH_2C_6H_5]Cl$ [35], and $[N(CH_3)_4]SCN$ [54]. With NaF [17, 21] and NaCN [21], complex formation is incomplete; with CsN_3 a 1:1 complex [61]; and with KCN [21] and $[N(CH_3)_4]CN$ [57], 1:4 complexes are produced. LiF [21] and CsCl [17, 21] do not react with $Be(C_2H_5)_2$ in ether.

$Be(C_2H_5)_2$ reacts with CH_3OH [10], C_2H_5OH [2,10], and i-C_3H_7OH [10] at −80°C to give the corresponding $Be(OR)_2$ and C_2H_6 [10]. Addition of $(CF_3)_3COH$ to $Be(C_2H_5)_2$ in C_6H_{14} gives $(Be(OC(CF_3)_3)_2)_2$ in 65% yield [60]; and with (t-$C_4H_9)_2CHOH$ in a 1:2 ratio in ether, $Be(OCH(C_4H_9\text{-}t)_2)_2$ is obtained [58]. Equimolar amounts of $(C_2H_5)_3COH$ in ether solution convert $Be(C_2H_5)_2$ into $(Be(C_2H_5)OC(C_2H_5)_3)_3$. With two equivalents of the alcohol, $(BeOC_4H_9\text{-}t)_2$ is obtained [40]. C_6H_5OH, 4-$CH_3C_6H_4OH$, 4-ClC_6H_4OH, and 2,4-$(NO_2)_2C_6H_3OH$ in ether at −50 to −20°C react with $Be(C_2H_5)_2$ to give the corresponding beryllium bisphenolates $Be(OR)_2$ and C_2H_6 [24]. The product of the reaction of $CH_3OCH_2CH_2OH$ with $Be(C_2H_5)_2$ in ether/C_6H_6 is $Be(OCH_2CH_2OCH_3)_2$, which is supposed to have a polymeric structure, and the analogous reaction with $N(CH_3)_2CH_2CH_2OH$ gives $Be(OCH_2CH_2N(CH_3)_2)_2$, which is oligomeric in C_6H_6 solution (n ≈ 10). During the reactions, C_2H_6 evolves [32]. With an equimolar amount of $CH_3OCH_2CH_2OH$ in ether, $Be(C_2H_5)OCH_2CH_2OCH_3$ is formed [40].

An equimolar mixture of $LiOC_4H_9$-t and $Be(C_2H_5)_2$ in ether reacts to yield liquid $Be(C_2H_5)OC_4H_9$-t in 36% yield as isolated product, but MOC_4H_9-t (M = Na, K) gives $(M[Be(C_2H_5)_2OC_4H_9\text{-}t])_2$. With M = Rb, the composition of the product was $Rb_2Be_2(C_2H_5)_3(OC_4H_9\text{-}t)_3$ [59].

$Be(C_2H_5)_2$ adds to the carbonyl group of $(C_2H_5)_2CO$ to form $(Be(C_2H_5)OC(C_2H_5)_3)_3$ [40]. With (t-$C_4H_9)_2CO$ in C_6H_{14}, however, C_2H_4 is evolved, and $Be(C_2H_5)OCH(C_4H_9\text{-}t)_2$ is formed in high yield. Ethene is also generated with t-C_4H_9CHO, but the residue was not characterized [58]. Benzophenone is reduced to $(C_6H_5)_2CHOH$ [6], and only C_2H_4 is evolved [58]. Michler's ketone gives a positive color test [6].

Equimolar quantities of CH_3SH, C_2H_5SH, i-C_3H_7SH, and t-C_4H_9SH give $(Be(C_2H_5)SR)_4$ upon reaction with $Be(C_2H_5)_2$. Two equivalents of C_2H_5SH give $Be(SC_2H_5)_2$. The latter compound reacts with $Be(C_2H_5)_2$ to form $(Be(C_2H_5)SC_2H_5)_4$, which is also accessible from $C_2H_5SSC_2H_5$ and $Be(C_2H_5)_2$, with $C_2H_5SC_2H_5$ as a byproduct [41]. $N(CH_3)_2CH_2CH_2SH$ and $Be(C_2H_5)_2$ (ratio 2:1) in ether/C_6H_6 give 2 mol of C_2H_6 and $Be(SCH_2CH_2N(CH_3)_2)_2$, for which a chelate structure is proposed [32].

$Se_2(C_2H_5)_2$ converts $Be(C_2H_5)_2$ (mole ratio 1:1) into $Be(SeC_2H_5)_2$. Only in the presence of pyridine could the complex $Be(C_2H_5)SeC_2H_5\cdot 2NC_5H_5$ be isolated. $Be(C_2H_5)_2$ and $SeHC_6H_5$ (mole ratio 1:1) react in ether to form $(Be(C_2H_5)SeC_6H_5\cdot O(C_2H_5)_2)_2$. With a 1:2 mole ratio of the reactants, $Be(SeC_6H_5)_2$ is formed [41].

$NH(CH_3)_2$ reacts with $Be(C_2H_5)_2$ in ether to form an unstable adduct which decomposes to generate C_2H_6 and $(Be(C_2H_5)N(CH_3)_2)_3$ [37]. Excess $NH(CH_3)_2$ yields $(Be(N(CH_3)_2)_2)_3$. The preparation can be carried out in C_7H_{16} or in the absence of a solvent [30]. Reactions of $Be(C_2H_5)_2$ with $NH(C_2H_5)_2$ and $NH(C_6H_5)_2$ in ether in the mole ratio 1:1 give $(Be(C_2H_5)N(C_2H_5)_2)_2$ and $(Be(C_2H_5)N(C_6H_5)_2)_2$, respectively [37]. The reaction with excess $NH(C_6H_5)_2$ in C_6H_6 yields $Be(N(C_6H_5)_2)_2$ [16]. The evolution of C_2H_6 was observed with $NH(C_2H_5)_2$ [10] and with $NH(C_6H_5)_2$ [16].

References on pp. 33/4

The reactions with substituted hydrazines NRR'NR"R"' occur in three stages: 1) adduct formation, 2) monosubstitution of C_2H_5, and 3) disubstitution of C_2H_5, if sufficiently vigorous conditions are employed. Steric effects in the hydrazine may determine the number of C_2H_5 groups in $Be(C_2H_5)_2$ which are replaced. $Be(C_2H_5)_2$ and $NH_2N(CH_3)_2$ (mole ratio 1:2.8) give at 50°C in C_8H_{18} $Be(NHN(CH_3)_2)_2$. In the reaction of $Be(C_2H_5)_2$ with $NH(CH_3)N(CH_3)_2$ (mole ratio 1:2.35), 1.5 equivalents of C_2H_6 are evolved at room temperature within 2 days. The elemental analysis of the product is in agreement with the formula $Be(C_2H_5)N(CH_3)NH_2$. If the reaction is carried out at 65°C in C_9H_{20}, increased formation of $Be(N(CH_3)NH_2)_2$ is observed. A 1:1 mixture of $Be(C_2H_5)_2$ and $NH(CH_3)NH(CH_3)$ in C_8H_{18} gives at 75°C $Be(N(CH_3)NHCH_3)_2$. The product obtained by reaction of $Be(C_2H_5)_2$ with an equivalent of $NH(CH_3)N(CH_3)_2$ in C_8H_{18} consists of $Be(C_2H_5)_2 \cdot NH(CH_3)N(CH_3)_2$ (15%) and $Be(C_2H_5)N(CH_3)N(CH_3)_2$. If the reaction is performed at 55°C for 4 h in C_7H_{16} in a mole ratio of 1:2, the main product is still $Be(C_2H_5)N(CH_3)N(CH_3)_2$, while $Be(N(CH_3)N(CH_3)_2)_2$ is formed only as a side product; with $N(CH_3)_2N(CH_3)_2$ as a reactant (1:1 mole ratio) in C_8H_{18}, $(Be(C_2H_5)_2 \cdot 0.5N(CH_3)_2N(CH_3)_2)_2$ is formed [26].

$C_6H_5N{=}CH_2$ in ether is reduced in an extremely violent reaction to yield $C_6H_5NHC_3H_7$ [6]. $C_6H_5CH{=}NCH_3$ and $C_6H_5CH{=}NC_6H_5$ react with $Be(C_2H_5)_2$ to form the compounds $(Be(C_2H_5)N(CH_3)CH(C_2H_5)C_6H_5)_2$ (73% yield) and $(Be(C_2H_5)N(C_6H_5)CH(C_2H_5)C_6H_5)_2$ (28% yield) with evolution of C_2H_6 and C_2H_4; $C_6H_5CH{=}NC_4H_9$-t affords the ortho-metallation product I in 50% yield with C_2H_6 being the only gas evolved [58].

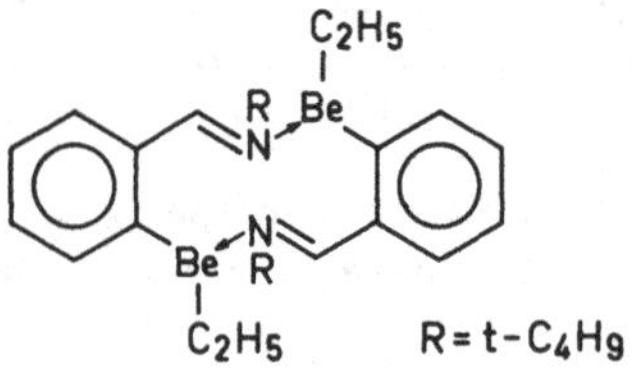

I

Adducts are formed with the following organic N compounds: $N(CH_3)_3$ gives an unstable 2:1 and a stable 1:1 complex [29], $N(C_2H_5)_3$ is only loosely bound to $Be(C_2H_5)_2$ [10]; $N(CH_3)_2N{=}NN(CH_3)_2$ forms a chelated 1:1 adduct [33], as does 2,2'-bipyridine [19, 23]. Pyridine gives $Be(C_2H_5)_2 \cdot 2NC_5H_5$ [23]; see pp. 35/7.

In reactions with $B(C_3H_7)_3$, $B(C_4H_9\text{-i})_3$, or $B(CH_2C_4H_9\text{-t})_3$, ligand exchange occurs, and the corresponding Be dialkyls are obtained in good yields. $B(CH_2Si(CH_3)_3)_3$ gives $(Be(CH_2Si(CH_3)_3)_2)_2$, but with $B(C_4H_9\text{-s})_3$ the insoluble $Be_2(C_2H_5)_3H$ and C_4H_8 are generated [51]. A similar exchange reaction occurs with $B(CH_2CH{=}CH_2)_3$; however, with the exclusion of $B(C_2H_5)_3$, no definite products could be isolated in the absence of stabilizing donors. Oligomeric organoberyllium compounds are formed [56]. $B(C_6H_5)_3$ gives $Be(C_6H_5)_2$ (88% yield) and $B(C_2H_5)_3$. This reaction was extended to other BR_3 compounds, e.g., R = 4-ClC_6H_4, 2-$CH_3C_6H_4$, 3-$CH_3C_6H_4$, 2,4-$(CH_3)_2C_6H_3$, and 1-naphthyl, but no formation of $B(C_2H_5)_3$ was observed with BR_3 (R = 2,4,6-$(CH_3)_3C_6H_2$, c-C_6H_{11}, c-C_5H_9, i-C_3H_7, and C_6F_5) at temperatures up to 110°C. Attempts to prepare ether-free $Be(CH_2C_6H_5)_2$ by reactions of $Be(C_2H_5)_2$ with $B(CH_2C_6H_5)_3$, $Hg(CH_2C_6H_5)_2$, or $(B(CH_2C_6H_5)_2)_2O$ failed [52]. Reaction of $Be(C_2H_5)_2$ with $BeCl_2$ and $Na[B(C_2H_5)_3H]$ in ether gives $Be(C_2H_5)H$ as an etherate, but no details are described [36, 42]. With closo-1,2-$C_2B_9H_{13}$ in ether/C_6H_6 gives 3,1,2-$BeC_2B_9H_{11}$ [63, 64]. If this reaction is performed with $Be(C_2H_5)_2 \cdot 0.33O(C_2H_5)_2$ in C_6H_6 instead of $Be(C_2H_5)_2 \cdot 2O(C_2H_5)_2$, also a polymeric $(3,1,2\text{-}BeC_2B_9H_{11})_n$ is formed [64], see Chapter 2, p. 152.

The products observed by reacting $Be(C_2H_5)_2$ with $Si(CH_3)_3N_3$ at 80°C are $Be(C_2H_5)N_3 \cdot Be(C_2H_5)_2$ and $Si(CH_3)_3C_2H_5$ (Section 4.2, pp. 200/1) [61]. Ligand exchange is also observed with $Be(C_4H_9\text{-}t)_2$ in ether. The primary product $(Be(C_2H_5)C_4H_9\text{-}t)_2$ reacts further with excess $Be(C_2H_5)_2$ on heating to 140°C; a hydride, $Be_2(C_2H_5)_3H$, is formed [55]. $Be(C_2H_5)_2$ reacts with equimolar amounts of $Be(C_5H_5)_2$ to give $Be(C_5H_5)C_2H_5$ [68]. With $Hg(C_6H_5)_2$ in xylene at 90°C, $Be(C_6H_5)_2$ is formed in 78% yield besides $Hg(C_2H_5)_2$ [52].

Uses. $Be(C_2H_5)_2$ has been shown to be a clean and controllable source of Be for p-type doping in the growth of GaAs layers [65, 67] and of $Al_xGa_{1-x}As$ layers [66] by organometallic vapor phase epitaxy.

$Be(C_2H_5)_2/TiCl_3$ are catalysts for the polymerization of C_3H_6. The intermediate formation of $Be(C_2H_5)Cl$ is assumed [20].

References:

[1] Cahours, A. (Ann. Chim. [Paris] [3] **58** [1860] 5/90, 22).
[2] Cahours, A. (Compt. Rend. **76** [1873] 1383/7).
[3] Lavroff, D. (J. Russ. Phys. Chem. Soc. **16** [1884] 93/100).
[4] Krause, E., Wendt, B. (Ber. Deut. Chem. Ges. **56** [1923] 466/7).
[5] Gilman, H. (J. Am. Chem. Soc. **45** [1923] 2693/5).
[6] Gilman, H., Schulze, F. (J. Chem. Soc. **1927** 2663/9).
[7] Gilman, H., Schulze, F. (J. Am. Chem. Soc. **49** [1927] 2904/7).
[8] Perkins, T. R. (J. Chem. Soc. **1929** 1687/91).
[9] Paneth, F., Loleit, H. (J. Chem. Soc. **1935** 366/71).
[10] Goubeau, J., Rodewald, B. (Z. Anorg. Allgem. Chem. **258** [1949] 162/79).

[11] Barbaras, G. D., Dillard, C., Finholt, A. E., Wartik, T., Wilzbach, K. E., Schlesinger, H. I. (J. Am. Chem. Soc. **73** [1951] 4585/90).
[12] Strohmeier, W., Hümpfner, K. (Z. Elektrochem. **60** [1956] 1111/4).
[13] Bähr, G., Thiele, K. H. (Chem. Ber. **90** [1957] 1578/86).
[14] Mazzanti, G., Longi, P. (Rend. Ist. Lomb. Sci. Lettere Cl. Sci. Mat. Nat. [2] **91** [1957] 775).
[15] Strohmeier, W., Hümpfner, K., Miltenberger, K., Seifert, F. (Z. Elektrochem. **63** [1959] 537/40).
[16] Longi, P., Mazzanti, G., Bernardini, F. (Gazz. Chim. Ital. **90** [1960] 180/8).
[17] Strohmeier, W., Gernert, F. (Z. Naturforsch. **16b** [1961] 760).
[18] Strohmeier, W., Gernert, F. (Z. Naturforsch. **17b** [1961] 128).
[19] Coates, G. E., Green, S. I. E. (Proc. Chem. Soc. **1961** 376).
[20] Firsov, A. P., Sandomirskaya, N. D., Tsvetkova, V. I., Chirkov, N. M. (Vysokomol. Soedin. **3** [1961] 1352/7; Polym. Sci. [USSR] **3** [1962] 943/9; C.A. **56** [1962] 10373).

[21] Strohmeier, W., Gernert, F. (Chem. Ber. **95** [1962] 1420/7).
[22] Coates, G. E., Cox, G. F. (Chem. Ind. [London] **1962** 269).
[23] Coates, G. E., Green, S. I. E. (J. Chem. Soc. **1962** 3340/8).
[24] Funk, H., Masthoff, R. (J. Prakt. Chem [4] **22** [1963] 250/4).
[25] Strohmeier, W., Seifert, F. (Z. Elektrochem. **63** [1959] 683/8).
[26] Fetter, N. R. (Can. J. Chem. **42** [1964] 861/6).
[27] Masthoff, R., Vieroth, C. (Z. Chem. [Leipzig] **5** [1965] 142).
[28] Strohmeier, W., Gernert, F. (Z. Naturforsch. **20b** [1965] 829/31).
[29] Peters, F. M. (J. Organometal. Chem. **3** [1965] 334/6).
[30] Fetter, N. R., Peters, F. M (Can. J. Chem. **43** [1965] 1884/6).

[31] Bell, N. A., Coates, G. E. (J. Chem. Soc. **1965** 692/9).
[32] Bell, N. A. (J. Chem. Soc. A **1966** 542/4).
[33] Fetter, N. R. (J. Chem. Soc. A **1966** 711/3).

[34] Bell, N. A., Coates, G. E. (J. Chem. Soc. A **1966** 1069/73).
[35] Strohmeier, W., Haecker, W., Popp, G. (Chem. Ber. **100** [1967] 405/11).
[36] Coates, G. E., Tranah, M. (J. Chem. Soc. A **1967** 615/7).
[37] Coates, G. E., Fishwick, A. H. (J. Chem. Soc. A **1967** 1199/204).
[38] Bell, N. A. (J. Organometal. Chem. **13** [1968] 513/5).
[39] Ashby, E. C., Arnott, R. C. (J. Organometal. Chem. **14** [1968] 1/11).
[40] Coates, G. E., Fishwick, A. H. (J. Chem. Soc. A **1968** 477/83).

[41] Coates, G. E., Fishwick, A. H. (J. Chem. Soc. A **1968** 635/40).
[42] Bell, N. A., Coates, G. E. (J. Chem. Soc. A **1968** 628/31).
[43] Coates, G. E., Roberts, P. D. (J. Chem. Soc. A **1968** 2651/5).
[44] Sanders, J. R., Ashby E. C., Carter J. H. (J. Am. Chem. Soc. **90** [1968] 6385/90).
[45] Fetter, N. R. (Organometal. Chem. Rev. A **3** [1968] 1/34).
[46] Chambers, D. B., Coates, G. E., Glockling, F. (Discussions Faraday Soc. No. 47 [1969] 157/64).
[47] Shepherd, L. H., Ter Haar, G. L., Marlett, E. M. (Inorg. Chem. **8** [1969] 976/9).
[48] Coates, G. E., Pendlebury, R. E. (J. Chem. Soc. A **1970** 156/60).
[49] Chambers, D. B., Coates, G. E., Glockling, F. (J. Chem. Soc. A **1970** 741/8).
[50] Coates, G. E., Francis, B. R. (J. Chem. Soc. A **1971** 1305/8).

[51] Coates, G. E., Francis, B. R. (J. Chem. Soc. A **1971** 1308/10).
[52] Coates, G. E., Srivastava, R. C. (J. Chem. Soc. Dalton Trans. **1972** 1541/4).
[53] Blindheim, U., Coates, G. E., Srivastava, R. C. (J. Chem. Soc. Dalton Trans. **1972** 2302/5).
[54] Atam, N., Müller, H., Dehnicke, K. (J. Organometal. Chem. **37** [1972] 15/23).
[55] Coates, G. E., Smith, D. L., Srivastava, R. C. (J. Chem. Soc. Dalton Trans. **1973** 618/22).
[56] Wiegand. G., Thiele, K.-H. (Z. Anorg. Allgem. Chem. **405** [1974] 101/8).
[57] Dehnicke, K., Atam, N. (Chimia [Aarau] **28** [1974] 663/4).
[58] Andersen, R. A., Coates, G. E. (J. Chem. Soc. Dalton Trans. **1974** 1171/80).
[59] Andersen, R. A., Coates, G. E. (J. Chem. Soc. Dalton Trans. **1974** 1729/36).
[60] Andersen, R. A., Coates, G. E. (J. Chem. Soc. Dalton Trans. **1975** 1244/5).

[61] Atam, N., Dehnicke, K. (Z. Anorg. Allgem. Chem. **427** [1976] 193/9).
[62] Strohmeier, W., Popp, G. (Z. Naturforsch. **23b** [1968] 38/41).
[63] Popp, G., Hawthorne, M. F. (J. Am. Chem. Soc. **90** [1968] 6553/4).
[64] Popp, G., Hawthorne, M. F. (Inorg. Chem. **10** [1971] 331/3).
[65] Parsons, J. D., Krajenbrink, F. G. (J. Electrochem. Soc. **130** [1983] 1782/3).
[66] Bottka, N., Sillman, R. S., Tseng, W. F. (J. Cryst. Growth **68** [1984] 54/9).
[67] Parsons, J. D., Lichtmann, L. S., Krajenbrink, F. G. (J. Cryst. Growth **77** [1986] 32/6).
[68] Bartke, T. C. (Diss. Univ. Wyoming 1975; Diss. Abstr. Intern. B **36** [1976] 6141).

1.2.1.4 Adducts of $Be(C_2H_5)_2$

$Be(C_2H_5)_2 \cdot O(C_2H_5)_2$

No etherate of $Be(C_2H_5)_2$ of a definite composition was isolated, but it is generally accepted that solutions of $Be(C_2H_5)_2$ mainly contain the 1:1 adduct. The adduct is not very stable, as the ether can be removed by prolonged heating at 50°C/10^{-3} Torr. After 24 h, no ether remains [11]. See the parent compound for reactions of $Be(C_2H_5)_2$ in ether, pp. 28/33.

$Be(C_2H_5)_2 \cdot O_2C_4H_8$ ($O_2C_4H_8$=1,4-dioxane)

The dioxane adduct of $Be(C_2H_5)_2$ has not been isolated, but molecular mass and dipole moment (μ=4.3 D) determinations show the presence of a 1:1 adduct with a highly polar structure. A chelating interaction appears to be most reasonable (Formula I) [2, 3].

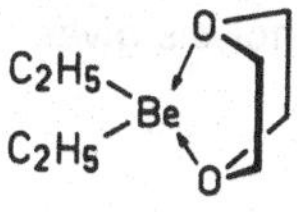

I

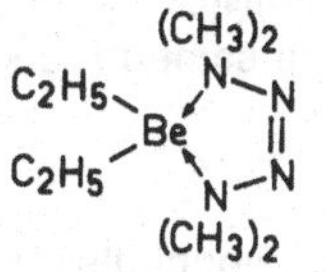

II

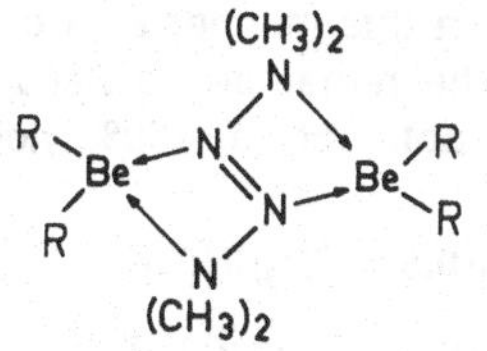

III

$Be(C_2H_5)_2 \cdot NH_3$

A colorless solid is formed if NH_3 is condensed onto an ether solution of $Be(C_2H_5)_2$ at −80°C followed by addition of cold C_6H_{14}. No C_2H_6 is formed under these conditions. Gas is evolved only above −60 to −50°C, leading eventually above 0 to 10°C to the precipitation of a polymer, presumably $(Be(C_2H_5)NH_2)_n$ [8].

$Be(C_2H_5)_2 \cdot N(CH_3)_3$

The 1:1 adduct is prepared from equimolar quantities of the components on a vacuum line [5, 7]. The 1:2 adduct decomposes at room temperature in a vacuum to the 1:1 adduct [7]. Reaction of $N(CH_3)_3$ with $Be_2(C_2H_5)_3H$ also presumably gives the adduct [13].

It is a colorless liquid, b.p. 65 to 66°C/10^{-3} Torr. It is monomeric in c-C_6H_{12} (by cryoscopy). The ^1H NMR spectrum has been registered, but the values are not given [7].

The adduct combines with an excess of $N(CH_3)_3$ at −78°C to form the 1:2 complex, see below [7]. With $NH(CH_3)N(CH_3)_2$ in C_8H_{18} at 25°C, ~50% $Be(C_2H_5)_2 \cdot NH(CH_3)N(CH_3)_2$ and ~50% polymeric $Be(C_2H_5)N(CH_3)N(CH_3)_2$ are obtained. $NH(CH_3)NH_2$ affords polymeric $Be(C_2H_5)N(CH_3)NH_2$ under analogous conditions [6]. See the 1:2 adduct below for the equilibrium between the 1:1 and 1:2 adduct.

$Be(C_2H_5)_2 \cdot 2N(CH_3)_3$

$Be(C_2H_5)_2$ or its 1:1 adduct with $N(CH_3)_3$ combine with an excess of $N(CH_3)_3$ at −78°C to form a 1:2 complex. After removal of the uncomplexed amine in vacuum at −65°C, the 1:2 ratio can be confirmed by tensimetric measurements.

The vapor pressure of the residue is 1.1, 6.8, 11, and 20.5 Torr at −33, −22.6, −15.5, and −7.6°C, respectively. At room temperature one of the amines can be removed in vacuum, leaving the more stable 1:1 adduct described above. Solutions of $Be(C_2H_5)_2$ and $N(CH_3)_3$ in c-C_5H_{10} in the mole ratio 1:2 show only one ^1H NMR signal for CH_3N without splitting from −90°C to room temperature. This result is explained by the equilibrium $Be(C_2H_5)_2 \cdot N(CH_3)_3 + N(CH_3)_3 \rightleftharpoons Be(C_2H_5)_2 \cdot 2N(CH_3)_3$. The spectral data are not given [7].

$Be(C_2H_5)_2 \cdot nN(C_2H_5)_3$

There is no discernible reaction between the components at low temperatures and on warming, but attempts to remove the amine from the reaction mixture meet with difficulties. Complex formation is likely to take place, but no further information is available [1].

References on p. 37

$Be(C_2H_5)_2 \cdot NH(CH_3)N(CH_3)_2$

For the preparation, $NH(CH_3)N(CH_3)_2$ is transferred to a solution of $Be(C_2H_5)_2$ in C_8H_{18} at −196°C. The mixture is warmed to room temperature and allowed to stand for another 2 h. Some C_2H_6 is evolved. After removal of the solvent in vacuum, the crude product is vacuum-transferred at 25°C/0.05 Torr onto a −196°C cold surface. Approximately 15% of the product is collected in this manner as a clear mobile liquid and analysis is in agreement with the 1:1 adduct. The remainder is $Be(C_2H_5)N(CH_3)N(CH_3)_2$. Starting with $Be(C_2H_5)_2 \cdot N(CH_3)_3$, the 1:1 adduct is obtained in ~50% yield by the same procedure. Methanolysis gives C_2H_6 [6].

$Be(C_2H_5)_2 \cdot 0.5\,N(CH_3)_2N(CH_3)_2$

This adduct is synthesized from the components (mole ratio 1:1) in C_8H_{18} on warming from −196°C to room temperature and then standing overnight. A white crystalline solid remains after removal of the solvent, m.p. 84 to 87°C. In C_6H_6 a degree of association of 1.24 was found for the 2:1 adduct (by cryoscopy). The ^1H NMR spectrum in C_6H_6 shows $\delta = 1.89$ ppm (s, CH_3N) and a triplet/quartet pattern for the C_2H_5 groups. Methanolysis gives C_2H_6 [6].

$Be(C_2H_5)_2 \cdot N(CH_3)_2CH_2CH_2N(CH_3)_2$

This adduct is mentioned to react with $C_6H_5C{\equiv}CH$ at 80°C for 24 days in a sealed tube to give $Be(C{\equiv}CC_6H_5)_2 \cdot N(CH_3)_2CH_2CH_2N(CH_3)_2$ in a 69% yield [12].

$Be(C_2H_5)_2 \cdot N(CH_3)_2N{=}NN(CH_3)_2$

The adduct is prepared from the components in C_7H_{16} at 25°C. The product is vacuum-sublimed at 25°C/10^{-2} Torr.

The melting point is 34 to 38°C. The ^1H NMR spectrum in c-C_5H_{10} shows resonances at $\delta = 4.59$ (s, CH_3N), 5.60 (t, CH_3C), and 7.08 ppm (q, CH_2; J(H,H) = 8 Hz) relative to C_6H_6. The low temperature spectrum is largely unchanged at −75°C. A symmetrical chelate structure (Formula II, p. 35) is proposed.

The adduct dissolves in C_6H_6 as a monomer. Vacuum pyrolysis at 70 to 90°C leads to the 2:1 complex and $N(CH_3)_2N{=}NN(CH_3)_2$ [9].

$Be(C_2H_5)_2 \cdot 0.5\,N(CH_3)_2N{=}NN(CH_3)_2$

It is prepared by vacuum pyrolysis of the 1:1 complex. The colorless viscous material melts at ~20°C and can be distilled at 110 to 120°C in a vacuum. The ^1H NMR spectrum has the following resonances in c-C_5H_{10}: $\delta = 4.35$ (s, CH_3N), 5.68 (t, CH_3C), 7.00 ppm (q, CH_2; J(H,H) = 8 Hz) relative to C_6H_6. A symmetrical structure (Formula III, p. 35, $R = C_2H_5$) is suggested.

The adduct is slightly associated in C_6H_6 (by cryoscopy). Pyrolysis at 200 or 220°C leads to a series of low molecular weight polymers of the approximate compositions $C_{16}H_{16}BeN_4$ to $C_{18}H_{21}BeN_2$; C_2H_6, C_2H_4, and H_2 are evolved [9].

$Be(C_2H_5)_2 \cdot 2\,NC_5H_5$

The pyridine adduct is obtained from the components in ether. It forms orange-yellow needles, which darken gradually from 90 to 100°C, and decompose rapidly above 140°C without melting [5].

The adduct is a monomer in C_6H_6 (by cryoscopy) [5]. Electrolysis in pyridine yields the radical $BeC_2H_5 \cdot NC_5H_5$, which can be isolated as a highly viscous black material [10].

$Be(C_2H_5)_2 \cdot N_2C_{10}H_8$ ($N_2C_{10}H_8$ = 2,2'-bipyridine)

The adduct precipitates from the reaction mixture of the components in ether as red needles. The decomposition temperature is 90°C. Sublimation is carried out at 100°C/15 × 10^{-4} Torr with only slight decomposition [5]. The UV absorption maximum is at λ = 461 nm (ε = 3700 $L \cdot mol^{-1} \cdot cm^{-1}$) [4, 5].

It dissolves in C_6H_6 as a monomer (by cryoscopy) but is only moderately soluble in ether. It becomes white in air, but does not appear to be affected by dry air. There is no reaction with $(C_6H_5)_2CO$ in boiling ether [5].

References:

[1] Goubeau, J., Rodewald, B. (Z. Anorg. Allgem. Chem. **258** [1949] 162/79).
[2] Strohmeier, W., Hümpfner, K. (Z. Elektrochem. **60** [1956] 1111/4).
[3] Strohmeier, W., Hümpfner, K., Miltenberger, K., Seifert, F. (Z. Elektrochem. **63** [1959] 537/40).
[4] Coates, G. E., Green, S. I. E. (Proc. Chem. Soc. **1961** 376).
[5] Coates, G. E., Green, S. I. E. (J. Chem. Soc. **1962** 3340/8).
[6] Fetter, N. R. (Can. J. Chem. **42** [1964] 861/6).
[7] Peters, F. M. (J. Organometal. Chem. **3** [1965] 334/6).
[8] Masthoff, R., Vieroth, C. (Z. Chem. [Leipzig] **5** [1965] 142).
[9] Fetter, N. R. (J. Chem. Soc. A **1966** 711/3).
[10] Strohmeier, W., Popp, G. (Z. Naturforsch. **22b** [1967] 891).

[11] Coates, G. E., Roberts, P. D. (J. Chem. Soc. A **1968** 2651/5).
[12] Coates, G. E., Francis, B. R. (J. Chem. Soc. A **1971** 160/4).
[13] Coates, G. E., Francis, B. R. (J. Chem. Soc. A **1971** 1308/10).

1.2.1.5 Dipropylberyllium Compounds and Their Adducts

The first preparation of dipropylberyllium was achieved by the reaction of Be metal with $Hg(C_3H_7)_2$ for several hours at 130 to 135°C. It was isolated by repeated distillation at 240 to 260°C.

The compound is obtained as a colorless liquid, b.p. 244 to 246°C. It is viscous at −17°C, but not yet solidified.

It fumes in air and is vigorously decomposed by H_2O. No information is available as to whether the prepared compound was the n- or iso-isomer [12].

$Be(C_3H_7)_2$

The compound is prepared from anhydrous $BeCl_2$ and $Mg(C_3H_7)Br$ in ether at reflux temperature. Decantation and distillation at 85°C/0.1 Torr gives the adduct $Be(C_3H_7)_2 \cdot 2O(C_2H_5)_2$. The ether in the adduct is removed by prolonged heating for ~24 h at 60°C (oil bath)/10^{-3} Torr (condenser −30°C) and subsequent distillation [5]. The compound is also formed by reaction of $Be(C_2H_5)_2$ with $B(C_3H_7)_3$ (mole ratio ~1:1) at room temperature for 8 days in a sealed tube. $B(C_2H_5)_3$ is collected in a trap, and $Be(C_3H_7)_2$ is obtained by distillation at 50°C/~10^{-3} Torr [10].

$Be(C_3H_7)_2$ is a colorless liquid, b.p. 40°C/10^{-3} Torr. It is a dimer in C_6H_6 solution as shown by cryoscopy [5]. The ionization potential of $Be(C_3H_7)_2$ is 8.71 ± 0.06 eV, determined from the mass spectra [6, 8].

References on p. 40

Electron impact of gaseous $Be(C_3H_7)_2$ produces ions which could all be identified by their mass spectra. Ion abundances clearly show that only dimers and monomers are present in the gas phase. The dinuclear ions $[Be_2C_4H_{11}]^+$, $[Be_2(C_3H_7)_2H]^+$, and $[Be_2(C_3H_7)H_2]^+$ were observed at 70 eV and 35 to 45°C and 210 to 215°C source temperature; $[BeC_3H_7]^+$ and $[BeC_3H_6]^+$ are the most abundant Be-containing ions in the spectra taken at 70 eV, 35 to 45°C, and 210 to 215°C, and at 12 eV, 210 to 215°C. The appearance potentials for these ions are 10.81 ± 0.05, and 9.86 ± 0.05 eV, respectively, and the BeC bond dissociation energy is 42.7 ± 3.2 kcal/mol [6, 8].

The compound adds one equivalent of NaH from an ethereal suspension of NaH to form $Na_2[Be(C_3H_7)_2H]_2$ [7]. Adducts (mole ratio Be : donor) are formed with ether (1:2), pyridine (1:2), and 2,2′-bipyridine (1:1). The adducts are described below [5].

$Be(C_3H_7)_2 \cdot 2O(C_2H_5)_2$

The adduct is obtained from an ethereal solution of $Be(C_3H_7)_2$ after vacuum distillation at 85°C (bath temperature)/0.1 Torr.

It is largely dissociated in C_6H_6 (by cryoscopy). The ether can be removed on prolonged heating under reflux at 60°C/10^{-3} Torr. Addition of pyridine gives the adduct described below. With $NH(CH_3)CH_2CH_2N(CH_3)_2$ (mole ratio 1:1) in C_6H_{14}, $(Be(C_3H_7)N(CH_3)CH_2CH_2N(CH_3)_2)_2$ is obtained [5].

$Be(C_3H_7)_2 \cdot 2NC_5H_5$

The adduct is obtained by addition of pyridine to the diether complex as a yellow crystalline solid melting at ca. 10°C. It reacts with 2,2′-bipyridine to yield the following complex [5].

$Be(C_3H_7)_2 \cdot N_2C_8H_6$ ($C_8H_6N_2$ = 2,2′-bipyridine)

The preceding adduct reacts in a 1:1 mole ratio with 2,2′-bipyridine in ether. After removal of the solvent the product crystallizes from C_5H_{12} as blood-red crystals, which decompose >90°C; 2,2′-bipyridine acts as a chelating ligand. The adduct is a monomer in C_6H_6 (by cryoscopy) [5].

$Be(C_3H_7\text{-i})_2$

The compound is prepared from anhydrous $BeCl_2$ and $Mg(C_3H_7\text{-i})Br$ in ether at room temperature. An ether-containing product can be removed in vacuum from the reaction mixture. Redistillation in vacuum still yields only an etherate, but boiling in vacuum in a special apparatus for 16 h gives a pure material [1].

$Be(C_3H_7\text{-i})_2$ is a mobile liquid, m.p. −9.5°C, b.p. 280°C (extrapolated), 40°C/0.53 Torr, 20°C/0.17 Torr. The vapor pressure curve can be described by the equation $\log p = 7.01 - 2280\ T^{-1}$ (p in Torr, T in K) for the temperature range 35 to 60°C. The heat of vaporization is 10.48 kcal/mol, and Trouton's constant is $\Delta H_v/T_v = 19.1\ cal \cdot mol^{-1} \cdot K^{-1}$ [1].

The ^1H NMR spectrum in $C_6D_5CD_3$ shows, at 60 to 80°C, a septet due to CH, but the resolution becomes poor as the temperature is reduced. Fine structure is absent at 33°C and does not reappear as the temperature is reduced further. The CH_3 resonances also change with temperature. Above 33°C these resonances exist of the usual sharp doublet, but considerable broadening is apparent at 20°C, and very little detail is present at −10°C. However, at −30°C at least five peaks are evident for CH_3, the appearance being similar at −45°C: δ = 0.81, 0.90, 1.00, 1.10, and 1.23 ppm. The other values are not reported. The behavior may be due to a distinction between protons contained in bridging groups and those contained in terminal groups being lost at higher temperatures owing to exchange [5].

The ionization potential is determined as 8.80 ± 0.02 eV from mass spectroscopic data [6, 8].

According to vapor pressure determinations of C_6H_6 solutions, the compound dissolves as a dimer in this solvent [1]. Electron impact of gaseous $Be(C_3H_7\text{-}i)_2$ produced ions which could all be identified by their mass spectra. At 70 eV and 55 to 75°C source temperature, the dimeric ions found are $[Be_2(C_3H_7)_4]^+$, $[Be_2(C_3H_7)_3]^+$, $[Be_2(C_3H_7)_2H]^+$, $[Be_2C_5H_{11}]^+$, $[Be_2C_4H_{11}]^+$, and $[Be_2(C_3H_7)H_2]^+$. The most abundant Be-containing monomers are $[BeC_3H_7]^+$ and $[BeC_3H_6]^+$ at 70 eV, 55 to 75°C and 215 to 220°C, and at 12 eV and 215 to 220°C source temperatures. Appearance potentials for the latter ions are 10.65 ± 0.01 and 9.60 ± 0.01 eV. The BeC bond dissociation energy is 48.7 ± 3.2 kcal/mol [6, 8].

On heating to 200°C for 8 h in a sealed bulb, C_3H_6 and $Be(C_3H_7\text{-}i)H$ are formed [1]. The compound fumes strongly on exposure to air, but does not inflame. The reaction with H_2O is of explosive violence unless controlled by cooling. Boiling a suspension of NaH and $Be(C_3H_7\text{-}i)_2$ in ether for 2 days yields $Na_2[Be(C_3H_7\text{-}i)_2H]_2$, which probably contains a hydride-bridged dimeric anion [7]. In the reaction with CH_3OH, C_3H_8 is evolved, and $Be(C_3H_7\text{-}i)OCH_3$ can be isolated in high yield [1, 11]. $Be(C_3H_7\text{-}i)_2$ reacts with C_2H_5SH in ether in the mole ratio 1:1 to give the adduct $(Be(C_3H_7\text{-}i)SC_2H_5 \cdot O(C_2H_5)_2)_2$, in which ether can be displaced by pyridine. The reaction of $i\text{-}C_3H_7SH$ with $Be(C_3H_7\text{-}i)_2$ in ether yields a viscous liquid, which on addition of pyridine gives the adduct $Be(C_3H_7\text{-}i)SC_3H_7\text{-}i \cdot 2NC_5H_5$. With $t\text{-}C_4H_9SH$ in ether in the mole ratio 1:1, the disproportionation products $Be(SC_4H_9\text{-}t)_2$ and $Be(C_3H_7\text{-}i)_2$ are obtained [4].

The reaction with $NH(CH_3)_2$ at room temperature yields C_3H_8, and $(Be(C_3H_7\text{-}i)N(CH_3)_2)_3$ is formed as the primary product [1, 3]. On heating the mixture to 100 to 220°C, C_3H_6 is evolved leaving $Be(H)N(CH_3)_2$ and some $Be(N(CH_3)_2)_2$. The latter compound is the main product if an excess of $NH(CH_3)_2$ is employed [1]. $(t\text{-}C_4H_9)_2C{=}NH$ reacts with $Be(C_3H_7\text{-}i)_2$ in ether in the mole ratio 2:1 to afford $(Be(N{=}C(C_4H_9\text{-}t)_2)_2)_2$ and C_3H_8 [9]. $NH(CH_3)CH_2CH_2N(CH_3)_2$ yields the chelate complex $(Be(C_3H_7\text{-}i)N(CH_3)CH_2CH_2N(CH_3)_2)_2$, and C_3H_8 is evolved [5]. Adducts are formed (mole ratio Be: donor) with $N(CH_3)_3$ (1:1) [1]; pyridine (1:2) [4]; $N(CH_3)_2CH_2CH_2N(CH_3)_2$ (1:1) [5]; and $N(CH_3)_2N{=}NN(CH_3)_2$ (1:1) [2]; see below. The stoichiometry of the adduct with ether is not known [1].

$Be(C_3H_7\text{-}i)_2 \cdot nO(C_2H_5)_2$

The adduct, whose stoichiometry is not known, is obtained during the synthesis of $Be(C_3H_7\text{-}i)_2$ (see p. 38). The ether is removed from the clear liquid by prolonged heating at 40 to 50°C in a vacuum [1].

$Be(C_3H_7\text{-}i)_2 \cdot N(CH_3)_3$

The adduct is obtained from equimolar quantities of the components in a vacuum system as a clear colorless liquid.

From the vapor pressure curve, $\log p = 7.95 - 2535\,T^{-1}$ (p in Torr, T in K) in the range 20 to 100°C is derived; the heat of vaporization is calculated as 11.6 kcal/mol. The extrapolated boiling point is 227°C at 760 Torr, and Trouton's constant is $\Delta H_v/T_v = 23.2$ cal·mol^{-1}·K^{-1}.

The adduct is a monomer in C_6H_6 (from vapor pressure). At 100°C thermal decomposition starts and is rapid at 200°C; C_3H_6 and $N(CH_3)_3$ are evolved. The residue consists of $(Be(C_3H_7\text{-}i)H)_n$ [1].

$Be(C_3H_7\text{-}i)_2 \cdot N(CH_3)_2CH_2CH_2N(CH_3)_2$

The components react in ether in the mole ratio 1:1. The residue left after evaporation of the solvent crystallizes from C_6H_{14} as colorless prisms, m.p. 76 to 78°C.

References on p. 40

The ^1H NMR spectrum in C_6H_6 shows resonances at $\delta = 1.50$ (d, CH_3C), 1.75 (s, CH_2), and 1.98 ppm (s, CH_3N). The spectrum is unchanged in $C_6D_5CD_3$ at −80°C. The diamine acts as a chelating ligand. The compound is soluble in C_6H_6 as a monomer (by cryoscopy) [5].

$Be(C_3H_7\text{-}i)_2 \cdot N(CH_3)_2N{=}NN(CH_3)_2$

This adduct with a chelating tetrazene is prepared like the previous adduct in C_7H_{16} and purified by vacuum sublimation at 25°C, m.p. 25 to 30°C.

The yellow compound decomposes on heating in a vacuum to 70 to 90°C to yield the following adduct and free $N(CH_3)_2N{=}NN(CH_3)_2$ [2].

$Be(C_3H_7\text{-}i)_2 \cdot 0.5\,N(CH_3)_2N{=}NN(CH_3)_2$

The 2:1 adduct is a residue of the vacuum pyrolysis of the 1:1 complex at 70 to 90°C. The compound is a colorless viscous liquid that crystallizes at 0 to 5°C. The structure given in Formula III, p. 35 with $R = i\text{-}C_3H_7$ is suggested.

The 2:1 adduct is monomeric in C_6H_6 (by cryoscopy). Thermal decomposition at 190°C gives largely C_3H_8, C_3H_6, N_2, H_2, and low molecular polymers depending on the experimental conditions [2].

$Be(C_3H_7\text{-}i)_2 \cdot 2\,NC_5H_5$

The adduct forms by addition of excess pyridine to $Be(C_3H_7\text{-}i)_2$. The golden-yellow needles (from C_6H_6/C_6H_{14} [1:1]) melt at 111 to 112°C. The compound is monomeric in C_6H_6 (by cryoscopy) [4].

References:

[1] Coates, G. E., Glockling, F. (J. Chem. Soc. **1954** 22/7).
[2] Fetter, N. R. (J. Chem. Soc. A **1966** 711/3).
[3] Coates, G. E., Fishwick, A. H. (J. Chem. Soc. A **1967** 1199/204).
[4] Coates, G. E., Fishwick, A. H. (J. Chem. Soc. A **1968** 635/40).
[5] Coates, G. E., Roberts, P. D. (J. Chem. Soc. A **1968** 2651/5).
[6] Chambers, D. B., Coates, G. E., Glockling, F. (Discussions Faraday Soc. No. 47 [1969] 157/64).
[7] Coates, G. E., Pendlebury, R. E. (J. Chem. Soc. A **1970** 156/60).
[8] Chambers, D. B., Coates, G. E., Glockling, F. (J. Chem. Soc. A **1970** 741/8).
[9] Farmer, J. B., Shearer, H. M. M., Sowerby, J. D., Wade, K. (J. Chem. Soc. Chem. Commun. **1976** 160/1).
[10] Coates, G. E., Francis, B. R. (J. Chem. Soc. A **1971** 1308/10).

[11] Coates, G. E., Fishwick, A. H. (J. Chem. Soc. A **1968** 477/83).
[12] Cahours, A. (Compt. Rend. **76** [1873] 1383/90).

1.2.1.6 Dibutylberyllium Compounds and Their Adducts

$Be(C_4H_9)_2$

The compound is best prepared from anhydrous $BeCl_2$ with an ether-free Grignard solution of $Mg(C_4H_9)I$ in C_6H_6 in 70% yield (41% after distillation) [8]. If the reaction is carried out in ether, the resulting etherate has to be refluxed in vacuum for ~24 h and subsequently vacuum-distilled to obtain an ether-free product [2, 12]. It is also prepared by the reaction of LiC_4H_9 with anhydrous $BeCl_2$ (mole ratio 2:1) in C_6H_6. After stirring overnight the solution is filtered, and the solvent is evaporated [35]. The compound is also formed in a reaction of Be metal with $Hg(C_4H_9)_2$ in the presence of an $HgCl_2$ catalyst, but no details are given [1 to 3]. No reaction occurs without a catalyst [2].

For the colorless viscous liquid [2] boiling points are given as ~110°C/0.008 Torr [8], 95°C/10^{-3} Torr [12], 105°C/10^{-3} Torr [35], and ~170°C/25 Torr [2].

The ^1H NMR spectrum in C_6H_6 shows a normal pattern for $CH_3CH_2CH_2$ (not given) and a multiplet for the α-CH_2 at $\delta = 0.38$ ppm consisting of at least eight lines at room temperature. Addition of an excess of ether transforms the multiplet into a triplet. The spectrum in C_6H_6 is assumed to arise from nonequivalent α-CH_2 groups in a dimer, which is present in C_6H_6, as determined by cryoscopy [8].

Pyrolysis at 200°C yields impure BeH_2 [32]. The compound is not spontaneously inflammable when exposed to air, but it oxidizes rapidly with the evolution of heat. The mercaptan-like odor thereby changes to that of C_4H_9OH. A positive test with Michler's ketone is observed. $C_6H_5N{=}CH_2$ is reduced to $C_6H_5NHC_5H_{11}$ after hydrolysis [2]. It does not react with NaH in ether at 20°C [34]. $Be(C_5H_5)C_4H_9$ is formed in the reaction with $Be(C_5H_5)_2$ [35].

$Be(C_4H_9\text{-}i)_2$

The compound is prepared via the Grignard method from $BeCl_2$ and $Mg(C_4H_9\text{-}i)Cl$ in ether. Decantation, removal of ether, and distillation at 70°C (oil bath)/10^{-3} Torr gives an ether adduct, $Be(C_4H_9\text{-}i)_2 \cdot 0.5O(C_2H_5)_2$. On heating under vacuum at 50°C/10^{-3} Torr (condenser −20°C) for 24 h and subsequent distillation, the ether-free product is obtained [12]. Another method of synthesis is the reaction of $Be(C_2H_5)_2$ and $B(C_4H_9\text{-}i)_3$ at 20°C. After 12 days, $Be(C_4H_9\text{-}i)_2$ and $B(C_2H_5)_3$ can be separated by distillation at 50 to 60°C/~10^{-2} Torr [20]. $Be(C_4H_9\text{-}i)_2$ forms by slow isomerization of $Be(C_4H_9\text{-}t)_2$ at room temperature [15].

The colorless liquid boils at 45°C/10^{-3} Torr. The ^1H NMR spectrum in $C_6D_5CD_3$ shows a broadening of all resonances as the temperature is lowered from 100 to −80°C, but no details are given [12]. An ionization potential of 8.74 ± 0.05 eV was determined in a mass spectroscopic study [14, 17].

The compound dissolves as a dimer in C_6H_6 (by cryoscopy) [12]. Pyrolysis at 85°C gives $Be(C_4H_9\text{-}i)H$ and $(CH_3)_2C{=}CH_2$ [15]. Heating of the compound in $C_{12}H_{26}$ at 195°C gives BeH_2 [32]. The same products are observed with the etherate [15, 32]. Mass spectra of gaseous $Be(C_4H_9\text{-}i)_2$ were taken at 70 eV, 75 to 90°C and 235 to 240°C source temperatures and at 12 eV and 240°C. All spectra show the molecular ion; the most abundant Be-containing species are $[BeC_4H_9]^+$ and $[BeC_4H_8]^+$ for which appearance potentials of 10.00 ± 0.05 and 9.41 ± 0.03 eV were determind. A BeC bond dissociation energy of 29.1 ± 3.2 kcal/mol was calculated. There is no evidence for a high concentration of dimers in the gaseous state. This is due to rapid thermal decomposition with formation of C_nH_m.

The compound reacts with LiH and NaH in boiling ether to form $M[Be(C_4H_9\text{-}i)_2H]$. The anion is presumably a hydride-bridged dimer. Excess $Be(C_4H_9\text{-}i)_2$ reacts further with the Na salt in a reversible reaction to give $Na_2[Be_3(C_4H_9\text{-}i)_6H_2]$ [18]; $t\text{-}C_4H_9CHO$ and the ether adduct of $Be(C_4H_9\text{-}i)_2$ react in C_6H_{14} in a 1:1 ratio to give $Be(C_4H_9\text{-}i)OCH_2C_4H_9\text{-}t$ and $(CH_3)_2C{=}CH_2$. Additional aldehyde gives $Be(OCH_2C_4H_9\text{-}t)_2$. Similarly, $(t\text{-}C_4H_9)_2CO$ gives $(Be(C_4H_9\text{-}i)\text{-}OCH(C_4H_9\text{-}t)_2)_2$ and $(CH_3)_2C{=}CH_2$, but excess ketone gives no further reaction [28]. C_6H_5CN is reduced by $Be(C_4H_9\text{-}i)_2$ with evolution of $(CH_3)_2C{=}CH_2$. The products formed after acid hydrolysis are C_6H_5CHO and $C_6H_5(i\text{-}C_4H_9)CO$. The aldehyde dominates at a reaction temperature of 67°C, whereas the ketone is mainly formed at 0, 30, and 47°C. $Be(C_4H_9\text{-}i)N{=}CHC_6H_5$ is assumed

References on p. 57

to be the product before hydrolysis [22]. Pyridine forms a 2:1 complex in ether [18]. A rapid ligand exchange with the optically active (+)(S)-$Zn(CH_2CH(CH_3)C_2H_5)_2$ to give (+)(R)-$Be(CH_2CH(CH_3)C_2H_5)_2$ is observed at room temperature [23].

$Be(C_4H_9\text{-}i)_2 \cdot 0.5\,O(C_2H_5)_2$

The semietherate is obtained by distillation of ethereal solutions of $Be(C_4H_9\text{-}i)_2$ at 70°C (bath temperature)/10^{-3} Torr. The ether is completely removed by boiling under reflux at 50°C (oil bath)/10^{-3} Torr (condenser −20°C) for 24 h [12]. See the parent compound, pp. 41/42, for further reactions.

$Be(C_4H_9\text{-}i)_2 \cdot 2\,NC_5H_5$ (NC_5H_5 = pyridine)

The adduct is formed from the components in the mole ratio 1:2 in ether. The pale yellow irregular hexagonal plates obtained by recrystallization from C_6H_{14} melt at 65°C [18].

$Be(C_4H_9\text{-}s)_2$ and **$Be(C_4H_9\text{-}s)_2 \cdot n\,O(C_2H_5)_2$**

The compound is mentioned to be dimeric in C_6H_6 in [12] with reference to an earlier paper [6]. But the quotation is presumably erroneous, since the compound is not mentioned in [6]. An attempted synthesis from $Be(C_2H_5)_2$ and $B(C_4H_9\text{-}s)_3$ failed [20]. An ether complex of unknown constitution is obtained from the reaction of LiC_4H_9-s with $BeCl_2$ (mole ratio 1:1) in boiling ether. After 3 h reflux and standing overnight, the solution is decanted, concentrated, and distilled at 10^{-3} Torr (bath temperature 70 to 75°C) [24].

An attempt to remove the ether by refluxing at 10^{-3} Torr for 24 h results in loss of C_4H_8, and $Be(C_4H_9\text{-}s)H$ as an etherate is formed [24].

$Be(C_4H_9\text{-}t)_2$

Preparation

The compound is prepared from anhydrous $BeCl_2$ and $Mg(C_4H_9\text{-}t)Cl$ in ether at room temperature. The ether solvent is removed by pumping to leave an etherate. Ether-free $Be(C_4H_9\text{-}t)_2$ cannot be obtained by distillation without decomposition [4, 5]. The etherate is treated with anhydrous $BeCl_2$, and free $Be(C_4H_9\text{-}t)_2$ can be distilled from this mixture into a cold trap [5, 10]. The yield is around 40% based on the etherate [5]. The compound is also prepared by stirring overnight a mixture of LiC_4H_9-t in excess and anhydrous $BeCl_2$ in C_5H_{12}. After filtration the pentane is removed in a vacuum, the colorless viscous residue is distilled at 25 to 30°C/10^{-2} Torr, and the second cut is retained [35].

The deuterated compound is prepared from t-C_4D_9Cl by the same method. It was characterized by its vapor pressure (not given) and by the IR spectrum (see below) [10].

Molecular Parameters and Physical Properties

The ether-free material is a clear, mobile liquid with a density of 0.65 g/cm³ at 25°C [5, 9], m.p. −16°C [5] or -16.8 ± 0.2°C [19]. It is monomeric in the vapor phase, as shown by molecular weight measurements. The vapor pressure between −5 and 25°C can be represented by the equation $\log p = 7.4 - 1745/T$ [10] or by $\log p = 7.496 - 1755/T$ (p in Torr, T in K), as determined between 0 and 30°C [19]. At 25°C, values of p = 40.80 Torr [19] or p = 35 Torr [5] are given. The heat of vaporization is determined as $\Delta H_v = 7.98 \pm 0.10$ kcal/mol [10] or 8.030 ± 0.180 kcal/mol [19]. The extrapolated boiling point is 115°C/760 Torr [10] or 107 ± 5°C/760 Torr [19]. Trouton's constant is $\Delta H_v/T_v = 20.6\ cal \cdot mol^{-1} \cdot K^{-1}$ [10].

The ^1H NMR spectrum of the neat compound shows a singlet at $\delta = 0.93$ ppm [10, 19] and in C_6H_6 at $\delta = 0.99$ ppm [35].

The vibrational spectra of neat $Be(C_4H_9\text{-}t)_2$ were studied and assigned in the gas, liquid [10, 21], and solid state (Tables 6 and 7) [21]. The IR spectrum of $Be(C_4D_9\text{-}t)_2$ in the gas phase was measured for comparison (Table 8, p. 44) [10]. The assignments given in [10] are partially revised in [21]. The spectra in the gaseous and liquid state (Table 6) correspond with a linear C–Be–C structure of D_{3d} symmetry [10, 21] with free rotation of the t-C_4H_9 groups [10]. A loss of symmetry is observed in the solid state, the symmetry D_3 being more probable than C_{2v} (Table 7). There are no dimers in the solid state [21]. The force constant k(BeC) $\sim 1.2 \times 10^5$ dyn/cm was calculated [10].

Table 6
IR and Raman Vibrations (in cm^{-1}) of $Be(C_4H_9\text{-}t)_2$ in the Gaseous and Liquid State [21].

IR gas	IR liquid	Raman liquid	p*)	assignment
2975(s)				
2950(s)	2950(w)			
2940(s)	2940(s)			
	2920(w)	2920(ms)	0.20	$\nu_{as}(CH)$
2900(m)				
	2880(m)			
2860(s)	2850(s)	2840(ms)	0.15	$\nu_s(CH)$
2798(w)				1390 + 1410
2738(m)				1490 + 1250
1495(s)	1485(s)			
1490(m)	1480(m)	1469(s)	1.90	$\delta_{as}(CH)$
1420(m)	1460(w)	1446(s)	1.50	
1390(ms)	1400(w)			$\delta_s(CH)$
	1380(w)			
1287(w)	1290(w)			
1250(s)	1245(s)			$\nu_{as}(CC)+\varrho(CH_3)(E_u)$
1200(vw)		1205(s)	0.41	$\varrho(CH_3)(A_{1g})$
1170(vw)	1170(vw)	1171(ms)	0.90	$\nu_{as}(CC)$
1050(ms)	1050(ms)			$\varrho(CH_3)(A_{2u})$
970(vs)	990(s)	995(w)	0.90	$\varrho(CH_3)(E_g)$
	970(s)			$\varrho(CH_3)(E_u)$
940(s)	955(s)			$\varrho(CH_3)(E_u) + \nu_{as}(CC)$
	930(s)	936(m)	1.40	$\varrho(CH_3)(E_g)$
870(m)	870(m)			$\nu_s(CC)$
		813(s)	0.60	$\nu_s(CC)(A_{1g})$
580(m)	580(m)			$\nu_{as}(BeC)$
		549(vs)	0.10	$\nu_s(BeC)(A_{1g})$
490(w)	490(m)			$\delta_{as}(CC_3)$
465(vw)	460(vs)	450(vw)	1.70	$\delta_{as}(CC_3)(E_g)$
		373(s)		
		270(w)	1.20	$\varrho(CC_3)(E_g)$
		220(ms)	0.30	$\delta_s(CC_3)(A_{1g})$

*) degree of polarization

References on p. 57

Table 7
IR and Raman Vibrations (in cm^{-1}) of Solid $Be(C_4H_9\text{-}t)_2$ at −180°C [21].

IR	Raman	assignment
2910(m)	2910(m)	$\nu(CH)$
2860(m)		
2820(m)	2830(m)	
1480(m)	1469(s)	$\delta_{as}(CH)$
1460(m)	1450(w)	
1390(m)	1385(vw)	$\delta_s(CH)$
1380(m)	1377(w)	
1235(s)	1235(w)	$\nu_{as}(CC) + \varrho_{as}(CH_3)$
1190(ms)	1188(vs)	$\varrho_{as}(CH)_3$
1170(s)	1159(s)	$\gamma_{as}(CC) + \varrho_{as}(CH_3)$
1020(vs)	1008(ms)	$\varrho_s(CH_3)$
990(w)		
930(sh)	930(s)	$\varrho_{as}(CH_3) + \nu_{as}(CC)$
910(s)	910	
860(w)		
810(m)	806(w)	$\nu_s(CC)$
790(w)	785(s)	
668(s)	660(w)	$\nu_{as}(BeC)$
565(s)	545(s)	$\nu_s(BeC)$
475(w)	474(w)	$\delta_{as}(CC_3)$
	440(w)	
	380(w)	$\delta_s(CC_3)$

Table 8
IR Vibrations (in cm^{-1}) of Gaseous $Be(C_4D_9\text{-}t)_2$ [10].

IR	assignment
2908(w, R)	C_4D_9CH impurity
2894(w, Q)	
2880(w, P)	
2225(m, sh)	$\nu(CD)$
2213(s, sh)	
2202(vs)	
2176(vs)	
2102(m)	
2057(vs)	
2025(m, sh)	
1398(vw, br)	
1319(w)	
1253(m)	
1219(s)	$\nu_{11}(e_u)$
1166(w)	$\nu_5(a_{2u})$
1121(vw)	
1083(w, sh)	$\delta_{as}(CD_3)$
1069(s, sh)	
1064(s)	
1018(m)	$\delta_s(CD_3)$
866(s)	$\varrho(CD_3)$ (e_u)
806(m)	$\varrho(CD_3)$ (e_u)
680(w, br)	$\varrho(CD_3)$ (a_{2u})
490(w, vbr)	impurity?
459(s, sh)	ν_{12} (e_u)
450(vs)	ν_6 (a_{2u})

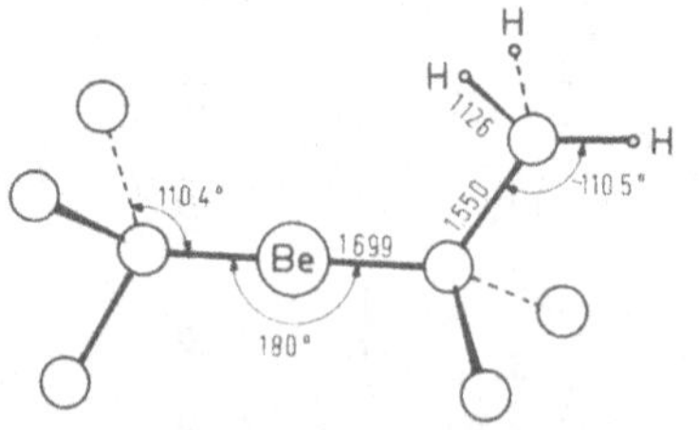

Fig. 6. Molecular structure of $Be(C_4H_9\text{-}t)_2$ gas in the staggered conformation (only 3 H atoms are shown) [11].

An electron diffraction study of $Be(C_4H_9\text{-}t)_2$ gas gave the following mean vibrational amplitudes u (= $\langle u^2 \rangle^{1/2}$): 0.048 ± 0.002 (BeC), 0.054 ± 0.001 (CC), and 0.079 ± 0.001 (CH) Å. The electron scattering pattern is consistent with a monomeric species which has a CBeC angle of 180°. The t-C_4H_9 groups appear to undergo virtually nonhindered rotation, which means that the barrier of rotation should be ≦0.6 kcal/mol. See **Fig. 6** for the structure, the calculated bond distances, and bond angles [11].

Chemical Reactions

$Be(C_4H_9\text{-}t)_2$ dissolves as a monomer in C_6H_6 (by cryoscopy) [10].

Ether-free $Be(C_4H_9\text{-}t)_2$, stored in a clean glass and protected from air and moisture, slowly isomerizes to $Be(C_4H_9\text{-}i)_2$ at room temperature. After 24 days the iso : tertiary ratio was 4.2 : 1; after 13.5 weeks it was 12 : 1; and after 40 weeks only $Be(C_4H_9\text{-}i)_2$ was present, as shown by ^1H NMR spectra [15]. Another report claims that ether-free $Be(C_4H_9\text{-}t)_2$ isomerizes almost completely to $Be(C_4H_9\text{-}i)C_4H_9\text{-}t$ within 3 h at 68°C, but no details are given. The etherate is unchanged after many months at room temperature [32]. Pyrolysis at 105 to 110°C causes elimination of one molar portion of $(CH_3)_2{=}CH_2$ and gives a mixture of $Be(C_4H_9\text{-}t)H$ and $Be(C_4H_9\text{-}i)H$. Thermal decomposition in refluxing $C_6H_5CH_3$ similarly gave 18% $Be(C_4H_9\text{-}t)H$ and 82% $Be(C_4H_9\text{-}i)H$ [15]. Thermal decomposition of ether-free or ether-containing $Be(C_4H_9\text{-}t)_2$ at 150°C [4, 5], 200°C [5, 16, 32], or 210°C [4] yields BeH_2 (up to 80% [5, 32]) and gaseous hydrocarbons, mainly $(CH_3)_2CHCH_3$ and $(CH_3)_2C{=}CH_2$. Optimum purity BeH_2 (90 to 98 wt%) is secured at a pyrolysis temperature of 200 ± 5°C in $C_{12}H_{16}$ from the mono-etherate [32]. Pyrolysis at 240 to 290°C causes increasing liberation of H_2 which becomes very rapid at about 300°C [4].

Mass spectra were obtained at 70 eV, 55 to 75 and 240°C, and at 12 eV and 240°C source temperature [17]. The parent ion derived from the monomer is observed, and there is no evidence for the existence of oligomers [10, 17]. The BeC bond cleavage to give $[C_4H_9]^+$ and $Be(C_4H_9)\cdot$ is an important process. The parent ion also eliminates C_4H_{10}, and $[BeC_4H_8]^+$ is a major Be-containing species. Loss of $(CH_3)_2C{=}CH_2$ from the parent ion is also observed. A complete list of the ions is given [17].

The compound reacts rapidly with H_2O to form $Be(OH)_2$, $(CH_3)_2CHCH_3$, $(CH_3)_2C{=}CH_2$, and traces of H_2. The same products are obtained with an ether-containing sample [5]. A slurry of $Be(C_4H_9\text{-}t)_2$ and NaH in ether form $Na_2[Be_2(C_4H_9\text{-}t)_4H_2]_2\cdot 4\,O(C_2H_5)_2$ at reflux temperature. The coordinated ether can be removed in vacuum [18]. Ligand redistribution reactions are observed with $Be(C_4H_9\text{-}t)_2$ etherate and $BeCl_2$ or $BeBr_2$ to form $Be(C_4H_9\text{-}t)X\cdot O(C_2H_5)_2$ (X = Cl [5, 12, 28], Br [12]). Ether-free $Be(C_4H_9\text{-}t)_2$ reacts with $BeCl_2$ in C_6H_6 to give $Be(C_4H_9\text{-}t)Cl$ [25]. $BeCl_2$ and LiH react with $Be(C_4H_9\text{-}t)_2$ in ether at 60°C to give $Be(C_4H_9\text{-}t)H$, which is isolated as adduct with $N(CH_3)_2CH_2CH_2N(CH_3)_2$. The analogous reaction with $BeBr_2$ gives only $Be(C_4H_9\text{-}t)Br$ etherate [24].

The reactions of ether-containing samples with CH_3OH [13], t-C_4H_9OH [13], or $(C_6H_5)_3COH$ [28] yield $(Be(C_4H_9\text{-}t)OR)_2$ (R = CH_3, t-C_4H_9, $(C_6H_5)_3C$), $(CH_3)_2CH{=}CH_2$, and ether. $(CH_3)_2NCH_2CH_2OH$ and $(CH_3)_2NCH_2CH_2SH$ yield $Be(C_4H_9\text{-}t)XCH_2CH_2N(CH_3)_2$ (X = O, S, respectively); both have a chelate structure [12]. Reactions with $O(CH_3)_2$, $O(C_2H_5)_2$, THF, $S(CH_3)_2$, and $S(C_2H_5)_2$ give adducts which are described on pp. 48/50.

$Be(C_4H_9\text{-}t)_2$ and t-C_4H_9CHO are converted into $(Be(C_4H_9\text{-}t)OCH_2C_4H_9\text{-}t)_3$ in ether at − 78°C by way of an orange-red color. Reactions with $(CH_3)_2CO$, $(t\text{-}C_4H_9)_2CO$, or $(C_6H_5)_2CO$ in a 1 : 1 mole ratio in ether give $(Be(C_4H_9\text{-}t)OCHR_2)_2$ (R = CH_3, C_4H_9-t, C_6H_5) and $(CH_3)_2C{=}CH_2$ at room temperature. With $(t\text{-}C_4H_9)_2CO$, in the mole ratio 1 : 2, $Be(OCH(C_4H_9\text{-}t)_2)_2$ is formed. $(Be(C_4H_9\text{-}t)OC(C_4H_9\text{-}t)_2)_2$ is also formed if $N(CH_3)_2CH_2CH_2N(CH_3)_2$ is added to the reaction

References on p. 57

solution; but in the presence of quinuclidine (1-azabicyclo[2.2.2]octane), $Be(C_4H_9\text{-t})\text{-}OC(C_4H_9\text{-t})_2 \cdot NC_7H_{13}$ is formed [28].

$NH(CH_3)_2$ forms with $Be(C_4H_9\text{-t})_2$ in ether a 1:1 adduct at room temperature. Even at 70°C $(CH_3)_2CHCH_3$ elimination is imperfect. In the presence of $N(CH_3)_3$, $(CH_3)_2CHCH_3$ evolution is much faster, and some $Be(C_4H_9\text{-t})N(CH_3)_2$ is formed. With $NH(CH_3)CH_2CH_2N(CH_3)_2$ at −100°C, the adduct precipitates, then dissolves again upon warming to −60°C; elimination of $(CH_3)_3CH$ is complete at room temperature, and $Be(C_4H_9\text{-t})N(CH_3)CH_2CH_2N(CH_3)_2$ is formed, presumably with a chelate structure [9]. With $N(CH_3)_3$, $N(C_2H_5)_3$, $N(CH_3)_2CH_2CH_2N(CH_3)_2$, and $N(CH_3)_2N{=}NN(CH_3)_2$, 1:1 adducts are formed which are described on pp. 52/4.

$Be(C_4H_9\text{-t})_2 \cdot O(C_2H_5)_2$ reacts with $t\text{-}C_4H_9CN$ in a 1:1 mole ratio in C_6H_{14} to give $Be(C_4H_9\text{-t})_2 \cdot NCC_4H_9\text{-t}$ [28]. With C_6H_5CN in C_6H_{14} in a 1:2 mole ratio, $Be(C_4H_9\text{-t})_2 \cdot 2NCC_6H_5$ is formed at room temperature [29]. $C_6H_5CH{=}NCH_3$ and $C_6H_5CH{=}NC_4H_9\text{-t}$ react with $Be(C_4H_9\text{-t})_2$ etherate in C_6H_{14} at room temperature to form the products $(Be(C_4H_9\text{-t})N(CH_3)CH_2C_6H_5)_2$ and the compound with the structure I ($R = t\text{-}C_4H_9$), respectively. $C_6H_5CH{=}CH\text{–}CH{=}NC_6H_5$ in ether affords $(Be(C_4H_9\text{-t})N(C_6H_5)CH_2CH{=}CHC_6H_5)_2$. $C_6H_5CH{=}NC_6H_5$, $(CH_3)_2CH\text{–}CH{=}NCH_3$, and quinuclidine (1-azabicyclo[2.2.2]octane) in ether or C_6H_{14} give 1:1 adducts with $Be(C_4H_9\text{-t})_2$ etherate [28].

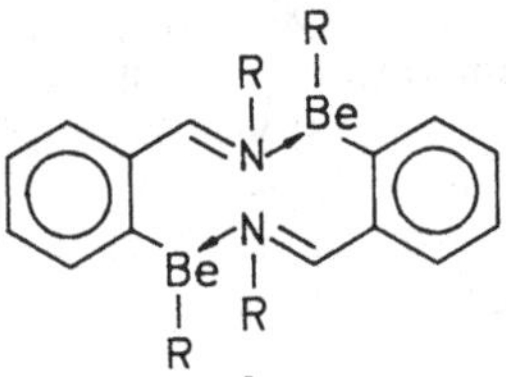

I

With $P(CH_3)_3$ and $P(C_2H_5)_3$, adducts are formed, see pp. 55/7.

$Be(CH_3)_2$, $Be(C_2H_5)_2$, $Be(C_5H_5)_2$, $Be(C_6H_4CH_3\text{-}2)_2$, $Be(C_6H_3(CH_3)_2\text{-}2,5)_2$, and $Be(C_6F_5)_2$ afford $(Be(C_4H_9\text{-t})CH_3)_3$, $(Be(C_4H_9\text{-t})R)_2$ ($R = C_2H_5$, $C_6H_4CH_3\text{-}2$, $C_6H_3(CH_3)\text{-}2,5$), and $Be(C_4H_9\text{-t})R$ ($R = C_5H_5$, C_6F_5), respectively, by reaction in C_6H_6 (with $Be(C_5H_5)_2$ also without a solvent [35]) at room temperature. No reaction was observed with $Be(C_6H_5)_2$ and $Be(C_{10}H_7)_2$ ($C_{10}H_7$ = 1-naphthyl). With $Be(C_6H_4CH_3\text{-}3)_2$, the exchange reaction is accompanied by $(CH_3)_2C{=}CH_2$ elimination; $Be(C_6H_4CH_3\text{-}3)H$, or its $N(CH_3)_3$ adduct, is isolated [25].

$Be(C_4H_9\text{-t})_2\text{–}O(C_2H_5)_2$ System

Liquid-solid equilibrium and Tammann diagrams of the system $Be(C_4H_9\text{-t})_2\text{–}O(C_2H_5)_2$ **(Fig. 7)** show that $Be(C_4H_9\text{-t})_2 \cdot O(C_2H_5)_2$ is a discrete phase (50 mol% ether), melting congruently at −4.4°C, with no evidence for dissociation in the solid-liquid transition. In the region $Be(C_4H_9\text{-t})_2 - Be(C_4H_9\text{-t})_2 \cdot O(C_2H_5)_2$ (0 to 50 mol% ether), a stable eutectic point is observed at 37±1 mol% ether and −33.2±0.2°C. The region $Be(C_4H_9\text{-t})_2 \cdot O(C_2H_5)_2\text{–}O(C_2H_5)_2$ (50 to 100 mol% ether) is more complicated and can only be interpreted by the existence of an unstable dietherate, $Be(C_4H_9\text{-t})_2 \cdot 2O(C_2H_5)_2$ (66.6 mol% ether) with a transition (decomposition) temperature of only −103.6°C (86.0±0.5 mol% ether) and an extrapolated melting point of −98°C. It gives rise to a stable eutectic point at 92.0±0.5 mol% ether with a melting point of −118.0±0.2°C. A metastable eutectic point for $Be(C_4H_9)_2 \cdot O(C_2H_5)_2\text{–}O(C_2H_5)_2$ is detected at 90.6±0.5 mol% ether and a melting point at −119.5±0.5°C.

The vapor of $Be(C_4H_9\text{-t})_2 \cdot O(C_2H_5)_2$ also consists of undissociated monomers. The results of vapor pressure measurements of the two systems $Be(C_4H_9\text{-t})_2\text{–}Be(C_4H_9\text{-t})_2 \cdot O(C_2H_5)_2$ and $Be(C_4H_9\text{-t})_2 \cdot O(C_2H_5)_2\text{–}O(C_2H_5)_2$ are shown in the isotherms given in **Fig. 8** a and Fig. 8 b. Both curves deviate positively from an ideal behavior. The activities of the components were calculated from the total pressure values [19].

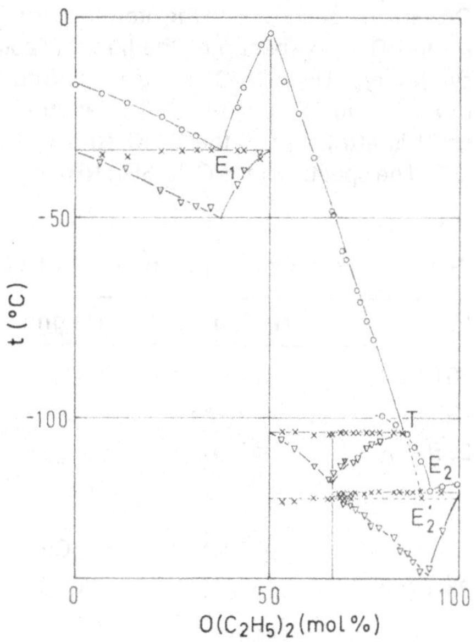

Fig. 7. Liquid-solid phase and Tammann diagrams of the system $Be(C_4H_9\text{-}t)_2$–$O(C_2H_5)_2$ (o = liquidus curve, x = solidus curve (eutectic), ∇ = Tammann curve, E = eutecticum, T = peritecticum) [19].

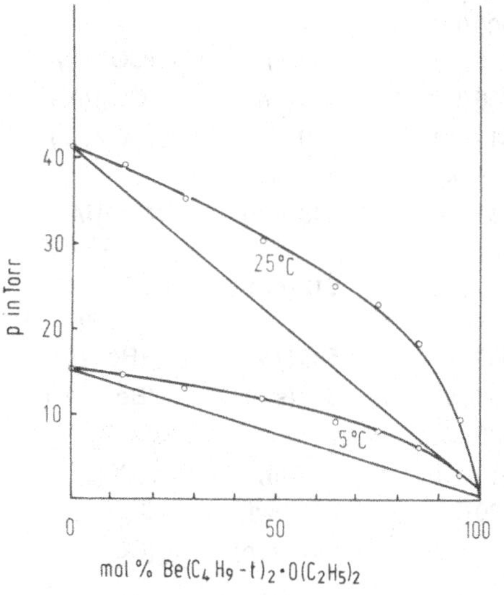

Fig. 8 a. Vapor pressure isotherms of the system $Be(C_4H_9\text{-}t)_2$–$Be(C_4H_9\text{-}t)_2 \cdot O(C_2H_5)_2$ at two different temperatures [19].

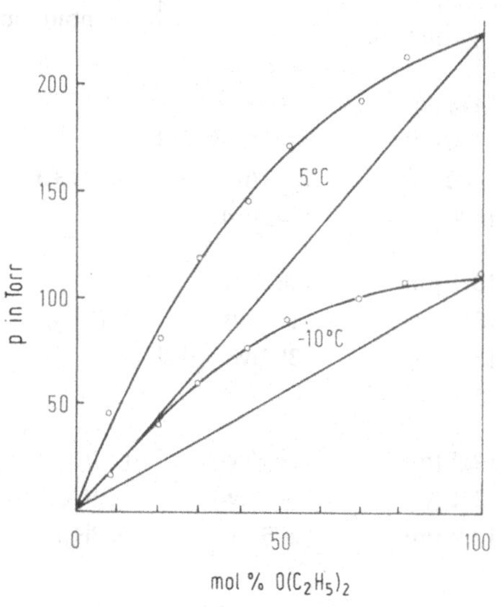

Fig. 8 b. Vapor pressure isotherms of the system $Be(C_4H_9\text{-}t)_2 \cdot O(C_2H_5)_2$–$O(C_2H_5)_2$ at two different temperatures [19].

References on p. 57

$Be(C_4H_9\text{-}t)_2 \cdot O(CH_3)_2$

The adduct is presumably obtained by distillation of an ethereal solution of $Be(C_4H_9\text{-}t)_2$ in $O(CH_3)_2$, as described in the literature for the following adduct. It has been characterized by the IR and Raman spectra of the liquid (Table 9). The spectra are assigned on the basis of a C_s symmetry. The Be–C stretching frequency is significantly lowered by complexation ($\Delta\nu \sim 50$ cm^{-1}), as is the C–O stretching mode ($\Delta\nu \sim 35$ cm^{-1}). A strong IR absorption at 650 cm^{-1} is attributed to the Be–O stretching mode. The corresponding Raman line is polarized [27]. The spectra of the OR_2, SR_2, NR_3 and PR_3 ($R = CH_3$, C_2H_5) adducts are compared in [31].

Table 9
IR and Raman Vibrations (in cm^{-1}) of Liquid $Be(C_4H_9\text{-}t)_2 \cdot O(CH_3)_2$ [27].

IR	Raman	assignment
3010(m)		
2990(s)	2980(vw)	
2950(s)	2957(w)	$\nu_{as}(CH_3)$, $\nu_s(CH_3)$
	2930(w)	
	2902(w)	
2870(m)	2863(w)	
	2835(w)	
2815(s)	2818(vs)	
2760(w)		combination
2690(w)		
1475(s)		
1460(m)	1465(s)	$\delta_{as}(CH_3)$
1448(m)	1449(m)	
1430(m)	1442(m)	
1375(m)	1385(w)	$\delta_s(CH_3)$
1350(w)	1375(w)	
1330(w)	1350(w)	
1310(w)		
1255(m)	1260(vw)	$r(CH_3)$
1240(w)	1240(vw)	$\nu_{as}(CC_3)$
1204(m)	1205(s)	$\varrho(CH_3)$
1185(w)	1184(vs, p)	$\nu'_{as}(CC_3)(A')$
	1172(s)	$\nu''_{as}(CC_3)(A'')$
1161(vw)		$r(CH_3)$
1153(vw)	1150(sh)	$r(CH_3) + \nu_{as}(CO)$
1089(vw)		
1062(s)	1060(w)	$\nu_{as}(CO) + \tau(CH_3)$
1005(vw)	1000(w)	$\varrho(CH_3)$
962(w)	955(w)	$\varrho(CH_3)$
930(vw)	929(s)	$\varrho(CH_3)$
900(m)		
885(m)	893(s)	$\nu_s(CO)(A')$
860(sh)	860(vw)	$\nu_s(CC_3)(A'')$
811(m)	810	$\nu_s(CC_3)(A')$
749(w)	752(w)	
653(vs)	648(s, p)	$\nu(BeO)(A') + \delta(CO)(A')$
	612(m, p)	$\delta(COC)(A') + \nu(BeO)(A')$
537(m)	530(vw)	$\nu_{as}(BeC)(A'')$
500(w)	503(s,p)	$\nu_s(BeC)(A')$
440(s)		$\delta_{as}(CC_3)$
360(w)	381(m)	$\delta_{as}(CC_3)$
280(w)	281(vw)	$\varrho(CC_3)$
	231(m, p)	$\delta_s(CC_3)$

$Be(C_4H_9\text{-}t)_2 \cdot O(C_2H_5)_2$

A material with a composition close to 1:1 stoichiometry is obtained from ethereal solutions of $Be(C_4H_9\text{-}t)_2$ by distillation [4, 32]. Thus, a vacuum distillation at 60°C for 2 days and at 75°C for 1 day, followed by a reflux distillation at 35 to 40°C in a vacuum for 4 h [4], or a distillation at 55 to 65°C/1 Torr [32] gives a colorless liquid with a mole ratio of 1:0.8 [4], 1.0:1.0 [19, 32], or 1:0.94 [5].

Densities of d = 0.78 g/cm³ at room temperature [5] and 0.7815 ± 0.0003 g/cm³ at 25°C [19] were measured. The melting point is −4.4 ± 0.2°C [19]. The vapor pressure at 25°C is p = 0.59 ± 0.21 Torr. The vapor pressure in the temperature range between 55 and 90°C follows the equation log p = 7.222 − 2222/T (p in Torr, T in K), the standard error being ± 0.21 Torr [19]; over the temperature range 30 to 70°C, it can be expressed by log p = 3.390 − 522.9/(t + 100) (p in Torr, t in °C) [32]. The heat of vaporization is $\Delta H_v = 10.2 \pm 0.6$ kcal/mol; the extrapolated boiling point is 238 ± 30°C at 760 Torr; and Trouton's constant is $\Delta H_v/T_v = 20$ cal·mol⁻¹·K⁻¹ [19]. A thermodynamic study of the system $Be(C_4H_9\text{-}t)_2$–$O(C_2H_5)_2$ is summarized on pp. 46/7 [19].

The ^1H NMR spectrum of the neat compound shows chemical shifts at δ = 0.808 (s, C_4H_9), 1.383 (t, CH_3C), and 3.925 ppm (q, CH_2) [19]; in $C_6D_5CD_3$ δ = 0.82 ppm (s, C_4H_9) [10].

The IR and Raman spectra of the liquid are assigned on the basis of a C_s symmetry. The results are summarized in Table 10. The Be–O stretching vibration is found at 650 cm⁻¹, ν(CO) at 790 and 1045 cm⁻¹. The assignment ν(BeC) for the vibrations at 520 and 540 cm⁻¹ are tentative [27]. The spectra of the OR_2, SR_2, NR_3, and PR_3 (R = CH_3, C_2H_5) adducts are compared in [31].

Table 10
IR and Raman Vibrations (in cm⁻¹) of Liquid $Be(C_4H_9\text{-}t)_2 \cdot O(C_2H_5)_2$ [27].

IR	Raman	assignment
3000 (w)		$\nu_{as}(CH_2)$
2960 (s)	2980 (m)	
2940 (s)	2930 (s)	
2920 (s)	2913 (m)	ν'_s, $\nu_{as}(CH_3)$
2880 (m)	2885 (w)	$\nu_s(CH_3)$
2830 (s)	2840 (w)	$\nu_s(CH_2)$
	2820 (vs)	
2785 (w)		
2705 (w)		
1480 (s)	1470	
1460 (m)	1460	$\delta_{as}(CH_3)$
	1440	
1370 (w)	1380	$\delta_s(CH_3)$
1320 (w)	1350	
	1310	$\omega(CH_2)$
1290 (w)	1290	$t(CH_2)$
1275 (w)		
1240 (w)	1240 (w)	$\nu_{as}(CC_3)$
	1205 (m)	$\varrho(CH_3)$
1190 (s)	1188 (vs, p)	$\nu'_{as}(CC_3)A'$ + $r(CH_3)$
1175 (m)	1170 (m)	$\nu''_{as}(CC_3)A''$
1148 (w)		$r(CH_3)$
	1105 (w)	$r(CH_3)$
1095 (w)	1095 (m)	$r(CH_3)$ + $\nu_{as}(CO)A''$
1083 (m)		$r(CH_3)$
1044 (vs)	1050 (w)	$\nu_{as}(CO)$ + $r(CH_3)$
1010 (vw)	1005 (w)	$\nu_s(CC_3)A'$
998 (w)		$\varrho(CH_3)$
960 (w)		$\varrho(CH_3)$
930 (w)	935 (m)	$\varrho(CH_3)$
910 (w)		
875 (vs)	884 (w)	$\nu_{as}(CC)A''$ + $\nu_s(CC_3)A''$
835 (w)	840 (m)	$r(CH_2)$
813 (m)	818 (s, p)	$\nu_s(CC_3)A'$
790 (w)	795 (w, p)	$\nu_s(CO)A'$
755 (w)		$r(CH_2)$
655 (s)	652 (s, p)	$\nu(BeO)A'$ + $\delta(COC)A'$
	547 (m, p)	δ (ether skeleton)
540 (m)	obscured	$\nu(BeC_2)A''$

References on p. 57

Table 10 (continued)

IR	Raman	assignment	IR	Raman	assignment
520 (w)	529 (m, p)	$\nu(BeC_2)A'$	445 (w)		$\delta_{as}(CC_3)$
	505 (vw)		385 (w)	385 (w)	$\delta_{as}(CC_3)$
475 (m)	480 (w, p)	δ (ether skeleton)		240 (m)	$\delta_s(CC_3)$

The reactions of the etherate are described together with the parent $Be(C_4H_9\text{-}t)_2$, pp. 45/46. The ether is replaced by addition of other donors, e.g., THF [25], $NH(CH_3)_2$ [9], $N(CH_3)_3$ [4], $N(CH_3)_2CH_2CH_2N(CH_3)_2$ [12], $N(CH_3)HCH_2CH_2N(CH_3)_2$ [9], t-C_4H_9CN [28], and C_6H_5CN [29]; see also the adducts described below.

$Be(C_4H_9\text{-}t)_2 \cdot 2O(C_2H_5)_2$

The 1:2 adduct is detected as an unstable phase in the system $Be(C_4H_9\text{-}t)_2$–$O(C_2H_5)_2$; see pp. 46/47 [19].

$Be(C_4H_9\text{-}t)_2 \cdot OC_4H_8$ (OC_4H_8 = tetrahydrofuran)

The THF adduct is prepared by addition of THF to the etherate followed by distillation under reduced pressure. The distillate solidifies at room temperature, m.p. 36°C. The compound sublimes at 25 to 30°C/10^{-3} Torr. It is a monomer in C_6H_6 (by cryoscopy) [25].

$Be(C_4H_9\text{-}t)_2 \cdot S(CH_3)_2$ and **$Be(C_4H_9\text{-}t)_2 \cdot S(C_2H_5)_2$**

The adducts are prepared by direct mixing of the components and subsequent distillation. The IR and Raman spectra of the liquids and the proposed assignments are given in Table 11. The assignments for the CH_3 groups are tentative, but the characteristic Be–S and Be–C stretching vibrations are easily located [26]. The spectra are compared with those of the OR_2, NR_3, and PR_3 (R = CH_3, C_2H_5) adducts in [31].

Table 11
IR and Raman Vibrations (in cm^{-1}) of Liquid $Be(C_4H_9\text{-}t)_2 \cdot S(CH_3)_2$ and $Be(C_4H_9\text{-}t)_2 \cdot S(C_2H_5)_2$ [26].

$Be(C_4H_9\text{-}t)_2 \cdot S(CH_3)_2$			$Be(C_4H_9\text{-}t)_2 \cdot S(C_2H_5)_2$		
IR	Raman	assignment	IR	Raman	assignment
3000 (w)	3000 (w)		2970 (w)	2970 (w)	
2920 (vs)	2929 (s)	$\nu(CH_3)$	2940 (s)	2934 (s)	
2890 (m)	2905 (w)		2922 (s)	2910 (w)	$\nu(CH_3)$
2850 (m)	2865 (w)		2880 (m)	2880 (w)	
2810 (s)	2820 (s)			2865 (w)	
1463 (m)	1458 (m)	$\delta_{as}(CH_3)$	2817 (s)	2820 (vs)	
1431 (m)	1440 (m)		2760	2750 (w)	$2\delta_s(CH_3)$
1378 (w)	1380 (w)		2690 (w)	2687 (w)	
1362 (w)	1370 (w)		1463 (vs)	1458 (vs)	$\delta_{as}(CH_3)$
1350 (w)	1350 (w)	$\delta_s(CH_3)$	1450 (m)	1442 (m)	
1335 (w)	1335 (w)		1422 (w)		

Table 11 (continued)

$Be(C_4H_9\text{-}t)_2 \cdot S(CH_3)_2$

IR	Raman	assignment
1240	1240	$\nu_{as}(CC)(A'+A'')$
1205 (m)	1205 (m)	$\varrho(CH_3)$
1179 (w)	1180 (s)	$\nu_{as}(CC)(A')$
1160 (vw)	1164	$\nu_{as}(CC)(A'')$
1029 (m)	1030 (w)	$r(CH_3)$
1004 (w)	1000 (w)	
990 (vw)	985 (w)	$\varrho(CH_3)$
970 (w)	960 (w)	$\varrho(CH_3)$
928 (w)	929 (m)	$\varrho(CH_3)$
909 (w)		$r(CH_3)$
890 (s)	889 (w)	$\nu_s(CC)(A'')$
860 (vw)	855 (vw)	
809 (m)	810 (s, p)	$\nu_s(CC)(A')$
	750 (w)	
730 (w)	736 (s)	$\nu_{as}(CS)$
682 (w)	683 (vs, p)	$\nu_s(CS)$
587 (s)	587 (s, p)	$\nu_s(BeS)$
488 (m)	490 (w)	$\nu_{as}(BeC)(A'')$
467 (m)	472 (w, p)	$\nu_s(BeC)(A')$
370 (w)	381 (w)	$\delta_{as}(CC_3)$
	285 (w)	$\delta_s(CSC)(A')$
	270 (w)	$\varrho(CC_3)$
	225 (w, p)	$\delta_s(CC_3)(A')$

$Be(C_4H_9\text{-}t)_2 \cdot S(C_2H_5)_2$

IR	Raman	assignment
1377 (m)	1380 (w)	
1360 (w)	1370 (w)	$\delta_s(CH_3)$
1348 (w)	1355 (w)	
1280 (w)		$\omega(CH_3)$
1257 (m)	1250 (vw)	
	1240 (vw)	$\nu_{as}(CC_3)(A'+A'')$
1205 (m)	1204 (m)	$\varrho(CH_3)$
1195 (vw)	1183 (s)	$\nu_{as}(CC_3)(A')$
1155 (vw)	1170 (m)	$\nu_{as}(CC_3)(A'')$
	1164 (m)	
1068 (w)	1076 (w)	$r(CH_3)$
1050 (vw)	1052	$r(CH_3)$
1002 (w)	1000 (w)	$\varrho(CH_3)$
980 (vw)	978 (w)	$\nu(CC)$
873		
931 (w)	930 (m)	$\varrho(CH_3)$
885 (s)	886 (w)	$\nu_s(CC_3)(A'')$
810 (m)	810 (m)	$\nu_s(CC_3)(A')$
778 (w)	780 (w)	$r(CH_2)$
	760 (w)	$\nu_{as}(CS)(A'')$
760	755 (w)	
687 (w)	683 (s, p)	
	646 (s, p)	$\nu_s(CS)(A')$
	629 (s, p)	
590 (vs)	591 (vs, p)	$\nu_s(BeS)(A')$
491 (m)	490 (w)	$\nu_{as}(BeC_2)(A'')$
467 (s)	470 (m, p)	$\nu_s(BeC_2)(A')$
	440 (vw)	$\delta_{as}(CC_3)$
365 (m)	380 (s)	$\delta(CCS) + \delta_{as}(CC_3)$
	330 (w, p)	$\delta(CCS)$
	310 (vw)	$\delta_s(CC_3)$
	229 (w, p)	$\delta_s(CC_3)$

$Be(C_4H_9\text{-}t)_2 \cdot NH(CH_3)_2$

$NH(CH_3)_2$ is condensed onto $Be(C_4H_9\text{-}t)_2 \cdot O(C_2H_5)_2$ in ether at −196°C (mole ratio 1:1). The mixture is warmed to room temperature and then briefly boiled. Hexane is added, and all volatile material is removed under reduced pressure, leaving the liquid adduct. Mixtures of $Be(C_4H_9\text{-}t)_2 \cdot O(C_2H_5)_2$, $NH(CH_3)_2$, and $N(CH_3)_3$ in C_5H_{12} give the adduct as a minor product together with $Be(C_4H_9\text{-}t)N(CH_3)_2$ [9].

References on p. 57

The ^1H NMR spectrum in C_6H_6 has a singlet at $\delta = 1.06$ ppm (C_4H_9) and a broad doublet at $\delta = 1.75$ ppm (CH_3N) [9]. The spectrum is presented in [33]. The IR spectrum of the liquid film has a ν(NH) absorption at 3289 cm^{-1}. Other absorptions are at 2907(s), 2801(s), 2747(m), 2717(w), 2684(w), 1466(s), 1410(w), 1381(m), 1355(w), 1259(m), 1219(m), 1197(m), 1132(w), 1115(m), 1034(s), 1019(s), 1000(m), 929(m, sh), 917(m), 892(s), 862(s), 813(s), 807(m, sh), 763(w, sh), 752(m), 695(s), 671(m, sh), and 544(s) cm^{-1} [9].

Only at ~70°C does gas evolve, and i-C_4H_{10} is formed, whereas neither $(CH_3)_2C{=}CH_2$ nor $NH(CH_3)_2$ could be detected [9].

Be(C_4H_9-t)$_2$·N(CH_3)$_3$

The adduct is prepared from Be(C_4H_9-t)$_2$ in ether and excess $N(CH_3)_3$ by standing overnight. Fractional distillation and condensations give a heterogeneous residue consisting of a liquid (probably Be(C_4H_9-t)$_2$ and its ether adduct) and a crystalline solid of melting point 42 to 47°C, which still contains some ether [4]. A cleaner product is obtained by the reaction of ether-free Be(C_4H_9-t)$_2$ with $N(CH_3)_3$ at −30°C. After removal of the excess $N(CH_3)_3$ at −30°C, the 1:1 adduct was sublimed at room temperature (diffusion pump vacuum) and condensed at −78°C, m. p. 45 to 46°C [12].

The vapor pressure between 25 and 70°C follows the equation $\log p = 5.77 - 1648\ T^{-1}$ (p in Torr, T in K). The ^1H NMR spectrum in C_6H_6 has two resonances at $\delta = 1.18$ (s, C_4H_9) and 1.89 ppm (s, CH_3N) [12]. The IR and Raman spectra of the crystalline solid at −180°C are summarized in Table 12. The assignments are based on the idealized model of C_s symmetry [30].

The adduct dissolves in C_6H_6 as a monomer (by cryoscopy). Noticeable decomposition occurs above 70°C [12].

The vibrational spectra of the adducts with OR_2, SR_2, NR_2, and PR_3 (R = CH_3, C_2H_5) are compared in [31].

Be(C_4H_9-t)$_2$·N(C_2H_5)$_3$

The adduct is obtained by addition of $N(C_2H_5)_3$ to Be(C_4H_9-t)$_2$ in stoichiometric amounts in the cold. It is characterized by the IR and Raman spectra of the crystalline solid at −180°C (Table 12). The assignments are tentative due to the complexity of the molecule [30]. See [31] for a comparison of the vibrational spectra of Be(C_4H_9-t)$_2$ adducts with OR_2, SR_2, NR_3, and PR_3 (R = CH_3, C_2H_5).

Be(C_4H_9-t)$_2$·N(CH_3)$_2$$CH_2CH_2$N($CH_3$)$_2$

The amine in c-$C_6H_{11}CH_3$ is added to Be(C_4H_9-t)$_2$·O(C_2H_5)$_2$ in c-$C_6H_{11}CH_3$ at −30°C. The adduct deposits as the solvent is evaporated, and it can be recrystallized from c-$C_6H_{11}CH_3$.

^1H NMR spectra were measured at 33, −10, −25, −50, and −90°C. At 33°C in c-$C_6D_{11}CD_3$ three singlets are observed at $\delta = 0.12$ (CH_3C), 2.42 (NCH_3), and 2.55 ppm (NCH_2); in $C_6D_5CD_3$, $\delta = 1.10$ (CH_3C) and 2.10 ppm (CH_2N and CH_3N coincide). The CH_3N resonance is split into two lines going from −25°C to lower temperatures; e. g., in $C_6D_5CD_3$ at −90°C $\delta = 1.25$ (s, CH_3C), 1.76 and 1.99 (s, CH_3N), and 2.23 ppm (s, CH_2) [12]. The spectra are reproduced as figures in [33]. The spectra are consistent with a constitution in which only one nitrogen is coordinated at a time, but in which rapid exchange takes place between complexed and free nitrogen [12].

Benzene solutions contain monomers (by cryoscopy). The solid decomposes at about 86°C to a yellow liquid [12]. The adduct reacts with (t-C_4H_9)$_2$CO in C_6H_{14} at room temperature to give (Be(C_4H_9-t)OCH(C_4H_9-t)$_2$)$_2$ [28].

Table 12
IR and Raman Vibrations (in cm^{-1}) of Solid $Be(C_4H_9\text{-}t)_2 \cdot N(CH_3)_3$ and $Be(C_4H_9\text{-}t)_2 \cdot N(C_2H_5)_3$ at −180°C [30].

$Be(C_4H_9\text{-}t)_2 \cdot N(CH_3)_3$ IR	Raman	assignment	$Be(C_4H_9\text{-}t)_2 \cdot N(C_2H_5)_3$ IR	Raman	assignment
2975(vs)	2973(w)		3000(w)		
2958(m)			2980(w)		
2944(m)	2940(w)	$\nu(CH_3)$	2960(m)	2960(m)	
2907(w)	2917(m)		2950(m)		$\nu(CH_3)$
2875(m)	2880(vw)		2900(m)		
	2869(m)		2860(m)		
1490(vs)			2810(s)	2814(m)	
1480(vs)	1471(m)		2750(vw)	2750(m)	
1460(s)			1485(m)	1482(ms)	
1452(m)	1446(m)		1475(m)		
1445(m)		$\delta(CH_3)$	1465(m)	1466(s)	$\delta_{as}(CH_3)$
1436(w)	1418(w)		1445(w)	1450(s)	
1405(w)			1438(vw)		
1365(m)	1371(vw)		1400(vw)		
1240(w)	1241(w)	$\nu_{as}(CC)(A'+A'')$	1390(m)	1385(w)	
		$+\nu_{as}(NC)$	1380(w)		$\delta_s(CH_3)+\delta(CH_2)$
1200(m)		$\varrho(CH_3$ of $C_4H_9)$	1350(w)	1350(w)	
1185(w)	1183(vs)	$\nu_{as}(CC)(A')$	1320(vw)	1315(w)	
1172(vw)			1270(vw)	1265(w)	$\tau(CH_2)$
1150(vw)	1157(m)	$\nu_{as}(CC)(A'')$	1250(vw)		$\nu_{as}(CC_3)(A'+A'')$
1100(vw)	1109(vw)	$r(CH_3$ of $NCH_3)$	1188(m)	1180(s)	$\nu_{as}(NC)$[a]
1050(vw)	1060(vw)	$\varrho(CH_3)$			$+\varrho(CH_3)$
1020(vw)	1020(vw)	$r(CH_3)$[a]	1170(vw)	1173(s)	$\nu_{as}(CC_3)(A')$
990(vw)	993(m)	$\varrho(CH_3)$	1155(m)	1160(m)	$\nu_{as}(CC_3)(A'')$
950(w)			1085(w)	1085(w)	
930(w)	928(m)	$\varrho(CH_3)$	1065(vw)		$\varrho(CH_3)+r(CH_3)$
880(m)	890(vw)	$\nu_s(CC)(A'')$	1045(vw)		
	841(w)		1030(m)	1035(w)	$r(CH_3)$[a]
820(s)	824(s)	$\nu_s(NC)$	1010(vw)	1010(m)	$\nu_s(CC$ of $C_2H_5)$
810(w)	809(s)	$\nu_s(CC)(A')$	990 (vw)	995(vw)	$r(CH_3)+\varrho(CH_3)$
	741(vw)		930(vw)	950(w)	$\varrho(CH_3)$
650(m)	658(s)	$\nu(BeN)(\nu_1)$	920(vw)	918(s)	$\nu_{as}(CC$ of $C_2H_5)$
620(w)			910(vw)	900(m)	$r(CH_3)$
575(w)	575(vw)	$\nu_{as}(BeC)(\nu_2)$	870(vw)	865(w)	$\nu_s(CC_3)(A'')$
540(m)	543(m)	$\nu_s(BeC)(\nu_3)$	835(s)	834(s)	$r(CH_2)$
495(w)	495(vw)	$\delta_{as}(CC_3)$	815(m)	816(s)	$\nu_s(CC_3)(A')$
		$+\delta_d(NC_3)$	785(vw)		$r(CH_2$ of $C_2H_5)$

References on p. 57

Table 12 (continued)

$Be(C_4H_9\text{-}t)_2\cdot N(CH_3)_3$ IR	Raman	assignment	$Be(C_4H_9\text{-}t)_2\cdot N(C_2H_5)_3$ IR	Raman	assignment
430 (w)	430 (w)	$\delta_s(NC_3)$	750 (m)		
	395 (w)	$\delta_{as}(CC_3)$	730 (w)	745 (m)	$\nu_s(NC)$
340 (w)	390(w)		660 (s)	650 (m)	$\nu_s(BeN)$
300 (w)		$\tau(CH_3)$	570 (vw)	570 (m)	$\delta_s(NC_3)$
		$\varrho(CC_3)$	550 (m)	555 (w)	$\nu_{as}(BeC)$
		$\delta_s(CC_3)$	510 (w)	515 (m)	$\nu_s(BeC)$
			485 (w)	490 (vw)	$\delta_{as}(CC_3)$
			415 (vw)	420 (s)	$\delta(NC_3)$
			390 (w)	380 (w)	$\delta_{as}(CC_3)$

[a] coupled vibrations

$Be(C_4H_9\text{-}t)_2\cdot N(CH_3)_2N{=}NN(CH_3)_2$

The adduct is obtained from the components in C_7H_{16} at room temperature. The solvent is removed via vacuum transfer to a cool trap. The remaining yellow liquid is vacuum-sublimed at 25°C onto a surface at −78°C.

The melting point is 44 to 48°C. It is a monomer in C_6H_6 (by cryoscopy). Thermal decomposition at 50°C in a sealed tube gives in a violent reaction an uncharacterized black char. Heating at 90°C in a vacuum system gives $N(CH_3)_2N{=}NN(CH_3)_2$, N_2, and $(CH_3)_2C{=}CH_2$. The composition of the residue corresponds to $Be(C_4H_9\text{-}t)_2\cdot N(CH_3)_2N{=}NN(CH_3)_2\cdot Be(C_4H_9\text{-}t)H$. The presence of a 2:1 adduct could not be proved [7].

$Be(C_4H_9\text{-}t)_2\cdot 0.5\,N(CH_3)_2N{=}NN(CH_3)_2$

The 2:1 adduct is assumed to be partly formed upon pyrolysis of the 1:1 adduct at 70 to 90°C. But the elemental analysis of the product agrees with a compound of the composition **$Be(C_4H_9\text{-}t)_2\cdot N(CH_3)_2N{=}NN(CH_3)_2\cdot Be(C_4H_9\text{-}t)H$**, m.p. 15 to 16°C. The molecular mass in C_6H_6 points to a higher mass (365, 350 by cryoscopy),or some association. Pyrolysis at 220°C gives $(CH_3)_2C{=}CH_2$ and a polymer of the composition $C_{10}H_{26}BeN_3$ [7].

$Be(C_4H_9\text{-}t)_2\cdot NC_7H_{13}$ (NC_7H_{13} = 1-azabicyclo[2.2.2]octane)

Quinuclidine (= 1-azabicyclo[2.2.2]octane) reacts with $Be(C_4H_9\text{-}t)_2\cdot O(C_2H_5)_2$ in C_6H_{14}. The adduct is recrystallized from C_6H_{14}. When heated, it shrinks at ~60°C and melts at 78 to 79°C. It dissolves as a monomer in C_6H_6 (by cryoscopy). Addition of $(t\text{-}C_4H_9)_2CO$ to a solution of the adduct in ether gives $Be(C_4H_9\text{-}t)OCH(C_4H_9\text{-}t)_2\cdot NC_7H_{13}$ at room temperature [28].

$Be(C_4H_9\text{-}t)_2\cdot N{\equiv}CC_4H_9\text{-}t$

The adduct is obtained from equimolar quantities of the nitrile and $Be(C_4H_9\text{-}t)_2\cdot O(C_2H_5)_2$ in C_6H_{14} at room temperature. Hexane is removed, and the white residue crystallizes as colorless needles from C_6H_{14} at 0°C, m.p. 27 to 28°C.

The IR spectrum in Nujol shows a band for $\nu(CN)$ at 2290 cm^{-1}. The complex is monomeric in C_6H_6 (by cryoscopy) and is unchanged after heating in C_6H_{14} for 8 h at 50°C and further at 65°C for 9 h [28].

$Be(C_4H_9\text{-}t)_2 \cdot 2\,N{\equiv}CC_6H_5$

$Be(C_4H_9\text{-}t)_2 \cdot O(C_2H_5)_2$ and the nitrile react in C_6H_{14} at room temperature in the mole ratio 1:2. A yellow precipitate forms. The mixture is gently warmed until the solid is redissolved, and standing overnight gives orange-red crystals, m. p. 64 to 65°C. In the IR spectrum in Nujol ν(CN) appears at 2278 cm^{-1}. It dissolves as a monomer in C_6H_6 (by cryoscopy) [29].

$Be(C_4H_9\text{-}t)_2 \cdot N(CH_3){=}CHC_3H_7\text{-}i$

Preparation occurs from $Be(C_4H_9\text{-}t)_2 \cdot O(C_2H_5)_2$ and $N(CH_3){=}CHC_3H_7\text{-}i$ (mole ratio 1:1) in C_6H_{14} within 1 h at room temperature. Recrystallization from C_6H_{14} gives colorless plates, m.p. 34°C. The IR spectrum in Nujol shows ν(C=N) at 1670 cm^{-1}. It dissolves as a monomer in C_6H_6 (by cryoscopy) [28].

$Be(C_4H_9\text{-}t)_2 \cdot N(C_6H_5){=}CHC_6H_5$

$C_6H_5N{=}CHC_6H_5$ reacts with $Be(C_4H_9\text{-}t)_2 \cdot O(C_2H_5)_2$ in ether within 30 min at room temperature. After removal of the ether the residue is crystallized from C_6H_{14}. The orange-red prisms melt at 128 to 130°C. IR (Nujol): 1605, 1595, and 1581 cm^{-1}. The adduct is a monomer in C_6H_6 (by cryoscopy) [28].

$Be(C_4H_9\text{-}t)_2 \cdot P(CH_3)_3$

The adduct is synthesized from the components in the absence of a solvent by three freeze-thaw cycles from −196 to −30°C in a vacuum system. The excess $P(CH_3)_3$ is removed at −30°C, and the adduct is sublimed at room temperature and condensed at −78°C, m. p. 44 to 46°C [12].

^1H NMR (C_6H_6): δ = 0.82 (d, CH_3P; J(P,H) = 3.5 Hz), 1.14 ppm (s, CH_3C) [12]. The IR and Raman spectra of the crystalline solid at −180°C are summarized in Table 13. Symmetry designations refer to a simplified C_s model [30]. A comparison of the vibrational spectra of adducts of $Be(C_4H_9\text{-}t)_2$ with OR_2, SR_2, NR_3, and PR_3 (R = CH_3, C_2H_5) is also made [31].

Solutions in C_6H_6 contain momomers (by cryoscopy). The adduct is readily decomposed with irreversible formation of hydrocarbons [12].

Table 13

IR and Raman Vibrations (in cm^{-1}) of Solid $Be(C_4H_9\text{-}t)_2 \cdot P(CH_3)_3$ and $Be(C_4H_9\text{-}t)_2 \cdot P(C_2H_5)_3$ at −180°C [30].

$Be(C_4H_9\text{-}t)_2 \cdot P(CH_3)_3$ IR	Raman	assignment	$Be(C_4H_9\text{-}t)_2 \cdot P(C_2H_5)_3$ IR	Raman	assignment
2990 (m)	2990 (w)	ν(CH_3)	2960 (m)	2966 (m)	ν(CH_3, CH_2)
	2980 (w)			2930 (m)	
			2925 (s)	2920 (m)	
2920 (s)	2920 (s)		2910		
			2880	2873 (m)	
2908 (s)			2860		
2870 (s)	2860 (w)		2810 (s)	2814 (m)	
2806 (s)	2810 (m)		2740 (w)	2746	

References on p. 57

Table 13 (continued)

$Be(C_4H_9\text{-}t)_2 \cdot P(CH_3)_3$ IR	Raman	assignment
1462(s)	1458(m)	
1435(w)	1441(w)	
1424(m)	1426(w)	
1390(w)	1380(vw)	$\delta(CH_3)$
1370(m)		
1355(w)		
1320(w)		
1310		
1250(vw)	1250(vw)	$\nu_{as}(CC)(A'+A'')$
1190(s)	1197(w)	$\varrho(CH_3)$
	1183(s)	$\nu_{as}(CC)(A')$
1170(m)	1163(s)	$\nu_{as}(CC)(A'')$
1155(m)		$r(CH_3)$
1050(w)		$\varrho(CH_3)$
1020(w)		?
1000(w)	1000(w)	$\varrho(CH_3)$
960(s)		$\varrho(CH_3) + r(CH_3)$
950(s)		$r(CH_3)$
945(s)		$\varrho(CH_3) + r(CH_3)$
930(m)	930(w)	$\varrho(CH_3)$
875(w)	878(vw)	$\nu_s(CC)(A'')$
840(w)	848(vw)	?
810(w)	815(m)	$\nu_s(CC)(A')$
730(m)	728(s)	$\nu_{as}(PC)$
	675(vs)	$\nu_s(PC)$
660(w)		free $P(CH_3)_3$
600(m)	598(m)	$\nu(BeP)$
510(m)	500(w)	$\nu_{as}(BeC)$
480(w)	489(w)	$\nu_s(BeC)$
450(w)		$\delta_{as}(CC_3)$
	389(w)	$\delta_{as}(CC_3)$
360(w)		
310(w)	320(w)	$\delta_{as}(PC_3)$
	272(w)	$\delta_s(PC_3) + r(CC_3)$
	220(w)	$\delta_s(CC_3)$

$Be(C_4H_9\text{-}t)_2 \cdot P(C_2H_5)_3$ IR	Raman	assignment
1462(vs)	1465(s)	
	1443(m)	
1425(w)		
1383(w)	1383(w)	$\delta(CH_3, CH_2)$
1370(w)		
1355(w)		
1310(vw)		
	1270(vw)	$\omega(CH_2)$
1255(vw)	1250(vw)	$\tau(CH_2)$
	1240(m)	$\nu_{as}(CC_3)(A'+A'')$
1190(m)	1194(vs)	$r(CH_3)$
	1182(vs)	$\nu_{as}(CC_3)(A')$
1172(m)	1170(m)	$\nu_{as}(CC_3)(A'')$
1065(m)	1057(m)	$\varrho(CH_3)$
1050(m)		
1042(s)	1045(w)	$\nu_s(CC)$
1030(w)		$r(CH_3)$
1010(w)		
	992(m)	$\varrho(CH_3)$
970(vw)	983(m)	$\nu_{as}(CC)$
940(vw)	935(m)	$\varrho(CH_3)$
890(s)		
870(vw)	869	$\nu_s(CC_3)(A')$
812(vw)	813(m)	$\nu_s(CC_3)(A'')$
768(s)	770(vw)	$r(CH_2)$
755(w)	755(vw)	$r(CH_2)$
725(w)	718(m)	$\nu_{as}(PC)$
705(w)	690(m)	
630(w)	636(s)	$\nu_s(PC)$
590(w)	590(w)	$\nu(BeP)$
520(w)	510(w)	$\nu_{as}(BeC)$
	485(m)	$\nu_s(BeC)$
455(vw)		$\delta_{as}(CC_3)$
	387(m)	
360(w)		$\delta_{as}(PC_3)$
330(w)	326(s)	$\delta_s(CCP)$
300(vw)		
	224(m)	$\delta_s(CC_3)$

$Be(C_4H_9\text{-}t)_2 \cdot P(C_2H_5)_3$

The adduct is said to be prepared from stoichiometric amounts of the components by the usual method. IR and Raman data of the solid are given (see Table 13) and assigned with a simplified C_s symmetry model [30]. A comparison of the spectra of $Be(C_4H_9\text{-}t)_2$ adducts with OR_2, SR_2, NR_3, and PR_3 ($R = CH_3$, C_2H_5) is made [31].

References:

[1] Lavroff, D. (J. Russ. Phys. Chem. Soc. **16** [1884] 93/100).
[2] Gilman, H., Schulze, F. (J. Chem. Soc. **1927** 2663/9).
[3] Gilman, H., Schulze, F. (J. Am. Chem. Soc. **49** [1927] 2904/7).
[4] Coates, G. E., Glockling, F. (J. Chem. Soc. **1954** 2526/9).
[5] Head, E. L., Holley, C. E., Rabideau, S. W. (J. Am. Chem. Soc. **79** [1957] 3687/9).
[6] Lardicci, L., Lucarini, L., Palagi, P., Pino, P. (J. Organometal. Chem. **4** [1965] 341/8).
[7] Fetter, N. R. (J. Chem. Soc. A **1966** 711/3).
[8] Glaze, W. H., Selman, C. M., Freeman, C. H. (Chem. Commun. **1966** 474).
[9] Coates, G. E., Fishwick, A. H. (J. Chem. Soc. A **1967** 1199/204).
[10] Coates, G. E., Roberts, P. D., Downs, A. J. (J. Chem. Soc. A **1967** 1085/91).

[11] Almenningen, A., Haaland, A., Nilson, J. E. (Acta Chem. Scand. **22** [1968] 972/6).
[12] Coates, G. E., Roberts, P. D. (J. Chem. Soc. A **1968** 2651/5).
[13] Coates, G. E., Fishwick, A. H. (J. Chem. Soc. A **1968** 477/83).
[14] Chambers, D. B., Coates, G. E., Glockling, F. (Discussions Faraday Soc. No. 47 [1969] 157/64).
[15] Coates, G. E., Roberts, P. D. (J. Chem. Soc. A **1969** 1008/12).
[16] Shepherd, L. H., Ter Haar, G. L., Marlett, E. M. (Inorg. Chem. **8** [1969] 976/9).
[17] Chambers, D. B., Coates, G. E., Glockling, F. (J. Chem. Soc. A **1970** 741/8).
[18] Coates, G. E., Pendlebury, R. E. (J. Chem. Soc. A **1970** 156/60).
[19] Mounier, J., Lacroix, R., Potier, A. (J. Organometal. Chem. **21** [1970] 9/19).
[20] Coates, G. E., Francis, B. R. (J. Chem. Soc. A **1971** 1308/10).

[21] Mounier, J. (J. Organometal. Chem. **38** [1972] 7/15).
[22] Giacomelli, G. P., Lardicci, L. (Chem. Ind. [London] **1972** 689/90).
[23] Lardicci, L., Giacomelli, G. P., De Bernardi, L. (J. Organometal. Chem. **39** [1972] 245/50).
[24] Blindheim, U., Coates, G. E., Srivastava, R. C. (J. Chem. Soc. Dalton Trans. **1972** 2302/5).
[25] Coates, G. E., Smith, D. L., Srivastava, R. C. (J. Chem. Soc. Dalton Trans. **1973** 618/22).
[26] Mounier, J. (J. Organometal. Chem. **56** [1973] 79/86).
[27] Mounier, J. (J. Organometal. Chem. **56** [1973] 67/77).
[28] Andersen, R. A., Coates, G. E. (J. Chem. Soc. Dalton Trans. **1974** 1171/80).
[29] Coates, G. E., Smith, D. L. (J. Chem. Soc. Dalton Trans. **1974** 1737/40).
[30] Mounier, J., Mula, B., Potier, A. (J. Organometal. Chem. **105** [1978] 289/301).

[31] Mounier, J., Mula, B., Potier, A. (J. Organometal. Chem. **107** [1976] 281/6).
[32] Baker, R. W., Brendel, G. J., Lowrance, B. R., Mangham, J. R., Marlett, E. M., Shepherd Jr., L. H. (J. Organometal. Chem. **159** [1978] 123/30).
[33] Coates, G. E. (Record Chem. Progr. **28** [1967] 3/23).
[34] Ashby, E. C., Prasad, H. S. (Inorg. Chem. **14** [1975] 2869/74).
[35] Bartke, T. C. (Diss. Univ. Wyoming 1975; Diss. Abstr. Intern. B **36** [1976] 6141).

1.2.1.7 Other Dialkylberyllium Compounds and Their Adducts

The preparation of BeR_2 (R = $-C_3H_6O(CH_3)-$, $-C_6H_4$-2-$CH_2O(CH_3)-$, $-C_3H_6N(CH_3)_2-$, and $-C_3H_6N(C_2H_5)_2-$) with chelating R groups was attempted without success from $BeCl_2$ and the corresponding Grignard reagent MgRCl [2].

$Be(C_5H_{11})_2 \cdot nO(C_2H_5)_2$ (n<1)

A solution of $Mg(C_5H_{11})Br$ in ether is added to $BeCl_2 \cdot 2\,O(C_2H_5)_2$ in ether during 20 min, and the mixture is boiled for another 45 min. After decantation, it is distilled at 50 to 105°C/10^{-5} Torr. Analysis of the distillate gives a ratio of Be : C_5H_{11} : $O(C_2H_5)_2$ = 1.00 : 2.00 : 0.675.

The compound reacts with $BeCl_2 \cdot 2O(C_2H_5)_2$ and LiH to give $Be(C_5H_{11})H$, which was isolated as an adduct with $N(CH_3)_2CH_2CH_2N(CH_3)_2$ [14].

$Be(CH_2C_4H_9\text{-t})_2$

The compound is prepared by heating a C_6H_6 solution of $Be(C_2H_5)_2$ and $B(CH_2C_4H_9\text{-t})_3$ to 70°C for 16 h. $B(C_2H_5)_3$ is distilled at room temperature/10^{-3} Torr, and the compound is then obtained at 50°C/10^{-3} Torr [8]. $Mg(CH_2C_4H_9\text{-t})Cl$ is added to $BeCl_2$ (mole ratio 2:1) in ether. Decantation of $MgCl_2$, concentration, and distillation at 50°C/10^{-3} Torr gives the monoetherate $Be(CH_2C_4H_9\text{-t})_2 \cdot O(C_2H_5)_2$ (see below) [7, 12].

The ether-free dialkyl has two singlets at δ = 0.43(CH_2) and 1.16 ppm (CH_3) in the ^{1}H NMR spectrum in C_6H_6 at 30°C [8]. The compound exists as a monomer–dimer equilibrium in C_6H_6. From the degrees of association measured cryoscopically in C_6H_6, namely 1.46, 1.38, and 1.34 at 1.42, 0.95, and 0.63 wt% concentration, the equilibrium constant for the reaction $(Be(CH_2C_4H_9\text{-t})_2)_2 \rightleftharpoons 2\,Be(CH_2C_4H_9\text{-t})_2$ was calculated as 0.041, 0.046, and 0.051 mol/L, respectively [8]. Pyrolysis of $Be(CH_2C_4H_9\text{-t})_2 \cdot O(C_2H_5)_2$ yields some $Be(CH_3)_2$, but no BeH_2 [12]. Ligand scrambling with $BeCl_2$, $BeBr_2$, or $Be(CH_3)_2$ affords $Be(CH_2C_4H_9\text{-t})X$ (X = Cl, Br, CH_3), respectively, which are isolated in the form of complexes with $N(CH_3)_2CH_2CH_2N(CH_3)_2$ [7]. Adducts with N-donors are prepared from the etherate (see below) [7].

$Be(CH_2C_4H_9\text{-t})_2 \cdot O(C_2H_5)_2$

The etherate is obtained as a colorless liquid by distillation of an ethereal solution of $Be(CH_2C_4H_9\text{-t})_2$ (prepared from $Mg(CH_2C_4H_9\text{-t})Cl$ and $BeCl_2$ in ether) at 50°C/10^{-3} Torr [7].

The ^{1}H NMR spectrum in C_6H_6 has resonances at δ = 0.19 (s, CH_2Be), 0.93 (t, CH_3CO), 1.30 (s, CH_3CCBe), and 4.69 ppm (q, CH_2O). The compound is monomeric in C_6H_6 (by cryoscopy). The adduct is very stable. The ether cannot be removed by heating under reflux at 10^{-3} Torr for 29 h. However, the ether is readily displaced by $N(CH_3)_2CH_2CH_2N(CH_3)_2$ and 2,2'-bipyridine (see below). Hydrolysis yields $C(CH_3)_4$ and $O(C_2H_5)_2$ [7]. Further reactions of the etherate are described together with the parent compound (see above).

$Be(CH_2C_4H_9\text{-t})_2 \cdot N(CH_3)_3$

The adduct is formed, unexpectedly, by reaction of $Be(CH_2C_4H_9\text{-t})Br$ and an excess of $N(CH_3)_3$ in C_6H_6 containing a little ether. When the mixture is warmed, the precipitate, formed at room temperature, dissolves at ~80°C. The cooled solution is filtered, and the filtrate is concentrated. Distillation of the residue at 50°C (bath temperature)/10^{-3} Torr produces the adduct as a colorless liquid.

The ^{1}H NMR spectrum in C_6H_6 has signals at δ = 0.26 (s, CH_2), 1.43 (s, CH_3C), and 1.83 ppm (s, CH_3N). The compound dissolves in C_6H_6 as a monomer (by cryoscopy). With equimolar amounts of $Be(CH_3)_2 \cdot N(CH_3)_3$ in c-C_5H_{10}, the mixed complex $Be(CH_3)CH_2C_4H_9\text{-t} \cdot N(CH_3)_3$ is formed. The equilibrium $Be(CH_3)_2 \cdot N(CH_3)_3 + Be(CH_2C_4H_9\text{-}t)_2 \cdot N(CH_3)_3 \rightleftharpoons 2\,Be(CH_3)CH_2C_4H_9\text{-}t \cdot N(CH_3)_3$ was followed by ^1H NMR at 30, −16, and −40°C. From the ratios of the CH_3Be and CH_2Be signals at −40°C, an equilibrium constant of K = 170 was obtained [7].

$Be(CH_2C_4H_9\text{-}t)_2 \cdot N(CH_3)_2CH_2CH_2N(CH_3)_2$

The diamine is added to $Be(CH_2C_4H_9\text{-}t)_2 \cdot O(C_2H_5)_2$ in C_6H_6. The solvent is evaporated, and the white residue is recrystallized from C_6H_{14} to give colorless crystals of the adduct, m.p. 78 to 80°C.

The ^1H NMR spectrum in C_6H_6 shows signals at δ = 0.08 (s, CH_2Be), 1.53 (s, CH_3C), 1.83 (s, CH_2N), and 2.00 ppm (s, CH_3N). The ^1H NMR spectrum in $C_6D_5CD_3$ does not change between room temperature and −80°C, indicating a four-coordinated Be with a chelating diamine.

The compound is a monomer in C_6H_6. Equimolar mixtures of $Be(CH_3)_2 \cdot N(CH_3)_2CH_2CH_2N(CH_3)_2$ and $Be(CH_2C_4H_9\text{-}t)_2 \cdot N(CH_3)_2CH_2CH_2N(CH_3)_2$ in c-C_5H_{10} give the mixed complex $Be(CH_3)CH_2C_4H_9\text{-}t \cdot N(CH_3)CH_2CH_2N(CH_3)_2$ with a chelating diamine [7].

$Be(CH_2C_4H_9\text{-}t)_2 \cdot N_2C_{10}H_8$ ($N_2C_{10}H_8$ = 2,2′-bipyridine)

$Be(CH_2C_4H_9\text{-}t)_2 \cdot O(C_2H_5)_2$ yields with 2,2′-bipyridine a deep red adduct which quickly decomposes at room temperature to a black tar [7].

(+)(R)-$Be(CH_2CH(CH_3)C_2H_5)_2$

The compound is prepared from $BeCl_2$ and (+)(R)-$LiCH_2CH(CH_3)C_2H_5$ in an ether/petroleum ether mixture. After 3 h reflux and cooling overnight, the product is isolated from the upper layer of the reaction mixture by distillation at 75 to 78°C/0.05 Torr in 40% yield [3]. It is also formed from (+)(S)-$Zn(CH_2CH(CH_3)C_2H_5)_2$ and $Be(C_4H_9\text{-}i)_2$ via ligand exchange in C_6H_6 at 25°C, but no isolation is described [10].

The compound has a boiling point of 79 to 80°C/0.02 to 0.03 Torr and a specific rotation of $[\alpha]_D^{25} = +35.47°$ in $C_6H_5CH_3$ [2] or +35.2° in xylene with an optical purity of 93% [17].

It dissolves in C_6H_6 as a dimer (by cryoscopy). It undergoes 61% racemization in $C_6H_5CH_3$ on heating to 70°C for 3 h, but no racemization occurs in dioxane within 8 h at 70°C. Thermal decomposition at higher temperatures yields $C_2H_5(CH_3)C{=}CH_2$ and $Be(CH_2CH(CH_3)C_2H_5)H$. CO_2 is easily absorbed by a solution of the compound in ether, and (+)(S)-$C_2H_5CH(CH_3)CH_2COOH$ is formed [3]. The reductions of $C_6H_5C(O)R$ (R = C_2H_5, C_3H_7-i, C_4H_9-t) yield, via stereospecific reactions, the optically active carbinols. The asymmetric induction is 14.8, 46.2, and 30.8%, respectively [6]. Treatment with an equimolar amount of C_6H_5CN at 47°C for 6 h gives (+)(S)-$C_6H_5(C_2H_5C(CH_3)CH_2)CO$ in 95 ± 2% optical purity and 32% yield after hydrolysis [17].

$Be(CH_2C_6H_5)_2 \cdot O(C_2H_5)_2$

The complex is synthesized from anhydrous $BeCl_2$ and $Mg(CH_2C_6H_5)Cl$ in ether and isolated by short-path distillation at 120 to 150°C/10^{-3} Torr in the form of colorless needles, m. p. 50 to 51°C. Attemps to prepare the ether-free compound failed, e. g., from $B(CH_2C_6H_5)_3$ or $(B(CH_2C_6H_5)_2)_2O$ and $Be(C_2H_5)_2$ or from $Hg(CH_2C_6H_5)_2$ with Be or $Be(C_2H_5)_2$. The adduct is a monomer in C_6H_6 [9].

$(Be(C_{14}H_{10})_2)_n$ (polymeric bianthrylberyllium)

The reaction of the radical anions generated in "lithium anthracene" (in excess of anthracene) with $BeCl_2$ in THF yields a yellow, nonvolatile, air-sensitive precipitate after several hours of

References on p. 62

stirring at room temperature. After Soxhlet extraction with THF for 18 h, a pale yellow insoluble material remains in ~20% yield, which is microcrystalline according to X-ray powder diffraction.

Elemental analysis gave only unsatisfactory results. Even strongly chelating solvents, such as $(CH_3)_2NCH_2CH_2N(CH_3)_2$ or $OP(N(CH_3)_2)_3$, do not dissolve the complex. Hydrolysis leads to compound I. The same reaction with D_2O gives the 10,10′-dideuterated compound I. The results support the polymeric structure shown in Formula II [15].

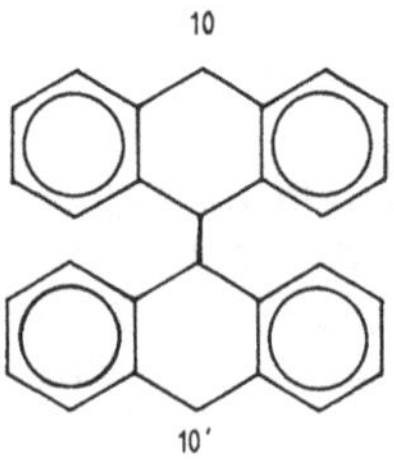

I

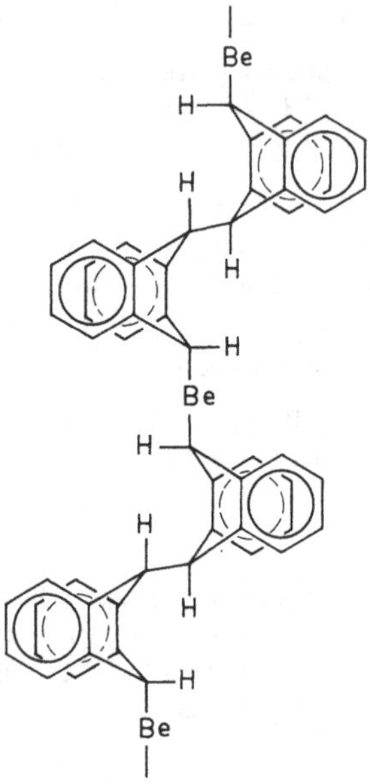

II

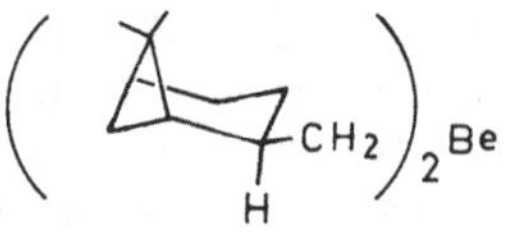

III

$Be(C_{10}H_{17})_2$ ($C_{10}H_{17}$ = [(1S,2S)-6,6-dimethylbicyclo[3.1.1]heptan-2-yl]methyl, Formula III)

The compound is prepared from cis-$C_{20}H_{18}MgCl$ and $BeCl_2$ (mole ratio 2:1) in ether at 0°C in 85% yield. Physical properties are not given.

The compound rapidly reduces prochiral ketones in moderate enantiomeric yields to the (R) alcohols. With i-$C_3H_7C(O)C_6H_5$ in ether, (+)i-$C_3H_7CH(OH)C_6H_5$ is obtained after hydrolysis with dilute H_2SO_4. For a conversion of 94%, the optical yield is 26% (R). Similar results are obtained with $(C_6H_5)_2CO$, t-$C_4H_9C(O)C_6H_5$, and t-$C_4H_9C(O)C{\equiv}CC_4H_9$ [18].

$Be(CH_2Si(CH_3)_3)_2$

$Be(C_2H_5)_2$ and $B(CH_2Si(CH_3)_3)_3$ are mixed and set aside for 8 h at room temperature. The mixture is separated by distillation at room temperature/10^{-3} Torr to remove $B(C_2H_5)_3$, and then at 65°C (bath temperature)/10^{-3} Torr to collect the compound [8].

The compound dissolves in C_6H_6 as a dimer. ^1H NMR spectrum (C_6H_6 at 30°C): δ = 0.14 (CH_3) and −0.17 ppm (CH_2). The spectrum of a solution in $C_6D_5CD_3$ changes quite extensively as the temperature is reduced, which is presumably due to a distinction between terminal and bridging $(CH_3)_3SiCH_2$ groups of the dimer. The chemical shift values are not given [8]. Adducts are formed with $N(CH_3)_2CH_2CH_2N(CH_3)_2$ [11] and 2,2′-bipyridine (see p. 61) [8]. An attempted reaction with $(t\text{-}C_4H_9)_2CO$ gave no definite product. Addition of $(C_6H_5)_2CO$ in C_6H_6 gives $Be(CH_2Si(CH_3)_3)OC(C_6H_5)_2CH_2Si(CH_3)_3$ [16].

$Be(CH_2Si(CH_3)_3)_2 \cdot N(CH_3)_2CH_2CH_2N(CH_3)_2$

This adduct is obtained from a slight excess of the diamine and $Be(CH_2Si(CH_3)_3)_2$ in C_6H_6. After evaporation of the solvent, it is recrystallized from C_6H_{14}, m. p. 96 to 97°C. It dissolves in C_6H_6 as a monomer (by cryoscopy) [11].

$Be(CH_2Si(CH_3)_3)_2 \cdot N_2C_{10}H_8$ ($N_2C_{10}H_8$ = 2,2'-bipyridine)

The adduct forms upon addition of 2,2'-bipyridine to $Be(CH_2Si(CH_3)_3)_2$ in C_6H_6. The solvent is removed, and the residue is crystallized from C_6H_{14}. The orange-red needles, m.p. 142°C, dissolve in C_6H_6 as monomers (by cryoscopy) [8].

$Be(CCl_3)_2$

It has been reported that anhydrous $BeCl_2$ reacts with $CHCl_3$ at 20°C with evolution of HCl. From a quantitative estimation of the content of Be and hydrolyzable Cl^-, it was concluded that products of the composition $Be(Cl)CCl_3$ and $Be(CCl_3)_2$ are present in solution. The products could not be isolated [1]. The experiment could not be repeated. Pure and dry $BeCl_2$ and $CHCl_3$ were found to give no reaction [4, 5].

$Be(-C_4H_8O(CH_3)-)_2$ (Formula IV)

The compound is prepared from anhydrous $BeCl_2$ and $Mg(C_4H_8OCH_3)Cl$ in ether. The mixture is warmed for 2 h, isooctane is added, and the ether is distilled. The isooctane solution is filtered, and the solvent is distilled. Vacuum distillation gives the compound in 45% yield. In a modified procedure, $BeCl_2 \cdot 2O(CH_3)C_4H_8Cl$ (prepared in situ) is added to Mg in ether. The mixture is boiled for 3 h, but the reaction is incomplete. Workup as before gives the compound in 22% yield. A polymeric byproduct is likely to be responsible for the low yields.

The compound is a colorless liquid, m.p. 7.6 to 7.8°C, b.p. 98 to 100°C/2.5 Torr, 108°C/4 Torr, 222 to 224°C/765 Torr.

In C_6H_6, only monomers are present (by cryoscopy). Hydrolysis yields $CH_3OC_4H_9$ and $Be(OH)_2$. No adduct is formed with an excess of ether [2].

$Be(-C_3H_6S(C_2H_5)-)_2$ (Formula V)

The compound is prepared from anhydrous $BeCl_2$ and $Mg(C_3H_6SC_2H_5)Cl$ as described for the previous compound, yield 62%. If the reaction is performed in $S(CH_3)_2$, the yield is only 35%.

The compound is a colorless liquid, m. p. 7 to 8°C, b. p. 122 to 123°C/2 Torr or 220 to 225°C/760 Torr (with partial decomposition).

In C_6H_6, only monomers are present (by cryoscopy). Hydrolysis yields $C_2H_5SC_3H_7$ and $Be(OH)_2$. No adduct is formed with an excess of ether [2].

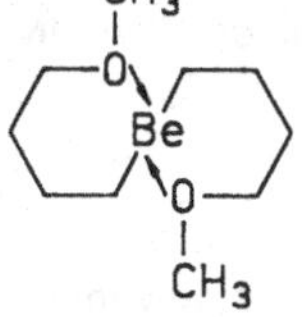

IV

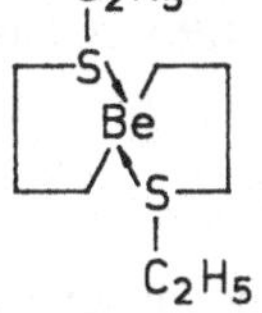

V

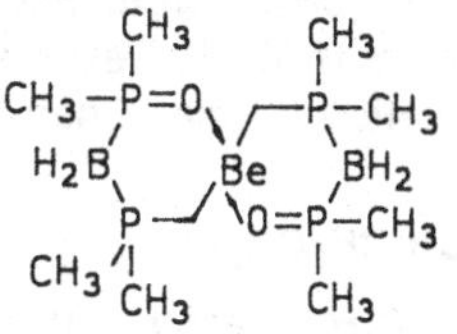

VI

References on p. 62

$Be(-CH_2P(CH_3)_2BH_2P(CH_3)_2O-)_2$ (Formula VI, p. 61)

The compound is prepared by addition of equivalent amounts of $BeCl_2$ to $Li(-CH_2P(CH_3)_2BH_2P(CH_3)_2O-)$ in THF at −78°C. After 16 h the solvent is evaporated at room temperature, and the product is crystallized from $C_6H_5CH_3$ and sublimed at 60 to 100°C/10^{-4} Torr to produce colorless crystals, m.p. 59°C. ^1H NMR(C_6D_6): $\delta = -0.13$ (d, CH_2; J(P,H) = 16.0 Hz), 1.11 (m, br, CH_3), and 1.27 ppm (d, CH_3; J(P,H) = 11.0 Hz). ^{13}C NMR(C_6D_6): $\delta = 6.7$ (br, CH_2), 15.6 (dd, CH_3P; $^1J(P,C) = 42$, $^3J(P,C) = 3$ Hz), 18.9 (d, CH_3P; $^1J(P,C) = 46$ Hz), and 20.8 ppm (d, br, CH_3P; $^1J(P,C) = 44$ Hz). ^{31}P NMR(C_6D_6): $\delta = -5.07$ (dq, PC_3; J(P,B) = 95, J(P,P) = 15 Hz), and 63.62 ppm (dq, PO; J(P,B) = 107, J(P,P) = 15 Hz). In agreement with Formula VI, the P atoms are not equivalent, nor are the two CH_3 groups on each P. The vibration at 1065 cm^{-1} for $\nu(P{=}O)$ in the IR spectrum indicates O–Be coordination.

The compound is moderately sensitive to air and moisture and soluble in polar aprotic solvents. The molecular ion is observed in the mass spectrum. On standing at room temperature, oligomerization is observed in the solid state [13].

References:

[1] Silber, P. (Ann. Chim. [Paris] [12] **7** [1952] 182/233, 218/29).
[2] Bähr, G., Thiele, K.-H. (Chem. Ber. **90** [1957] 1578/86).
[3] Lardicci, L., Lucarini, L., Palagi, P., Pino, P. (J. Organometal. Chem **4** [1965] 341/8).
[4] Andersen, R. A., Kovar, R. A., Morgan, G. L. (unpublished observations 1969 from [5]).
[5] Coates, G. E., Morgan, G. L. (Advan. Organometal. Chem. **9** [1970] 195/257, 220).
[6] Giacomelli, G. P., Menicagli, R., Lardicci, L. (Tetrahedron Letters **1971** 4135/8).
[7] Coates, G. E., Francis, B. R. (J. Chem. Soc. A **1971** 1305/8).
[8] Coates, G. E., Francis, B. R. (J. Chem. Soc. A **1971** 1308/10).
[9] Coates, G. E., Srivastava, R. C. (J. Chem. Soc. Dalton Trans. **1972** 1541/4).
[10] Lardicci, L., Giacomelli, G. P., De Bernardi, L. (J. Organometal. Chem. **39** [1972] 245/50).

[11] Coates, G. E., Smith, D. L., Srivastava, R. C. (J. Chem. Soc. Dalton Trans. **1973** 618/22).
[12] Baker, R. W., Brendel, G. J., Lowrance, B. R., Mangham, J. R., Marlett, E. M., Shepherd, L. H. (J. Organometal. Chem. **159** [1978] 123/30).
[13] Schmidbaur, H., Weiß, E. (Angew. Chem. **93** [1981] 300/2; Angew. Chem. Intern. Ed. Engl. **20** [1981] 283/4).
[14] Blindheim, U., Coates, G. E., Srivastava, R. C. (J. Chem. Soc. Dalton Trans. **1972** 2302/5).
[15] Berke, C. M, Streitwieser Jr., A. (J. Organometal. Chem. **197** [1980] 123/34).
[16] Andersen, R. A., Coates, G. E. (J. Chem. Soc. Dalton Trans. **1974** 1171/80).
[17] Giacomelli, G. P. Lardicci, L. (Chem. Ind. [London] **1972** 689/90).
[18] Giacomelli, G., Falorni, M, Lardicci, L. (Gazz. Chim. Ital. **115** [1985] 289/91).

1.2.2 Divinyl- and Diallylberyllium, and Adducts of $Be(CH_2CH{=}CH_2)_2$

A divinylberyllium compound could not be prepared up to now. Attempts to prepare **$Be(CH{=}C(CH_3)_2)_2$** from $BeCl_2$ and $Li[CH{=}C(CH_3)_2]$ resulted only in polymeric decomposition products [1, 3]. However, the parent **$Be(CH{=}CH_2)_2$** and its cation were the subject of an MNDO calculation [5]; see Chapter 5, p. 203.

$Be(CH_2CH{=}CH_2)_2$

$Be(C_2H_5)_2$ reacts with stoichiometric amounts of $B(CH_2CH{=}CH_2)_3$ without a solvent or in hydrocarbon solvents with a fast and quantitative exchange of the C_2H_5 and $CH_2CH{=}CH_2$

groups. $B(C_2H_5)_3$ can be evaporated, but $Be(CH_2CH{=}CH_2)_2$ cannot be isolated since it is transferred to oligomeric organoberyllium compounds, which separate as a white solid at a reaction temperature of 0 to 60°C or as a viscous mass below 0°C reaction temperature. The formation of oligomers is proved by hydrolysis, which yields terminally unsaturated hydrocarbons, e.g., $CH_2{=}CHCH_2CH(CH_3)CH_3$ and a nonadiene. The IR spectrum of the solid product in KBr shows an absorption at 1650 cm^{-1} for ν(C=C) and broad bands between 1300 and 400 cm^{-1}, characteristic for overlapping bands of BeC and CH vibrations.

The oligomerization reaction can be suppressed in the presence of donor molecules. In THF as solvent, the exchange reaction between $Be(C_2H_5)_2$ and $B(CH_2CH{=}CH_2)_3$ is slowed down and is less complete than in nonpolar solvents. Hydrolysis of the product gives C_2H_6, which results from unreacted $Be(C_2H_5)_2$. The product has the composition $Be(CH_2CH{=}CH_2)_2 \cdot 2OC_4H_8$ (OC_4H_8 = tetrahydrofuran). An adduct, probably $Be(CH_2CH{=}CH_2)_2 \cdot N_2C_{10}H_8$, is also obtained by addition of 2,2′-bipyridine to an ether solution of the product [4].

An AX_4 pattern in the ^1H NMR spectrum of $Be(CH_2CH{=}CH_2)_2$ in ether is mentioned (data not given) [2, 3].

$Be(CH_2CH{=}CH_2)_2 \cdot 2OC_4H_8$ (OC_4H_8 = tetrahydrofuran)

The adduct is described as a product in the reaction of $Be(C_2H_5)_2$ with $B(CH_2CH{=}CH_2)_3$ in THF. The composition of the very air-sensitive solid is deducted from an elemental analysis. Hydrolysis yields about 49% of the expected $CH_3CH{=}CH_2$ [4].

$Be(CH_2CH{=}CH_2)_2 \cdot N_2C_{10}H_8$ ($N_2C_{10}H_8$ = 2,2′-bipyridine)

Treatment of the products obtained from $Be(C_2H_5)_2$ and $B(CH_2CH{=}CH_2)_3$ in ether with 2,2′-bipyridine develops a deep red color ascribed to a 2,2′-bipyridine complex similar to those known from beryllium dialkyls [4].

References:

[1] Coates, G. E., Cox, G. F. (unpublished observations 1961 from [3]).
[2] Drew, D., Morgan, G. L. (unpublished observations 1969 from [3]).
[3] Coates, G. E., Morgan, G. L. (Advan. Organometal. Chem. **9** [1970] 195/257, 206/7).
[4] Wiegand, G., Thiele, K.-H. (Z. Anorg. Allgem. Chem. **405** [1974] 101/8).
[5] Glidewell, C. (J. Organometal. Chem. **217** [1981] 273/80).

1.2.3 Dialkynylberyllium Compounds and Their Adducts

General Remarks. Completely uncomplexed dialkynylberyllium compounds could not be prepared. The prepared complexes are obtained as etherates. $Be(C{\equiv}CC_6H_5)_2$ forms a 1:1 etherate, which probably exists as an equilibrium of monomers and dimers in solution [2]. From the etherates of $Be(C{\equiv}CCH_3)_2$ and $Be(C{\equiv}CC_4H_9\text{-}t)_2$, the ether can be largely removed in vacuum, whereby the compounds form electron-deficient polymers with bridging alkynyl groups [3].

The etherates are described below. They are the starting materials for adducts with other donor molecules. These adducts have 1:1 or 1:2 stoichiometries (determined from the molecular weight in C_6H_6 by cryoscopy) and are represented by the structures shown in Formulas I, II, III, and IV [2, 3]. The adducts are summarized in Table 14, pp. 65/7.

The phenyl derivatives are not very sensitive to air and moisture [2].

References on p. 68

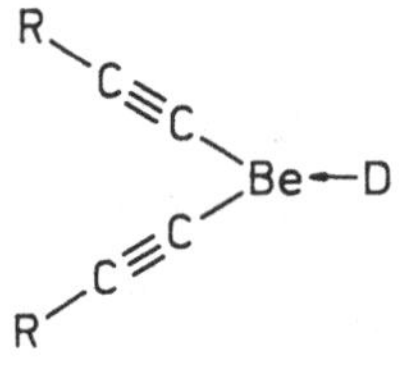

I

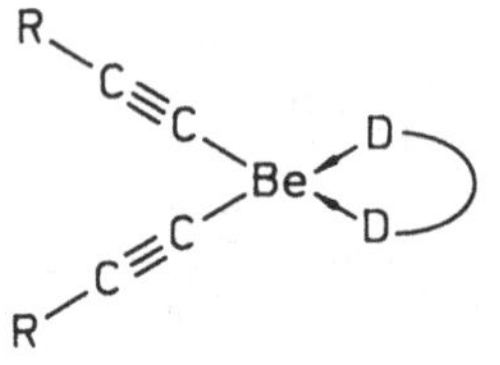

II

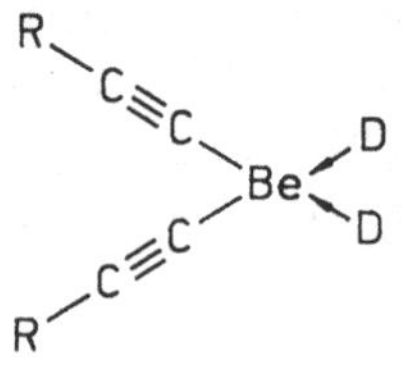

III

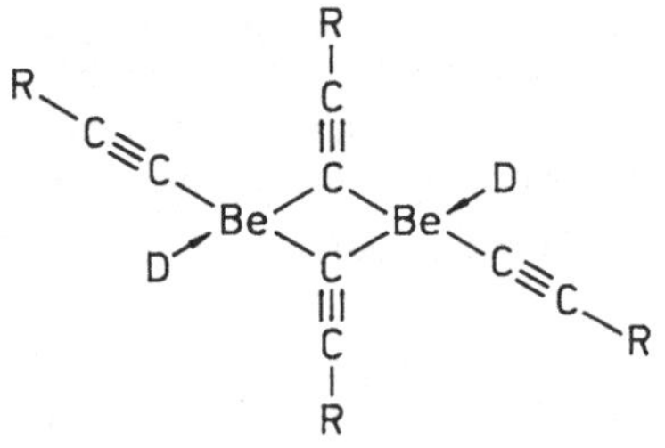

IV

The experimentally unknown parent compound of this series, **Be(C≡CH)$_2$** and its cation were investigated by an MNDO calculation [4]; see Chapter 5, p. 203.

Be(C≡CCH$_3$)$_2$·nO(C$_2$H$_5$)$_2$ (0<n≦1?)

$BeCl_2$ in ether is added to a suspension of LiC≡CCH$_3$ (prepared in situ from LiC_4H_9 and CH$_3$C≡CH) in C_6H_{14}/ether (mole ratio 1:2). Filtration and removal of the solvent gives a white etherate (n=1?). It is still soluble in C_6H_6, but not in C_6H_{14}, and is the starting material for the THF adduct (Table 14, No. 9, p. 66). When it is held under vacuum for 30 min at room temperature, it is no longer soluble even in hot C_6H_6. Analysis indicates a composition with n=0.37, and alcoholysis with $CH_3OCH_2CH_2OH$ is extremely vigorous even at −60°C. After additional 18 h at 25°C/10^{-3} Torr, a material with n=0.05 remains. It is believed to be a propynyl-bridged polymer. On account of the explosion risk, no melting point was determined. IR vibrations (Nujol) are found at 2119(s, νC≡C), 1360(w), 1032(w), 994(m), 976(m), 703(m), 494(s), and 473(s) cm^{-1}.

The compound does not dissolve in $S(CH_3)_2$. It is the starting material for the adducts Nos. 1, 10, 18 to 20 in Table 14, pp. 65/7 [3].

Be(C≡CC$_4$H$_9$-t)$_2$·nO(C$_2$H$_5$)$_2$ (0<n≦1?)

$BeCl_2$ in ether is added to LiC≡CC$_4$H$_9$-t in C_6H_6 (mole ratio 1:2). After filtration the solvent is evaporated until a small volume remains. Addition of C_6H_{14} gives a white solid (n=1?). If this solid is pumped at room temperature for 18 h under reduced pressure, a product that is insoluble in C_6H_6 is produced. The IR spectrum (Nujol) contains a medium-strong unsymmetrical IR vibration for ν(C≡C) at 2090 cm^{-1} with a weak shoulder at 2100 cm^{-1}. A polymeric structure with bridging butynyl groups is suggested. This compound is the starting material for the adducts Nos. 5 and 21 in Table 14, pp. 66/7 [3].

$Be(C{\equiv}CC_6H_5)_2 \cdot O(C_2H_5)_2$

$LiC{\equiv}CC_6H_5$ in ether (prepared in situ from LiC_4H_9 and $C_6H_5C{\equiv}CH$) is filtered into $BeCl_2$ (mole ratio 2:1) in ether. Concentration of the filtrate gives colorless needles, m.p. 149 to 150°C with decomposition. The IR spectrum in C_6H_{14} shows ν(C≡C) at 2105 cm^{-1}; ^1H NMR spectrum (C_6D_6): δ=1.10(t,CH_3), 4.07(q,CH_2), 7.09 and 7.71 ppm (m,C_6H_5). The degree of association was found to be ca. 1.6 to 1.7 in C_6H_6 (by cryoscopy). This is in agreement with an equilibrium between the monomer, Formula I (D = $O(C_2H_5)_2$) and the dimer, Formula IV (D = $O(C_2H_5)_2$).

The compound gradually decomposes at room temperature with the development of a yellow color. A red color develops above 120°C. Reaction with $Be(OC_4H_9\text{-}t)_2$ in a 1:1 mole ratio gives $(Be(C{\equiv}CC_6H_5)OC_4H_9\text{-}t)_n$; and in a 1:2 ratio, $Be_3(C{\equiv}CC_6H_5)_2(OC_4H_9\text{-}t)_4$ (Formula V, p. 67) results. The $O(C_2H_5)_2$ ligand is replaced by other donors, see Table 14, Nos. 2, 3, 7, 12, pp. 65/6 [2].

The adducts described in Table 14 are prepared by the following methods:

Method I: Nearly ether-free $Be(C{\equiv}CR)_2$ (R = CH_3, C_4H_9-t) or $Be(C{\equiv}CC_6H_5)_2 \cdot O(C_2H_5)_2$ reacts with the appropriate donor in C_6H_6 in a 1:1 ratio (Nos. 2, 3, 5, 7, 18 to 21) or with an excess of donor (Nos. 1, 9, 12). For Nos. 2, 19, and 21, $N(CH_3)_3$ is condensed onto $Be(C{\equiv}CR)_2$ in C_6H_6 at −196°C. No. 9 is prepared from the etherate of $Be(C{\equiv}CCH_3)_2$ dissolved in THF. Reaction occurs at room temperature, and the product precipitates directly (Nos. 2, 5, 7), by removal of the solvent (Nos. 1, 3), or by addition of C_6H_{14} (Nos. 9, 12). For Nos. 18 to 21, the mixture is heated until the precipitate is dissolved (40 to 70°C), then filtered, and the solvent is evaporated (Nos. 18, 19, 21), or the mixture is cooled (No. 20) [2, 3].

Method II: An excess of the donor is condensed onto $Be(C{\equiv}CR)_2 \cdot 2OC_4H_8$ in C_6H_6 at −196°C (Nos. 10, 13, 14) or the donor is added to $Be(C{\equiv}CR)_2 \cdot 2OC_4H_8$ in C_6H_6 (Nos. 4, 6, 10, 13 to 17), THF (No. 8), or without solvent (No. 7) in a 1:1 mole ratio (Nos. 4, 8, 15) or in excess (Nos. 6, 7, 16, 17). The products precipitate directly (Nos. 7, 15 to 17), after warming (Nos. 4, 10, 13, 14), or after addition of C_6H_{14} (No. 6) [2, 3].

Method III: The dimer $(Be(C{\equiv}CCH_3)_2 \cdot NC_5H_5)_2$ in C_6H_6 reacts with pyridine. No. 11 crystallizes as the solution is cooled [3].

Table 14

$Be(C{\equiv}CR)_2 \cdot nD$ (n = 1,2) and $(Be(C{\equiv}CR)_2 \cdot D)_2$ Adducts.

Further information on compounds preceded by an asterisk is given at the end of the table.
Explanations, abbreviations, and units on p. X.

No.	R, D method of preparation [Ref.]	properties and remarks [Ref.]
1:1 monomeric adducts (Formula I, p. 64)		
1	CH_3, $N(C_2H_5)_3$ I [3]	small prisms, m.p. 75 to 76° (from C_6H_6/C_6H_{14}) IR: ν(C≡C) at 2135 [3]
*2	C_6H_5, $N(CH_3)_3$ I [2]	small white needles, m.p. 186 to 188° (dec.) (from hot C_6H_6) IR (Nujol): ν(C≡C) at 2085(m) [2]
*3	C_6H_5, $N(C_2H_5)_3$ I [2]	large colorless prisms, m.p. 102 to 106° (from C_6H_6/C_6H_{14}) ^1H NMR (C_6D_6): 0.81(t,CH_3), 2.59(q,CH_2), 7.06(m,C_6H_5) IR (Nujol): ν(C≡C) at 2090(m) [2]

Table 14 (continued)

No.	R, D method of preparation [Ref.]	properties and remarks [Ref.]
1:1 monomeric adducts (Formula II, p. 64)		
4	CH_3, $N(CH_3)_2CH_2CH_2N(CH_3)_2$ II [3]	colorless needles, m.p. 196 to 201° (from C_6H_6 or sublimed) sublimation at 100°/10^{-3} Torr sample contains 1 mol C_6H_6, which is lost on heating IR (Nujol): ν(C≡C) at 2128(w) [3]
5	t-C_4H_9, $N_2C_{10}H_8$ (2,2′-bipyridine) I [3]	white solid, m.p. 292 to 296° (dec.) IR (Nujol?): ν(C≡C) at 2106(w) insoluble in boiling C_6H_6 darkens at 220° [3]
6	C_6H_5, $CH_3OCH_2CH_2OCH_3$ II [2]	white feathery crystals, m.p. 190° (dec.) cannot be redissolved in C_6H_6 [2]
*7	C_6H_5, $N(CH_3)_2CH_2CH_2N(CH_3)_2$ I, II, [2]	colorless needles, m.p. 168 to 170° (from C_6H_6/C_6H_{14}) ^{1}H NMR (C_6D_6): 1.91 (s, CH_2), 2.29 (s, CH_3), 7.15 and 7.70 (m, C_6H_5) [2]
8	C_6H_5, $N_2C_{10}H_8$ (2,2′-bipyridine) II [2]	white plates, m.p. 295 to 302° (dec.) [2]
1:2 monomeric adducts (Formula III, p. 64)		
9	CH_3, OC_4H_8 (THF) I [3]	colorless needles, m.p. 105 to 106° ^{1}H NMR (C_6D_6): 1.34 (t, H-3,4 in THF), 1.94 (s, CH_3), 3.95 (t, H-2,5 in THF) IR (Nujol): ν(C≡C) at 2137(w) starting material for adducts Nos. 4 and 10 [3]
10	CH_3, $N(CH_3)_3$ II [3]	colorless needles, m.p. 172 to 177° (dec.) IR (Nujol): ν(C≡C) at 2123(w) [3]
11	CH_3, NC_5H_5 III [3]	dec. at 130° IR (Nujol): ν(C≡C) at 2125(w) [3]
*12	C_6H_5, OC_4H_8 (THF) I [2]	colorless needles, m.p. 138 to 140° (dec.) (from THF/C_6H_{14} or C_6H_6/C_6H_{14}) ^{1}H NMR (C_6D_6): 1.41 (t, CH_2), 3.98 (t, CH_2), 7.14 and 7.70 (m, C_6H_5) starting material for adducts Nos. 6 to 8, 13 to 17 [2]
*13	C_6H_5, NH_2CH_3 II [2]	white precipitate, m.p. 86 to 87° (dec.) IR (Nujol?): ν(C≡C) at 2100(vw) [2]
14	C_6H_5, $NH(CH_3)_2$ II [2]	colorless needles, m.p. 151 to 153° (from hot C_6H_6) [2], 152 to 158° (dec.) [1] hardly soluble in C_6H_6 at 20° [2]
15	C_6H_5, $NH(C_6H_5)_2$ II [2]	white precipitate, softens at 204 to 207°, does not melt below 310° IR (Nujol): ν(C≡C) at 2114(vw) insoluble in C_6H_6 [2]

Table 14 (continued)

No.	R, D method of preparation [Ref.]	properties and remarks [Ref.]
16	C_6H_5, $N(CH_3)_3$ II [2]	white precipitate, m.p. 195 to 200° (dec.) heating in vacuum: see No. 2, p. 65 [2]
17	C_6H_5, NC_5H_5 II [2]	white precipitate, m.p. 226 to 230° (dec.) insoluble in $C_6H_5CH_3$ at 80°, C_6H_{14}, and THF [2]
1:1 dimeric adducts (Formula IV, p. 64)		
18	CH_3, OC_4H_8 (THF) I [3]	colorless prisms, m.p. 144 to 146° (dec.), from C_6H_6/C_6H_{14} ^{1}H NMR(C_6H_6): 1.18(m, H-3,4 in THF), 1.92(s, CH_3), 4.07(m, H-2,5 in THF) IR (Nujol): ν(C≡C) terminal at 2138(w), ν(C≡C) bridging at 2126 (ms) [3]
*19	CH_3, $N(CH_3)_3$ I [3]	colorless plates, m.p. 185 to 190° (dec.) (from C_6H_6/C_6H_{14}) ^{1}H NMR(C_6H_6 at 30°): 1.05(s, CH_3C), 1.79(s, CH_3N) IR (Nujol): ν(C≡C) terminal at 2134(w), ν(C≡C) bridging at 2115(s) [3]
20	CH_3, NC_5H_5 I [3]	crystals, dec. at ~130° hardly soluble in C_6H_6 IR (Nujol): ν(C≡C) terminial at 2136, ν(C≡C) bridging at 2120 [3]
*21	t-C_4H_9, $N(CH_3)_3$ I [3]	colorless prisms, m.p. 119° (from C_6H_{14}) ^{1}H NMR(C_6H_6 at 30°): 1.31(CH_3C), 2.59(CH_3N) IR (Nujol): ν(C≡C) terminal at 2111(w), ν(C≡C) bridging at 2091(s) [3]

* Further information:

$Be(C{\equiv}CC_6H_5)_2 \cdot N(CH_3)_3$ (Table **14**, No. **2**) is also obtained by heating the 1:2 adduct, No. 16, at 90 to 100°C/10^{-3} Torr for 24 h. It is only sparingly soluble in C_6H_6 at room temperature; therefore, the molecular weight was not determined. It may be a dimeric 1:1 complex of Formula IV, p. 64 [2].

$Be(C{\equiv}CC_6H_5)_2 \cdot N(C_2H_5)_3$ (Table **14**, No. **3**) reacts in a 1:1 ratio with t-C_4H_9OH in C_6H_6 to give $Be(C{\equiv}CC_6H_5)OC_4H_9$-t, whereas a slight excess of the alcohol (mole ratio 3:4) leads to $Be_3(C{\equiv}CC_6H_5)_2(OC_4H_9\text{-t})_4$ (Formula V) [2].

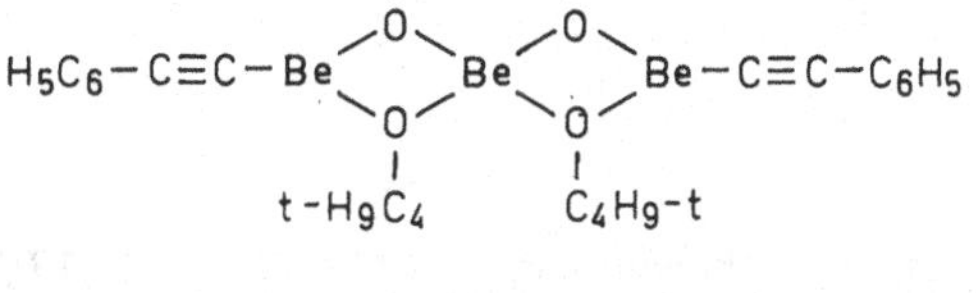

V

$Be(C{\equiv}CC_6H_5)_2 \cdot N(CH_3)_2CH_2CH_2N(CH_3)_2$ (Table **14**, No. **7**) is also obtained in 69% yield by reaction of $Be(C_2H_5)_2 \cdot N(CH_3)_2CH_2CH_2N(CH_3)_2$ with $C_6H_5C{\equiv}CH$ at 80°C for 24 days in a sealed tube. It can be sublimed at 110 to 120°C/10^{-3} Torr, but with appreciable loss of material [2].

$Be(C{\equiv}CC_6H_5)_2 \cdot 2OC_4H_8$ (OC_4H_8 = tetrahydrofuran, Table **14**, No. **12**) forms no stable adduct with NH_3 [1]. OC_4H_8 is also not displaced by $N(C_2H_5)_3$. It reacts with CH_3OH in C_6H_6 to give $Be(OCH_3)_2$, $C_6H_5C{\equiv}CH$, and C_4H_8O. With t-C_4H_9OH, C_6H_5OH, and t-C_4H_9SH, the adducts $Be(C{\equiv}CC_6H_5)R \cdot OC_4H_8$ (R = OC_4H_9-t, OC_6H_5, SC_4H_9-t) are obtained. Addition of $LiC{\equiv}CC_6H_5$ (mole ratio 1:1 or 1:2) yields $Li_2[Be(C{\equiv}CC_6H_5)_4]$ [2].

$Be(C{\equiv}CC_6H_5)_2 \cdot 2NH_2CH_3$ (Table **14**, No. **13**) eliminates $C_6H_5C{\equiv}CH$ only upon heating to 60 to 80°C [1, 2]. Heating in C_6H_6 to 80°C gives a product that is insoluble in C_6H_6 and does not melt when heated to 300°C. The compound was not further identified [2].

$Be(C{\equiv}CCH_3)_2 \cdot N(CH_3)_3$ and **$Be(C{\equiv}CC_4H_9\text{-t})_2 \cdot N(CH_3)_3$** (Table **14**, Nos. **19** and **21**) react with $Be(CH_3)_2 \cdot N(CH_3)_3$ in $C_6H_5CH_3$ to give $(Be(CH_3)C{\equiv}CR \cdot N(CH_3)_3)_2$ (R = CH_3, t-C_4H_9) [3].

It follows from an X-ray crystal structure determination that the crystals of No. 21 are triclinic with a = 9.115(1), b = 13.519(1), c = 8.384(1) Å, α = 102.8(1)°, β = 99.0(1)°, and γ = 94.8(1)°; space group $P\bar{1}$-C_i^1 (No. 2); Z = 4; R = 0.076 for 1493 independent reflections.

The unit cell contains two independent centrosymmetric dimers in which the alkynyl groups exhibit different types of interactions with the Be atoms. The two dimers are shown in **Fig. 9**a and Fig. 9b together with the important bond distances and bond angles. The Be atoms are in approximately tetrahedral environments in the two dimers. In the dimer shown in Fig. 9a only C(1) is involved in the bonding to both Be atoms, while the dimer in Fig. 9b exhibits nearly a side-on coordination of the C(1)≡C(2) and C(1')≡C(2') bond [5].

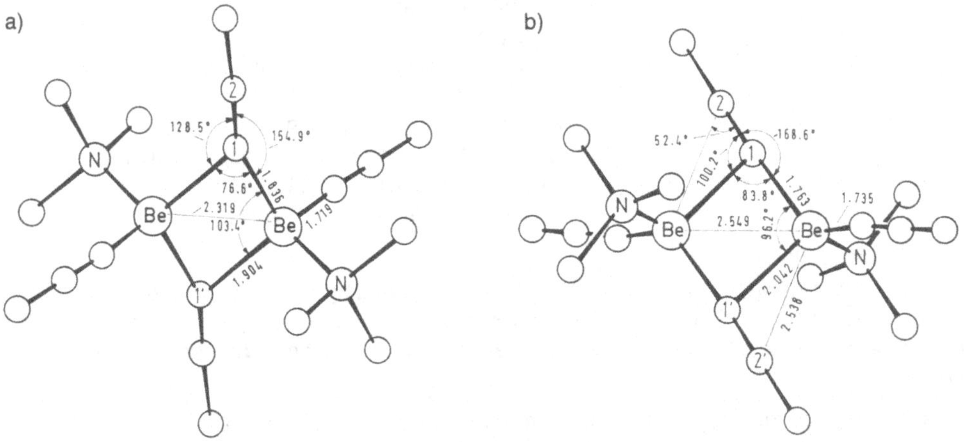

Fig. 9 a and b. Molecular structures of the two independent dimers of $Be(C{\equiv}CCH_3)_2 \cdot N(CH_3)_3$ [5].

References:

[1] Coates, G. E., Morgan, G. L. (Advan. Organometal. Chem. **9** [1970] 195/257, 210).
[2] Coates, G. E., Francis, B. R. (J. Chem. Soc. A **1971** 160/4).
[3] Coates, G. E., Francis, B. R. (J. Chem. Soc. A **1971** 474/7).
[4] Glidewell, C. (J. Organometal. Chem. **217** [1981] 273/80).
[5] Bell, N. A., Nowell, I. W., Shearer, H. M. M. (J. Chem. Soc. Chem. Commun. **1982** 147/8).

1.2.4 Diarylberyllium Compounds and Their Adducts

Diarylberyllium compounds have been less extensively studied than the dialkyl compounds. A number of donor-free species are analytically characterized, but little structural information is available. This also applies to the 1:1 and 1:2 adducts with O-, S-, N-, and P-containing donor molecules. In many ways, beryllium diaryls resemble the aluminium triaryls [15].

1.2.4.1 Diphenylberyllium, $Be(C_6H_5)_2$

Preparation

Donor-free $Be(C_6H_5)_2$ is readily prepared from $Hg(C_6H_5)_2$ and metallic Be in an inert high-boiling solvent. The earliest reports describe the preparation in sealed tubes in the absence of a solvent at 225°C in the presence of a trace of $HgCl_2$ [1], at 210 to 220°C [9], or at 170°C with $HgCl_2$ and $BeBr_2$ as catalysts [21]. It was later shown that the procedure can also be carried out in conventional inert gas protected glassware at temperatures up to 220°C [8]. The apparent method of choice is to carry out the reaction in dry o-xylene at 140 [13, 28] to 150°C [5, 6]. Metal activation by overnight contact with an ether solution of $Be(C_2H_5)_2$ [13, 16], or addition of a trace of $HgCl_2$ to the reaction mixture are advantageous for shorter reaction times (72 to 180 h at 140°C) and better yields [13,16,28]. The insoluble $Be(C_6H_5)_2$ is separated from the xylene by filtration. Then the compound, contaminated with Hg, is recrystallized from hot C_6H_6 [28].

Another method is the ligand exchange reaction of $Be(C_2H_5)_2$ and $B(C_6H_5)_3$ at room temperature in the absence of a solvent, which gives an 88% yield of $Be(C_6H_5)_2$, or the ligand exchange between $Be(C_2H_5)_2$ and $Hg(C_6H_5)_2$ in xylene at 90°C. After 15 h, the product crystallizes from the reaction mixture in 78% yield [16].

Phenylation of anhydrous $BeCl_2$ in ether with the Grignard reagent $Mg(C_6H_5)Br$ gives ethereal solutions of $Be(C_6H_5)_2$, which can be used for further reactions. After addition of dioxane $Be(C_6H_5)_2 \cdot 2O(C_2H_5)_2$ crystallizes. No ether-free material can be obtained from this adduct [13]. The phenylation of $BeCl_2$ with LiC_6H_5 [2, 14, 20] also gives ethereal solutions, from which no donor-free $Be(C_6H_5)_2$ can be obtained [5, 16]. See also p. 71.

Enthalpy of Formation

Caloric measurements of the acid hydrolysis of $Be(C_6H_5)_2$ lead to a standard enthalpy of formation of the solid of $\Delta H^\circ_{f(solid)} = 153.1 \pm 2.5$ kJ/mol. The enthalpy of formation of monomeric gaseous $Be(C_6H_5)_2$ of $\Delta H^\circ_{f(gas)} \approx 303$ kJ/mol was calculated from an estimated value for the heat of sublimation of the monomer of $\Delta H_s \approx 150$ kJ/mol [17].

Molecular Parameters and Physical Properties

$Be(C_6H_5)_2$ forms large transparent crystals from C_6H_6 [8], described as hexagonal prisms [5]. Melting points are given as 244 to 248°C (with dec.) [9], 245 to 250°C (with dec.) [28], 248 to 250°C (with dec.) [13,16], and 249 to 251°C [16].

It follows from a mass spectrometric study that the volatility of $Be(C_6H_5)_2$ is very low. The heat of sublimation of the monomer, obtained from measurements of the variation of pressure versus probe temperature in a mass spectrometer without applying an electron voltage, was roughly estimated to $\Delta H_s = 150$ to 200 kJ/mol. A combination of $\Delta H^\circ_f(Be(C_6H_5)_{2\,gas})$, $\Delta H^\circ_f(Be_{gas})$, and $\Delta H^\circ_f(C_6H_{5\,gas})$ leads to a Be–C bond dissociation energy of D(Be–C) ≈ 336 kJ/mol. An ionization potential of 9.2 ± 0.1 eV was obtained from a mass spectroscopic study [17].

References on p. 76

A characteristic IR band at 761 cm^{-1} for the out-of-plane CH vibration in Nujol is mentioned in a comparative study of substituted benzenes [7]. The ^{1}H NMR spectrum of the etherate is said to be as expected, but no details are published [14].

No structural information is available concerning pure solid $Be(C_6H_5)_2$, whose limited solubility in C_6H_6 may be due to a polymeric constitution in the crystalline state or to a very good crystal packing of an oligomer of high symmetry. No molecular mass was determined for solutions [16].

Calculations for monomeric and dimeric $Be(C_6H_5)_2$ were made by the PRDDO method. The monomeric $Be(C_6H_5)_2$, with perpendicular C_6H_5 rings, is calculated to be 0.4 kcal/mol more stable than the isomer with coplanar C_6H_5 rings. The calculated total energies for the monomers are: −474.3369 a.u. (perpendicular C_6H_5) and −474.3362 a.u. (coplanar C_6H_5). The dimeric structure is strongly favored. Only the dimers with perpendicular configuration of the bridging C_6H_5 groups relative to the BeC_2Be plane are considered, since the planar structure should be much higher in energy in comparison to $HBe(C_6H_5)_2BeH$. The dimerization energy is given as −47.4 kcal/mol for a planar conformation of the terminal C_6H_5 groups and −39.2 kcal/mol for a perpendicular conformation of the terminal C_6H_5 groups. The isomer with the planar terminal C_6H_5 groups is slightly favored. The calculated total energies are: −948.7493 a.u. (perpendicular terminal C_6H_5 groups) and −948.7349 a.u. (planar terminal C_6H_5 groups) [25].

Chemical Reactions

The compound is soluble in C_6H_6, xylene, ether, dioxane, and THF. Dipole moments are found as $\mu = 1.6$ and 4.3 D in C_6H_6 and dioxane, respectively. The high value in dioxane is taken as evidence for adduct formation in this solvent [8]. $Be(C_6H_5)_2$ dissolves in ether and dioxane with warming [5].

Pyrolysis of $Be(C_6H_5)_2$ at 240 to 250°C in vacuum gives C_6H_6 as the main volatile product. In an electron impact mass spectroscopic study, it is shown that, at a source temperature of 180 to 200°C and 70 eV, no Be-containing ions but only hydrocarbon ions are produced in the unsaturated vapor. At 200 to 240°C source temperature the spectrum changes dramatically and the Be-containing ions $[Be_3(C_6H_5)_6]^+$, $[Be(C_6H_5)_2]^+$, and $[BeC_6H_5]^+$ are produced, which disappear again above 250°C (base peak $[C_6H_6]^{\bullet+}$). The appearence potentials for $[Be(C_6H_5)_2]^+$ and for $[BeC_6H_5]^+$ are 9.2 ± 0.1 eV and 13.4 ± 0.2 eV. This leads to a dissociation energy for the Be–C bond of 412 ± 20 kJ/mol. A fragmentation scheme for the monomer $Be(C_6H_5)_2$ is proposed from the traces of metastable peaks [17]; see also [28].

An excess of dry HCl reacts with $Be(C_6H_5)_2$ to give $BeCl_2$ and C_6H_6 [13]. Acid hydrolysis with excess 1.16 M H_2SO_4 leads to $BeSO_4$ and C_6H_6. The reaction enthalpy is $\Delta H = -440.6 \pm 2.5$ kJ/mol [17]. LiH is added in ether to form the crystalline salt $Li[Be(C_6H_5)_2H]$ as a monoetherate [6]. LiC_6H_5 gives the analogous product $Li[Be(C_6H_5)_3]$ [3 to 5]; $Na[C(C_6H_5)_3]$ yields $Na[Be(C_6H_5)_2C(C_6H_5)_3]$ as a dietherate [5]; and $LiC_{13}H_9$ ($C_{13}H_9$ = fluorenyl) gives $Li[Be(C_6H_5)_2C_{13}H_9] \cdot 4O_2C_4H_8$ ($O_2C_4H_8$ = dioxane) upon addition of dioxane [5].

Rapid redistribution in the system $Be(C_6H_5)_2$–$BeBr_2$ in ether was observed by selected precipitation studies with dioxane, molecular association, and low-temperature NMR techniques. The equilibrium $Be(C_6H_5)_2 + BeBr_2 \rightleftharpoons 2Be(C_6H_5)Br$ in ether lies predominantly to the right [20]. The results are in contrast to an earlier study [21]. In isotopic labelling experiments with $^{7}BeBr_2$ and $Be(C_6H_5)_2$ in ether, it was shown, by precipitation of $BeBr_2$ as dietherate, that the ^{7}Be label is not statistically distributed between $Be(C_6H_5)_2$ and $BeBr_2$ after precipitation of the $BeBr_2$ [21]. Reactions with LiH, $BeBr_2$ [22], or $BeCl_2$ [24] in ether give $Be(C_6H_5)H$ [22, 24]. The latter product is also obtained by reaction with $Na[B(C_2H_5)_3H]$ and $BeCl_2$ [23]. The hydride obtained with $Sn(C_6H_5)_3H$ is contaminated with phenyltin compounds [23]. No reaction is observed with $Be(C_4H_9\text{-}t)_2 \cdot O(C_2H_5)_2$ in C_6H_6 [18] or with $Be(C_5H_5)_2$ [28].

$Be(C_6H_5)_2$ is converted into $(Be(C_6H_5)OCH_3)_4$ by the interaction with one equivalent of CH_3OH in C_6H_6 [8]. If $Be(C_6H_5)_2$ and CH_3OH react in the mole ratio 1:1 in C_6H_6 containing ether, the product contains complexed $O(C_2H_5)_2$ [13] in the ratio $Be:O(C_2H_5)_2 = 1:1$ [26]. $(Be(C_6H_5)OC(C_6H_5)_3)_2$ forms from the reaction with $(C_6H_5)_2CO$ in C_6H_6 [19]. The reaction of ether solutions of $Be(C_6H_5)_2$ with $(C_6H_5)_2CO$ yields $(C_6H_5)_3COH$ after hydrolysis [5]; with $RCH{=}CHC(O)C_6H_5$ ($R = C_6H_5$, $4\text{-}N(CH_3)_2C_6H_4$), the 1,4-addition products $R(C_6H_5)CHCH_2C(O)C_6H_5$ are obtained after hydrolysis [2]. Butadiene is converted into cis-$(C_6H_5)_3CCH_2CH{=}CHCH_3$ in low yield by a mixture of $Na[C(C_6H_5)_3]$ and $Be(C_6H_5)_2$ in ether, followed by hydrolysis. Similarly, isoprene yields $(C_6H_5)_3CCH_2CH{=}C(CH_3)_2$ [9]. 2,3-Dihydrobenzofuran gives with a mixture of $Na[C(C_6H_5)_3]$ and $Be(C_6H_5)_2$ in ether, after hydrolysis, $(C_6H_5)_3CCH_2CH_2C_6H_4OH\text{-}2$ as the main product [10]. The role of $Be(C_6H_5)_2$ is not well understood [9, 10].

With $NH(CH_3)_2$ a 1:1 adduct is formed initially, which is slowly transformed into $(Be(C_6H_5)N(CH_3)_2)_2$ at 40°C in ether/$C_6H_5CH_3$. $NH(CH_3)CH_2CH_2N(CH_3)_2$ yields $(Be(C_6H_5)N(CH_3)CH_2CH_2N(CH_3)_2)_2$, and $NH(C_6H_5)_2$ affords $(Be(C_6H_5)N(C_6H_5)_2)_2$. In all these reactions, C_6H_6 is eliminated [13].

A solution of 4,4'-bipyridine and ~5-fold excess of $Be(C_6H_5)_2$ in THF reacts with a K mirror to yield a radical cation $[C_6H_5BeNC_5H_4C_5H_4NBeC_6H_5]^{\bullet +}$ [27].

$Be(C_6H_5)_2$ or its etherate form a number of adducts, which are described in the following section.

1.2.4.2 Adducts of $Be(C_6H_5)_2$

$Be(C_6H_5)_2 \cdot 2O(CH_3)_2$

Tensimetric measurements of the reaction of $Be(C_6H_5)_2$ with $O(CH_3)_2$ at 0°C show that the vapor pressure is too low to be detected at a ratio of $Be:O(CH_3)_2 = 1:1.98$. The pressure rises rapidly as the ratio is increased to 1:2.00, 1:2.06, and 1:2.28. It is concluded that a 1:2 complex is present, but no attempts have been made to isolate and characterize the adduct [13].

$Be(C_6H_5)_2 \cdot nO(C_2H_5)_2$ ($n = 1$ to 2)

$Be(C_6H_5)_2$ is always obtained as an etherate, if the preparations described on p. 69 are carried out in ether solution or if the products are worked up with ether [5, 6, 13, 16]. $Be(C_6H_5)_2 \cdot 2O(C_2H_5)_2$ crystallizes from ether [6] in cubic colorless crystals, melting at 28 to 32°C to form a yellow liquid [13]. Heating to 130°C in vacuum causes liberation of ether and leaves a product that melts at 160 to 165°C with decomposition [6]. Others found that the 1:2 adduct obtained from ether/dioxane or ether at −78°C loses ether already very rapidly during the normal processes of separation and removal of the solvent to give $Be(C_6H_5)_2 \cdot nO(C_2H_5)_2$ with $1 < n < 2$. From $Be(C_6H_5)_2$, dissolved in ether, long needles deposit when the solution is cooled to −78°C. These crystals, when separated on a sintered filter disk and dried at room temperature under reduced pressure, have a melting point of 55 to 59°C and a stoichiometry of $Be(C_6H_5)_2 \cdot 1.8O(C_2H_5)_2$. Tensimetric titrations at 0°C show that the $Be(C_6H_5)_2$–ether system has a dissociation pressure of 60 Torr in most of the range from $n > 1$ to $n = 2$. Only for a ratio Be:ether ~1:1 is the vapor pressure close to 0 Torr, which would favor the existence of a 1:1 complex [13]. Other reactions of the etherates are described with the parent compound on pp. 70/1.

References on p. 76

$Be(C_6H_5)_2 \cdot CH_3OCH_2CH_2OCH_3$

Addition of $CH_3OCH_2CH_2OCH_3$ to a solution of $Be(C_6H_5)_2$ in ether and subsequent concentration gives the adduct as colorless needles, m.p. 143 to 145°C. It dissolves in C_6H_6 as a monomer (by cryoscopy), liquifies in air, and decomposes slowly with water [13].

$Be(C_6H_5)_2 \cdot O{=}C(C_4H_9\text{-t})_2$

The adduct is obtained by mixing the components in ether and evaporation of the solvent. It crystallizes from C_6H_{14} in colorless needles, containing some free ketone, m.p. 75 to 76°C. IR absorptions (in Nujol) for ν(C=O) are at 1649 and 1567 cm^{-1}.

It is a monomer in C_6H_6 (by cryoscopy) and can be boiled in $C_6H_5CH_3$ solution without decomposition. The donor is replaced by quinuclidine (1-azabicyclo[2.2.2]octane) (see below) [19].

$Be(C_6H_5)_2 \cdot nN(CH_3)_3$ ($n = 1$ to <2)

The amine is condensed onto a frozen mixture of $Be(C_6H_5)_2$ in C_6H_6. Warming to room temperature, stirring for 15 min, and addition of C_6H_{14} gives colorless needles, which are separated on a sintered filter disk and dried at reduced pressure. The remaining substance has a melting point of 65 to 67.5°C and the composition $Be(C_6H_5)_2 \cdot 1.27\,O(C_2H_5)_2$. Tensimetric titrations at 0°C show that the $Be(C_6H_5)_2$–$N(CH_3)_3$ system has an appreciable dissociation pressure at $n \sim 2$. Only at a ratio of Be:$N(CH_3)_3$ ~1:1 is the vapor pressure close to 0 [13].

$Be(C_6H_5)_2 \cdot 2NC_7H_{13}$ (NC_7H_{13} = 1-azabicyclo[2.2.2]octane)

The adduct is generated in a ligand displacement reaction of $Be(C_6H_5)_2 \cdot O{=}C(C_4H_9\text{-t})_2$ with quinuclidine in ether. Star-like needles crystallize from C_6H_6/C_6H_{14}, m.p. 136 to 137°C. The colorless material is dissociated in C_6H_6 (by cryoscopy) and also yields quinuclidine on heating in a vacuum [19].

$Be(C_6H_5)_2 \cdot N(CH_3)_2CH_2CH_2N(CH_3)_2$

The amine in C_6H_6 is added to $Be(C_6H_5)_2$ in a 1:1 mixture of C_6H_6/ether. Concentration gives a small amount of crystalline product. Addition of C_6H_{14} and boiling gives colorless plates as the solution cools, m. p. 150 to 156°C (dec.). It dissolves in C_6H_6 as a monomer (by cryoscopy) [13].

$Be(C_6H_5)_2 \cdot C_6H_4(N(CH_3)_2)_2$-1,2

1,2-$(N(CH_3)_2)_2C_6H_4$ in C_6H_6 is added to $Be(C_6H_5)_2$ in a 1:1 mixture of C_6H_6/ether. The adduct crystallizes when the solution is concentrated by evaporation, m.p. 260 to 261°C (dec.).

It is a monomer in C_6H_6 (by cryoscopy). The diamine is evolved when the adduct is heated at 70 to 75°C in a vacuum. Heating at 160 to 165°C results in a partial sublimation, and the diamine is removed [13].

$Be(C_6H_5)_2 \cdot N_2C_{10}H_8$ ($N_2C_{10}H_8$ = 2,2′-bipyridine)

The adduct precipitates immediately from a mixture of $Be(C_6H_5)_2$ and 2,2′-bipyridine in ether. It crystallizes as pale yellow needles after extraction with C_6H_6 [12].

The long wavelength UV absorption (no solvent given) is at $\lambda = 353$ nm ($\varepsilon = 500$ $L \cdot mol^{-1} \cdot cm^{-1}$) [11, 12]. The compound is only slowly hydrolyzed by cold water, but readily decomposed by warm dilute H_2SO_4 [12].

$Be(C_6H_5)_2 \cdot N(CH_3)$=$CHC_6H_5$

The adduct is obtained in 63% yield from a 1:1 ratio of $Be(C_6H_5)_2$ and C_6H_5CH=NCH_3 in ether after stirring for 1 h. The volatile materials are removed in vacuum, and recrystallization from C_6H_6/C_6H_{14} gives colorless needles, m.p. 122 to 124°C. IR absorptions (Nujol) are at 1630 and 1590 cm^{-1} for ν(C=N). It is a monomer in C_6H_6 (by cryoscopy) [19].

$Be(C_6H_5)_2 \cdot N(C_6H_4CH_3\text{-}4)$=$CHC_6H_5$

The adduct is obtained, as the previous compound, from C_6H_5CH=$NC_6H_4CH_3$-4 in 61% yield. Tan prisms are formed from C_6H_6/C_6H_{14}, which shrink at 100°C and melt at 134 to 135°C. IR absorptions (Nujol) are at 1615, 1590, and 1575 cm^{-1} for ν(C=N). It is a monomer in C_6H_6 (by cryoscopy) [19].

$Be(C_6H_5)_2 \cdot 2S(CH_3)_2$

$Be(C_6H_5)_2 \cdot nO(C_2H_5)_2$ and an excess of $S(CH_3)_2$ are stirred for 1 h at room temperature and then cooled overnight at −78°C, whereby the adduct crystallizes as colorless plates, which are dried at −15°C under reduced pressure, m.p. 111 to 113°C (dec.). A tensimetric study confirms the 1:2 stoichiometry. The adduct is partially dissociated in C_6H_6 (by cryoscopy) [13].

$Be(C_6H_5)_2 \cdot 1.5S(C_2H_5)_2$

In a tensimetric study of the system $Be(C_6H_5)_2$–ether at 0°C the vapor pressure is close to 0 Torr at a 1.53:1 ratio of $S(C_2H_5)_2$ and Be. No adduct has been isolated [13].

$Be(C_6H_5)_2 \cdot CH_3SCH_2CH_2SCH_3$

The components are mixed in C_6H_6 and boiled for 10 min with stirring. Addition of C_6H_{14} gives colorless needles overnight, m.p. 127 to 128°C. It is a monomer in C_6H_6 (by cryoscopy) [13].

$Be(C_6H_5)_2 \cdot 2P(CH_3)_3$

$Be(C_6H_5)_2 \cdot nO(C_2H_5)_2$ reacts with an excess of $P(CH_3)_3$ in $C_6H_5CH_3$ at room temperature. Addition of C_6H_{14} and cooling to −78°C results in colorless needles, which are washed with $C_6H_{14}/P(CH_3)_3$ and dried at −15°C under reduced pressure, m.p. 132 to 134°C (dec.). A tensimetric measurement shows a vapor pressure minimum at a Be:P ratio of 1:2. The adduct is partially dissociated in C_6H_6 (by cryoscopy) [13].

1.2.4.3 Other Diarylberyllium Compounds

The BeR_2 type compounds, with R = substituted phenyl or 1-naphthyl, are summarized in Table 15. They are prepared by the following reactions:

Method I: a) $BeCl_2 \cdot 2O(C_2H_5)_2$ is added to a stirred suspension of LiC_6F_5 in C_6H_6/ether (ratio C_6H_6:ether = 1:4) at −78°C. The mixture is allowed to warm to room temperature, stirring being continued for a further 2 h. Decantation and concentration under reduced pressure gives impure No. 1 [16].

b) For No. 6, $Mg(C_6H_2(CH_3)_3\text{-}2,4,6)Br$ in ether is added dropwise to $BeCl_2 \cdot 2O(C_2H_5)_2$ in ether. The mixture is refluxed for 5 h, decanted, and the solvent is evaporated. The white residue is sublimed at 140 to 150°C/10^{-3} Torr [16].

References on p. 76

Method II: $Be(C_2H_5)_2$ and BR_3 (ratio ~1.5:1) are mixed at room temperature and kept for 36 h (No. 2) or 72 h (Nos. 3 to 5) at 25°C. $B(C_2H_5)_3$ is evaporated, C_6H_6 (No. 2) or $C_6H_5CH_3$ (Nos. 3 to 5) is added and refluxed (for C_6H_6) or heated to 90°C for 10 h. No. 7 is obtained by stirring a mixture of $Be(C_2H_5)_2$ and $B(C_{10}H_7\text{-}1)_3$ in $C_6H_5CH_3$ at 100°C for 48 h. No. 2 separates upon cooling and is washed with C_6H_6; Nos. 3 to 5 and 7 remain after evaporation of all the volatile material. They are washed with C_6H_{14} (No. 7 with C_6H_6) [16].

The compounds are obtained in nearly quantitative yield. For Nos. 3 to 5, which are sufficiently soluble in C_6H_6, molecular weight determinations (by cryoscopy) show them to be dimeric. All compounds are sensitive to air and moisture [16]. Except for No. 2, 1:1 or 1:2 adducts with O- and/or N-donors are described [16, 18]. The etherates are mentioned with the parent compounds. The other adducts are summarized in Table 16.

Table 15

BeR_2 Type Compounds with R = Substituted Phenyl or 1-Naphthyl.
Explanations, abbreviations, and units on p. X.

No.	R method of preparation [Ref.]	properties and remarks [Ref.]
1	C_6F_5, as etherate ($Be:O(C_2H_5)_2:C_6F_5$ ~1:0.70:1.89) Ia [16]	light brown solid, m.p. ~90°, mild explosion at ~130° (bath temperature) when short-path distillation is attempted [16]
2	$C_6H_4Cl\text{-}4$ II [16]	dec. >300° insoluble in C_6H_6, dissolves slowly in ether [16]
3	$C_6H_4CH_3\text{-}2$ II [16]	m.p. 215 to 217° [16] dimeric in C_6H_6 (by cryoscopy) [16] reaction with LiH and $BeCl_2$ gives $Be(C_6H_4CH_3\text{-}2)H$ [24] addition of $LiC_6H_4CH_3\text{-}2$ gives $Li[Be(C_6H_4CH_3\text{-}2)_3]$ [16] ligand exchange with $Be(C_4H_9\text{-t})_2$ in C_6H_6 gives $(Be(C_6H_4CH_3\text{-}2)C_4H_9\text{-t})_2$ [18]
4	$C_6H_4CH_3\text{-}3$ II [16]	m.p. 168 to 170° (from C_6H_6/C_6H_{14}) dimeric in C_6H_6 (by cryoscopy) [16] reaction with LiH and $BeCl_2$ gives $Be(C_6H_4CH_3\text{-}3)H$ [18] reaction with $Be(C_4H_9\text{-t})_2$ gives $Be(C_6H_4CH_3\text{-}3)H$ and $(CH_3)_2C{=}CH_2$ after heating to 140°/10^{-3} Torr [18]
5	$C_6H_3(CH_3)_2\text{-}2,5$ II [16]	solidified glass, m.p. 158 to 160° dimeric in C_6H_6 (by cryoscopy) [16] ligand exchange with $Be(C_4H_9\text{-t})_2$ gives $(Be(C_6H_3(CH_3)_2\text{-}2,5)C_4H_9\text{-t})_2$ [18]
6	$C_6H_2(CH_3)_3\text{-}2,4,6$, as etherate (~1:0.7) Ib [16]	m.p. 175 to 178° heating at 30° for 3 days does not remove the ether [16]
7	$C_{10}H_7\text{-}1$ (1-naphthyl) II [16]	m.p. 255 to 258° (dec.) [16] hardly soluble in ether, C_6H_6, and other hydrocarbon solvents [16, 18] no reaction with $Be(C_4H_9\text{-t})_2$ [18]

1.2.4.4 Adducts of BeR_2 with R = Substituted Phenyl or 1-Naphthyl

Addition compounds of diarylberyllium with O- and N-donors are known. They are described in Table 16 except for the adducts with $O(C_2H_5)_2$, which are treated with the parent compounds on pp. 73/4.

The adducts of Table 16 are prepared by the following method:

> A slight excess of the donor is added to a solution of BeR_2 ($Be(C_6F_5)_2$ and $Be(C_6H_2(CH_3)_3$-2,4,6$)_2$ as etherate) in C_6H_6 (in ether for No. 1 [18]). The reaction is exothermic. After stirring for 15 to 30 min the volatile material is evaporated under reduced pressure, and the residue is recrystallized (Nos. 2, 4 to 12). Nos. 1, 3, and 13 crystallize upon concentration of the reaction mixture [16, 18].

All adducts are monomeric in C_6H_6, as determined by cryoscopy [16, 18].

Table 16
Adducts of the $BeR_2 \cdot D$, $BeR_2 \cdot D{-}D$, $BeR_2 \cdot 2D$ Types with R = Substituted Phenyl or 1-Naphthyl.
Method of preparation, see above.
Explanations, abbreviations, and units on p. X.

No.	adduct	properties and remarks [Ref.]
1	$Be(C_6F_5)_2 \cdot 2OC_4H_8$ (OC_4H_8 = THF)	needles, m.p. 134 to 136° reaction with $Be(C_5H_5)_2$ in C_6H_6 gives $Be(C_6F_5)C_5H_5$ [18]
2	$Be(C_6F_5)_2 \cdot N(CH_3)_2CH_2CH_2N(CH_3)_2$	small needles (from C_6H_6/C_6H_{14}), dec. at 123 to 126° to a black tar [16]
3	$Be(C_6F_5)_2 \cdot 2NC_5H_5$	changes color at 150° and dec. to a black liquid at 170° [16]
4	$Be(C_6H_4CH_3$-2$)_2 \cdot N(CH_3)_3$	small needles, m.p. 198 to 200° (from C_6H_6/C_6H_{14}) [16]
5	$Be(C_6H_4CH_3$-2$)_2 \cdot NC_7H_{13}$ (NC_7H_{13} = quinuclidine)	needles, m.p. 220 to 222° (from C_6H_6/C_6H_{14}) [18]
6	$Be(C_6H_4CH_3$-2$)_2 \cdot$ $N(CH_3)_2CH_2CH_2N(CH_3)_2$	m.p. 178 to 180° (from C_6H_6/C_6H_{14}) [16]
7	$Be(C_6H_4CH_3$-2$)_2 \cdot 2NC_5H_5$	plates, m.p. 171 to 173° (dec.)(from C_6H_6/C_6H_{14}) [16]
8	$Be(C_6H_4CH_3$-3$)_2 \cdot$ $CH_3OCH_2CH_2OCH_3$	needles, m.p. 130 to 131° (from C_6H_{14}) [16]
9	$Be(C_6H_4CH_3$-3$)_2 \cdot$ $N(CH_3)_2CH_2CH_2N(CH_3)_2$	m.p. 145 to 147° (from C_6H_6/C_6H_{14}) [16]
10	$Be(C_6H_4CH_3$-3$)_2 \cdot 2NC_5H_5$	m.p. 177 to 179° (dec.)(from C_6H_6/C_6H_{14}) [16]
11	$Be(C_6H_3(CH_3)_2$-2,5$)_2 \cdot$ $N(CH_3)_2CH_2CH_2N(CH_3)_2$	m.p. 177 to 179° (from C_6H_6/C_6H_{14}) [16]
12	$Be(C_6H_2(CH_3)_3$-2,4,6$)_2 \cdot N(CH_3)_3$	needles, m.p. 130 to 132° (from C_6H_6/C_6H_{14}) the melt solidifies to a glass [16]
13	$Be(C_{10}H_7)_2 \cdot NC_5H_5$ ($C_{10}H_7$ = 1-naphthyl)	shrinkage at 115°, m.p. 178 to 180° (dec.) [16]

References on p. 76

References:

[1] Gilman, H., Schulze, F. (J. Chem. Soc. **1927** 2663/9).
[2] Gilman, H., Kirby, R. H.(J. Am. Chem. Soc. **63** [1941] 2046/8).
[3] Wittig, G., Keicher, G. (Naturwissenschaften **34** [1947] 216).
[4] Wittig, G. (Angew. Chem. **62** [1950] 231/6).
[5] Wittig, G., Meyer, F. J., Lange, G. (Liebigs Ann. Chem. **571** [1951] 167/201).
[6] Wittig, G., Hornberger, P. (Liebigs Ann. Chem. **577** [1952] 11/25).
[7] Margoshes, M., Fassel, V. A. (Spectrochim. Acta **7** [1955/56] 14/24).
[8] Strohmeier, W., Hümpfner, K. (Z. Elektrochem. **60** [1956] 1111/4).
[9] Wittig, G., Wittenberg, D. (Liebigs Ann. Chem. **606** [1957] 1/23).
[10] Wittig, G., Kolb, G. (Chem. Ber. **93** [1960] 1469/76).

[11] Coates, G. E., Green, S. I. E. (Proc. Chem. Soc. **1961** 376).
[12] Coates, G. E., Green, S. I. E (J. Chem. Soc. **1962** 3340/8).
[13] Coates, G. E., Tranah, M. (J. Chem. Soc. A **1967** 236/9).
[14] Ashby, E. C., Arnott, R. C. (J. Organometal. Chem. **14** [1968] 1/11).
[15] Coates, G. E., Morgan, G. L. (Advan. Organometal. Chem. **9** [1970] 195/257, 204/5).
[16] Coates, G. E., Srivastava, R. C. (J. Chem. Soc. Dalton Trans. **1972** 1541/4).
[17] Glockling, F., Morrison, R. J., Wilson, J. W. (J. Chem. Soc. Dalton Trans. **1973** 94/6).
[18] Coates, G. E., Smith, D. L., Srivastava, R. C. (J. Chem. Soc. Dalton Trans. **1973** 618/22).
[19] Andersen, R. A., Coates, G. E. (J. Chem. Soc. Dalton Trans. **1974** 1171/80).
[20] Sanders Jr., J. R., Ashby, E. C., Carter II, J. H. (J. Am Chem. Soc. **90** [1968] 6385/90).

[21] Dessy, R. E. (J. Am. Chem. Soc. **82** [1960] 1580/2).
[22] Bell, N. A., Coates, G. E. (J. Chem. Soc. A **1966** 1069/73).
[23] Coates, G. E., Tranah, M. (J. Chem. Soc. A **1967** 615/7).
[24] Blindheim, U., Coates, G. E., Srivastava, R. C. (J. Chem. Soc. Dalton Trans. **1972** 2302/5).
[25] Marynick, D. S. (J. Am. Chem. Soc. **103** [1981] 1328/33).
[26] Coates, G. E., Fishwick, A. H. (J. Chem. Soc. A **1968** 477/83).
[27] Kaim, W. (J. Organometal. Chem. **241** [1983] 157/69).
[28] Bartke, T. C. (Diss. Univ. Wyoming 1975; Diss. Abstr. Intern. B **36** [1976] 6141).

1.2.5 BeRR′ Type Compounds and Their Adducts

This section describes the compounds in which two different R groups are bonded to Be. The prepared compounds are summarized in Table 17. Not all compounds of this type could be isolated without a donor. The donor-free examples are described as Nos. 1 to 4 in Table 17, followed by those containing an O- or N-donor (Nos. 5 to 16). Except for Nos. 5 and 6, for which n<1, 1:1 adducts are obtained. As far as determined, the complexes with the chelating ligand $N(CH_3)_2CH_2CH_2N(CH_3)_2$ are monomeric in C_6H_6, whereas those with $N(CH_3)_3$ are dimeric with bridging R groups, leading in all cases to four-coordinated Be. Nos. 15 and 16 are described to be of the type shown in Formula I and must therefore be treated as dinuclear species.

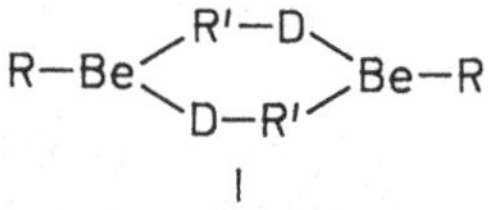

I

The compound **$Be(C_4H_9$-t)C_6F_5** is mentioned in the literature, but it follows from the text that this material is actually $Be(C_6F_5)C_5H_5$, which is treated in Section 3.4, p. 180 [4].

A detailed molecular orbital study using the PRDDO approximation was performed for the up to now experimentally unknown **$Be(CH_3)C_6H_5$**, see Chapter 5, p. 204 [6].

The following methods are used to prepare the compounds of Table 17:

Method I: Equimolar amounts of BeR_2 and BeR'_2 are mixed in C_6H_6 (Nos. 2 to 4) or without a solvent (No. 2). The solvent is evaporated under reduced pressure, and the residue is distilled (Nos. 1, 2) or recrystallized (Nos. 3, 4) [4].

Method II: The donor is added in excess to BeRR' in C_6H_6 (Nos. 7, 12, 13) or without a solvent (Nos. 5, 6). The volatile material is evaporated under reduced pressure. The adduct remains (Nos. 7, 12, 13), or the residue is distilled (Nos. 5, 6) [4].

Method III: $BeR_2 \cdot N(CH_3)_3$ and $BeR'_2 \cdot N(CH_3)_3$ (mole ratio 1:1) are dissolved in $C_6H_5CH_3$ (Nos. 10, 11) or in c-C_5H_{10} (No. 8). For Nos. 9 and 14, $BeR_2 \cdot O(C_2H_5)_2$ (R = $CH_2C_4H_9$-t or C_4H_9-t) in C_6H_6 is stirred with $Be(CH_3)_2$ or $Be(C_6F_5)_2 \cdot 0.7\,O(C_2H_5)_2$, respectively, and the diamine is subsequently added (after 2 days for No. 14) in excess. For workup, the volatile material is evaporated [1, 2, 4].

Method IV: $C_6H_5CH{=}NR''$ (R'' = C_4H_9-t, C_6H_5) in C_6H_{14} is added to $Be(C_2H_5)_2 \cdot O(C_2H_5)_2$ (mole ratio 1:1) in C_6H_{14} and stirred at room temperature for 4 h (No. 15) or 12 to 47 h (No. 16). The solvent is removed in vacuum. For No. 15 the resulting orange liquid is pumped in vacuum overnight at 10^{-2} Torr. No. 16 solidifies upon addition of C_6H_{14} [5].

Table 17

BeRR' Type Compounds and Their Adducts.

Further information on numbers preceded by an asterisk is given at the end of the table. Explanations, abbreviations, and units on p. X.

No.	compound method of preparation (yield in %) [Ref.]	properties and remarks [Ref.]
*1	$Be(CH_3)C_4H_9$-t I [4]	colorless liquid, b.p. 32°/10^{-3} Torr ^1H NMR($C_6D_5CD_3$): −0.09 (br, C_4H_9), 1.01 (s, CH_3) trimeric in C_6H_6 or c-C_6H_{12} (by cryoscopy) [4]
2	$Be(C_2H_5)C_4H_9$-t I [4]	colorless liquid, (from distillation at 60 to 65°/10^{-3} Torr) dimeric in C_6H_6 (by cryoscopy) air-sensitive pyrolysis of the pure liquid or a mixture with excess $Be(C_2H_5)_2$ at 140° for 1 h gives $Be_2(C_2H_5)_3H$ [4]
3	$Be(C_4H_9\text{-t})C_6H_4CH_3$-2 I [4]	prisms, m.p. 135° (dec.) (from C_6H_{14} at 5 to 6°) dimeric in C_6H_6 (by cryoscopy) thermal dec. in boiling xylene for 24 h gives $Be(C_6H_4CH_3\text{-2})H$ and $(CH_3)_2C{=}CH_2$ a 1:1 adduct is formed with $N(CH_3)_2CH_2CH_2N(CH_3)_2$, see No. 12 [4]
4	$Be(C_4H_9\text{-t})C_6H_3(CH_3)_2$-2,5 I [4]	prisms, m.p. 139 to 141° (dec.) (from C_6H_{14}/C_6H_6 10:1) dimeric in C_6H_6 (by cryoscopy) gives a 1:1 adduct with $N(CH_3)_2CH_2CH_2N(CH_3)_2$; see No. 13 [4]

References on p. 80

Table 17 (continued)

No.	compound method of preparation (yield in %) [Ref.]	properties and remarks [Ref.]
adducts of BeRR′		
5	$Be(CH_3)C_4H_9$-t·n$O(C_2H_5)_2$, n = 0.58 II [4]	colorless liquid (from distillation at room temperature/10^{-3} Torr) analysis gives Be:$O(C_2H_5)_2$ = 1:0.58 [4]
6	$Be(CH_3)C_4H_9$-t·n$N(CH_3)_3$, n = 0.55 II [4]	isolated by distillation above room temperature/10^{-3} Torr analysis gives Be:$N(CH_3)_3$ = 1:0.55 ^{1}H NMR(C_6D_6): −0.13(CH_3Be), 0.94(CH_3C), 1.87(CH_3N) [4]
7	$Be(CH_3)C_4H_9$-t· $N(CH_3)_2CH_2CH_2N(CH_3)_2$ II [4]	liquid [4]
8	$Be(CH_3)CH_2C_4H_9$-t·$N(CH_3)_3$ III [2]	identified by ^{1}H NMR in an equilibrium mixture of $Be(CH_3)_2 \cdot N(CH_3)_3$ and $Be(CH_2C_4H_9\text{-t})_2 \cdot N(CH_3)_3$ in c-C_5H_{10}, K = 170 at −40° ^{1}H NMR (c-C_5H_{10}): −1.00 ± 0.02(CH_3Be) from +30 to −40°, −0.18(CH_2) at 30°, −0.21(CH_2) at −16 to −40° [2]
*9	$Be(CH_3)CH_2C_4H_9$-t· $N(CH_3)_2CH_2CH_2N(CH_3)_2$ III [2]	colorless prisms, m.p. 41° (from C_5H_{12}) ^{1}H NMR (C_6H_6, 30°): −0.12(CH_2Be), 1.45(CH_3C), 1.93(s, CH_2N, m at −40°), 2.02(s, CH_3N, d at −40°) monomeric in C_6H_6 (by cryoscopy) [2]
*10	$Be(CH_3)C{\equiv}CCH_3 \cdot N(CH_3)_3$ III [1]	colorless needles, m.p. 98° (from C_6H_{14}) ^{1}H NMR (C_6H_6 at 30°): −0.44(s, CH_3Be), 1.76(s, CH_3C), 2.27(s, CH_3N) IR: ν(C≡C) at 2110 dimeric in C_6H_6 (by cryoscopy) [1]
11	$Be(CH_3)C{\equiv}CC_4H_9$-t·$N(CH_3)_3$ III [1]	small colorless prisms, m.p. 123 to 125° (from C_6H_{14}) ^{1}H NMR (C_6H_6): −0.50(s, CH_3Be), 1.26(s, CH_3C), 2.32(s, CH_3N) IR: ν(C≡C) at 2092 dimeric in C_6H_6 (by cryoscopy) same structure as for No. 10 is proposed [1]
12	$Be(C_4H_9\text{-t})C_6H_4CH_3$-2· $N(CH_3)_2CH_2CH_2N(CH_3)_2$ II [4]	prisms, m.p. 90 to 92° (from C_6H_{14}) monomeric in C_6H_6 (by cryoscopy) [4]
13	$Be(C_4H_9\text{-t})C_6H_3(CH_3)_2$-2,5· $N(CH_3)_2CH_2CH_2N(CH_3)_2$ II [4]	prisms, m.p. 105 to 107° (from C_6H_6/C_6H_{14} 1:10) monomeric in C_6H_6 (by cryoscopy) [4]
14	$Be(C_4H_9\text{-t})C_6F_5$· $N(CH_3)_2CH_2CH_2N(CH_3)_2$ III [4]	prisms, m.p. 95 to 98° (dec.) (from C_6H_6/C_6H_{14}) monomeric in C_6H_6 (by cryoscopy) [4]

Table 17 (continued)

No.	compound method of preparation (yield in %) [Ref.]	properties and remarks [Ref.]
*15	$Be(C_2H_5)C_6H_4(CH{=}NC_4H_9\text{-t})\text{-}2$ IV (50) [5]	yellow prisms that shrink at 120°, m.p. 122 to 124° (from C_6H_6/C_6H_{14} 1:2) IR: ν(C=N) at 1605(s); 1540(m) [5]
*16	$Be(C_4H_9\text{-t})C_6H_4(CH{=}NC_4H_9\text{-t})\text{-}2$ IV (27 to 50) [5]	orange-red needles that shrink at 120°, melt at 128 to 129°, and sublime at 70 to 75°/10^{-2} Torr (from C_6H_{14}/C_6H_6 3:1) IR: 1600(s), 1530(w) [5]

* Further information:

$Be(CH_3)C_4H_9$-t (Table **17**, No. **1**). The ^{1}H NMR spectrum is not changed even at −70°C. Thermal decomposition at 150°C in a N_2 atmosphere for 2 h and at 120°C/10^{-3} Torr for 10 h yields $Be(CH_3)H$, which is isolated as the $N(CH_3)_3$ adduct. The liquid is extremely air-sensitive. It reacts with t-C_4H_9OH to give $Be(OC_4H_9\text{-t})_2$; $CH(CH_3)_3$ and CH_4 are formed in an equimolar ratio. Adducts are generated by reaction with $O(C_2H_5)_2$, $N(CH_3)_3$, and $N(CH_3)_2CH_2CH_2N(CH_3)_2$; see Nos. 5 to 7 [4].

$Be(CH_3)CH_2C_4H_9\text{-t}\cdot N(CH_3)_2CH_2CH_2N(CH_3)_2$ (Table **17**, No. **9**). The appearance of only one CH_3N and CH_2N signal at ambient temperature suggests a chelated structure with rapid dissociation of the Be–N bonds to make the two CH_3 and CH_2 groups on N equivalent on the NMR time scale. At −40°C these signals are split [2].

$Be(CH_3)C{\equiv}CCH_3\cdot N(CH_3)_3$ (Table **17**, No. **10**). The crystal and molecular structure has been determined by single crystal X-ray diffraction at 20°C. The monoclinic crystals (a = 13.371, b = 9.290, c = 7.249 Å, β = 101.485°) belong to the space group $P2_1/n\text{-}C^5_{2h}$ (No. 14) with two dimeric molecules in the unit cell; d_c = 0.9201 g/cm³. For 1156 intensities, the final R = 0.049. The molecular structure is shown in **Fig. 10** together with the intramolecular distances and angles. The terminal triple-bonded carbon atoms of the two propynyl groups serve as bridging atoms between two Be atoms. The dimer is centrosymmetric, and the four-membered ring formed by the two Be atoms and the two bridging C atoms is planar [3].

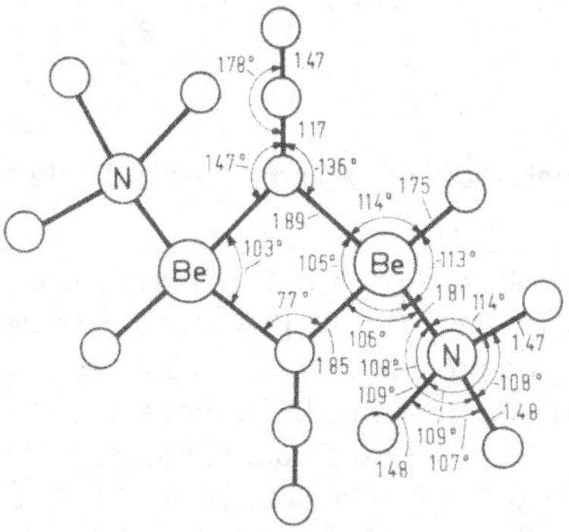

Fig. 10. Molecular structure of $(Be(CH_3)C{\equiv}CCH_3\cdot N(CH_3)_3)_2$ [3].

Alcoholysis of the compound with two molar equivalents of $CH_3OCH_2CH_2OH$ for every mole of dimer results in the fission of 40% of the CH_3Be groups as CH_4, and of 60% of the $CH_3C{\equiv}CBe$ groups as $CH_3C{\equiv}CH$ [1].

References on p. 80

Be(R)C$_6$H$_4$(CH=NC$_4$H$_9$-t)-2 (R = C_2H_5, C_4H_9-t, Table **17**, Nos. **15** and **16**). According to cryoscopic measurements in C_6H_6, No. 15 is dimeric, and the structure shown in Formula II (R = C_2H_5) is proposed. For No. 16, a degree of association of 1.5 to 1.65 was found, which suggests an equilibrium between Formula II (R = t-C_4H_9) and Formula III. Treatment of Nos. 15 and 16 with CO_2 in ether, followed by hydrolysis, gives 2-OHCC$_6$H$_4$COOH [5].

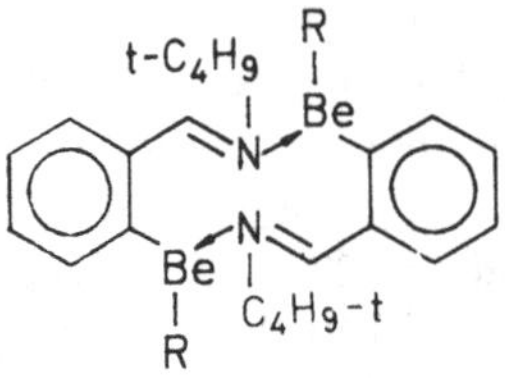

II

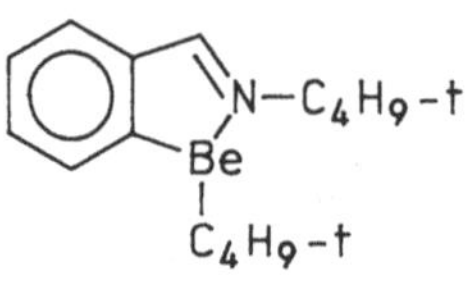

III

References:

[1] Coates, G. E., Francis, B. R. (J. Chem. Soc. A **1971** 474/7).
[2] Coates, G. E., Francis, B. R. (J. Chem. Soc. A **1971** 1305/8).
[3] Morosin, B., Howatson, J. (J. Organometal. Chem. **29** [1971] 7/14).
[4] Coates, G. E., Smith, D. L., Srivastava, R. C. (J. Chem. Soc. Dalton Trans. **1973** 618/22).
[5] Andersen, R. A., Coates, G. E. (J. Chem. Soc. Dalton Trans. **1974** 1171/80).
[6] Marynick, D. S. (J. Am. Chem. Soc. **103** [1981] 1328/33).

1.2.6 Diorganylberyllates

1.2.6.1 Hydrido(diorganyl)beryllates and Their Etherates

The compounds are addition products of alkali hydrides MH (M = Li, Na) with BeR_2: $MH + BeR_2 \rightarrow M[BeR_2H]$. Elemental analyses, IR spectra, and the X-ray structure of No. 13 show that the molecules have a dimeric structure with bridging hydrogen atoms (Formula I).

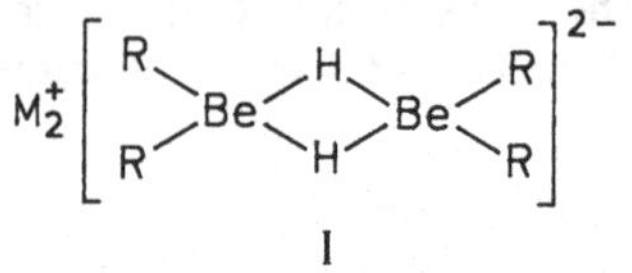

I

The compounds and their adducts are summarized in Table 18. They are prepared by the following methods:

Method I: Nos. 12 to 14 and 16 are prepared by reaction of BeR_2 with an excess of MH (M = Na, Li) in boiling ether (No. 13 [2, 4]) for 2 days (No. 13 [8]), 19 h (No. 14 [9]), or at 110°C, ether being allowed to distill off (No. 12 [8]). For No. 17, $Be(C_6H_5)_2 \cdot 2O(C_2H_5)_2$ and excess LiH are heated to 160 to 165°C and extracted 20 times with ether [1]. The mixtures are filtered, and the ether is evaporated at 0°C (No. 12 [8]), or the adducts crystallize upon concentration (No. 13 [5, 10], Nos. 14 and 16 [9]) and cooling (No. 17 [1]). No. 16 is filtered in a special apparatus to avoid removal of the ether [9].

Method II: Nos. 4 to 7 and 9 are prepared from their etherates by evaporation of the residues obtained in Method I in a vacuum (No. 5 [2, 4], No. 6 [4]) for 12 h (No. 7 [9]), 24 h/10^{-4} Torr (No. 9 [9]), or at 35 to 40°C (No. 4 [8]).

Method III: For Nos. 1 to 3, 8, and 10, no etherates are characterized, and the compounds are obtained directly by reaction of BeR_2 ($Be(CH_3)_2 \cdot N(CH_3)_3$ for No. 3 [6]) with an excess of MH (M = Li, Na) in boiling ether (2 days for Nos. 3, 8; 13 h for No. 10) [2, 4, 6, 9]. The solutions are filtered and the ether is evaporated [4, 6]. Nos. 8 and 10 are extracted several times with C_6H_{14}, followed by evaporation under reduced pressure [9]. For No. 4, reaction only occurs when the ether is distilled off until the mixture reaches 110°C and is kept for 12h at that temperature [2].

Method IV: Nos. 11 and 15 are prepared from the corresponding $Na_2[BeR_2H]_2$. For No. 11, $O(CH_3)_2$ is condensed onto ether-free $Na_2[Be(C_2H_5)_2H]_2$ at −78°C, and the mixture is agitated until all the solid has dissolved. The filtered solution is allowed to warm to −20°C, allowing the solvent to boil off. For No. 15, LiBr in ether is added to $Na_2[Be(C_4H_9\text{-i})_2H]_2$ in ether. The NaBr formed is separated and the filtrate concentrated [9].

Table 18

$M^+[BeR_2H]^-$ Type Compounds and Their Adducts.
Further information on numbers preceded by an asterisk is given at the end of the table.
Explanations, abbreviations, and units on p. X.

No.	compound method of preparation [Ref.]	properties and remarks [Ref.]
1	$Na[Be(CH_3)_2H]$ III [2, 4]	white, m.p. 195 to 196° [4] IR (Nujol): ν(BeH_2Be) at 1328(s), 1165(s) [4], 1325 [6], others at 1255(ms), 1189(s), 1138(vs), 1086(sh), 1018(vs), 789(vs), 754(m), 612(vw) [4] solubility in ether at 20°: 1.6 g/L [4] reaction with $BeCl_2$ in the presence of excess NaH gives a product of the composition $Be_3(CH_3)_4H_2$, see Section 4.1, p. 199 [4, 11].
2	$Na[Be(CH_3)_2D]$ III [4]	m.p. 196° IR (Nujol): ν(BeD_2Be) at 917(vs), 869(vs) [4], 920, 870 [6], others at 1261(m), 1199(ms), 1144(s), 1075(sh), 1015(vs), 797(vs), 777(s), 595(vw) [4]
3	$Na[Be(CD_3)_2H]$ III [6]	needles, m.p. 200 to 201° IR (Nujol): ν(BeH_2Be) at 1333, 1164, others at 1043, 934, 899, 713, 678, 600 [6]
4	$Li[Be(C_2H_5)_2H]$ II [8], III [2]	m.p. 227 to 229° IR (Nujol): ν(BeH_2Be) at 1580(s), 1390 insoluble in C_6H_6 [8] see No. 12 for reactions
*5	$Na[Be(C_2H_5)_2H]$ II [2]	colorless needles [2], m.p. 198° [2, 4] IR (Nujol): ν(BeH_2Be) at 1294(s), 1065(vs) [4] see No. 13 for reactions
6	$Na[Be(C_2H_5)_2D]$ II [4]	white needles, m.p. 200 to 201° IR (Nujol): ν(BeD_2B) at 951(s), 835(vs) [4]
7	$Na[Be(C_3H_7)_2H]$ II [9]	dec. at 203° [9]

References on p. 84

Table 18 (continued)

No.	compound method of preparation [Ref.]	properties and remarks [Ref.]
8	$Na[Be(C_3H_7\text{-}i)_2H]$ III (with ⅓ conversion) [9]	only characterized by elemental analysis (recrystallized from C_6H_6) thermal dec. at 90° for 24 h gives $Be(C_3H_7\text{-}i)_2$ and $CH_2{=}CHCH_3$ [9]
9	$Li[Be(C_4H_9\text{-}i)_2H]$ II [9]	only characterized by elemental analysis thermal dec. in mineral oil at 208° gives a precipitate of $Li[BeH_3]$ [9]
*10	$Na[Be(C_4H_9\text{-}i)_2H]$ III [9]	white powdery material [9]
adducts with ethers		
11	$Na[Be(C_2H_5)_2H]\cdot O(CH_3)_2$ IV [9]	colorless needles (from $O(CH_3)_2$ at −20°) dissociation pressure: 169 Torr/20.5° is handled under an atmosphere of $O(CH_3)_2$ at room temperature [9]
*12	$Li[Be(C_2H_5)_2H]\cdot O(C_2H_5)_2$ I [8]	m.p. 33 to 35° (from C_6H_{14} at −78°) IR (Nujol): ν(BeH_2Be) at 1575(m), 1390 dissociation pressures: 14 Torr/39°, 23 Torr/50° [8]
*13	$Na[Be(C_2H_5)_2H]\cdot O(C_2H_5)_2$ I [2 to 5, 8, 10]	needles [10] (grown from ether solutions) [5, 10] dissociation pressure: 17 Torr/20° [4, 10]
14	$Na[Be(C_3H_7)_2H]\cdot n\,O(C_2H_5)_2$ (n = 0.63) I [9]	needles (washed with cold ether) a stoichiometric adduct could not be obtained due to the low dissociation pressure of ether evaporation in a vacuum for 12 h gives the ether-free compound No. 7 [9]
15	$Li[Be(C_4H_9\text{-}i)_2H]\cdot n\,O(C_2H_5)_2$ (n = 0.93) IV [9]	needles 24 h at 10^{-4} Torr gives the ether-free compound No. 9 [9]
*16	$Na[Be(C_4H_9\text{-}t)_2H]\cdot 2\,O(C_2H_5)_2$ I [9]	crystals ^1H NMR (ether): 0.33(C_4H_9) vs. CH_3 of ether [9]
17	$Li[Be(C_6H_5)_2H]\cdot O(C_2H_5)_2$ I (82) [1]	colorless rhombic crystals, dried at 0° in vacuum dec. in air with occasional ignition with H_2O, H_2 evolves vigorously [1]

* Further information:

$Na[Be(C_2H_5)_2H]$ (Table **18**, No. **5**). The solubility in ether at 20°C is as high as 30 g/L [4] or 0.33 g-atom/L [2]. The needles inflame in air [2]. Pyrolysis with $BeCl_2$ at 180°C is reported to give $Na_2[BeH_4]$ [3]; see also the etherate, No. 13, for the reaction with $BeCl_2$. Very little reaction takes place for the ether-free $Na_2[Be(C_2H_5)_2H]_2$ with gaseous C_2H_4 in a sealed tube at 84°C; but in the presence of ether, an almost quantitative uptake is observed [7]. The compound combines with $Be(C_2H_5)_2$ in C_6H_{14} to give a product, whose composition corresponds to $Na[Be_3(C_2H_5)_6H]$. Thermal decomposition regenerates a condensate of $Be(C_2H_5)_2$ and leaves $Na_2[Be(C_2H_5)_2H]_2$ [9]. Adducts are formed with $O(CH_3)_2$ [9] and $O(C_2H_5)_2$ [5, 10], but no adducts could be crystallized with THF or $CH_3OCH_2CH_2OCH_3$ [9].

Na[Be(C_4H_9-i)$_2$H] (Table **18**, No. **10**) is only obtained with a residual amount of ether (Be:O(C_2H_5)$_2$ = 1:0.05) after repeated (5 times) addition of C_6H_{14} and evaporation. It is insoluble in hydrocarbons. Exposure to vacuum at 40°C results in some decomposition. Thermal decomposition in a vacuum at 95°C gives Be(C_4H_9-i)$_2$, Be(C_4H_9-i)H, CH_2=C(CH_3)$_2$, and other products. The result is similar in boiling C_7H_{16} (98°C). An adduct is formed with excess Be(C_4H_9-i)$_2$ in C_6H_{14}, which has the composition $Na_2[Be_3R_6H_2]$ [9].

Li[Be(C_2H_5)$_2$H]·O(C_2H_5)$_2$ (Table **18**, No. **12**) gives No. 4 when heated for 4 h at 35 to 40°C in a vacuum. The ether and Be(C_2H_5)$_2$ are lost when the adduct is heated for 4 h at 80°C. A similar reaction occurs in boiling C_6H_6. Reaction with $BeCl_2$ in ether, in a 4:1 ratio of Li:$BeCl_2$, gives $Li_2[BeH_4]$. If the ratio of Li:$BeCl_2$ is 3:1, a material with the approximate composition Li[BeH_3] is obtained, which is believed to consist of $Li_2[BeH_4]$ and BeH_2 [8]. At a ratio of Li:$BeCl_2$ = 2:1, a product of the composition $Be_3(C_2H_5)_4H_2$ is formed, which may be a mixture of Be(C_2H_5)$_2$ and Be(C_2H_5)H [2].

Na[Be(C_2H_5)$_2$H]·O(C_2H_5)$_2$ (Table **18**, No. **13**) is the only example in this section for which the crystal structure was determined. A preliminary X-ray structure is published in [5]. More exact values and details are given in [10]. The crystals belong to the monoclinic space group $P2_1/c-C^5_{2h}$ (No. 14) with a = 5.044(6), b = 11.17(2), c = 20.90(3) Å, β = 101.25(16)°; for Z = 2 follows d_c = 0.95 and d_m = 0.95(2) g/cm³; R = 0.117 for 1094 observed reflections. A perspective view of the molecule with some bond distances is shown in **Fig. 11**.

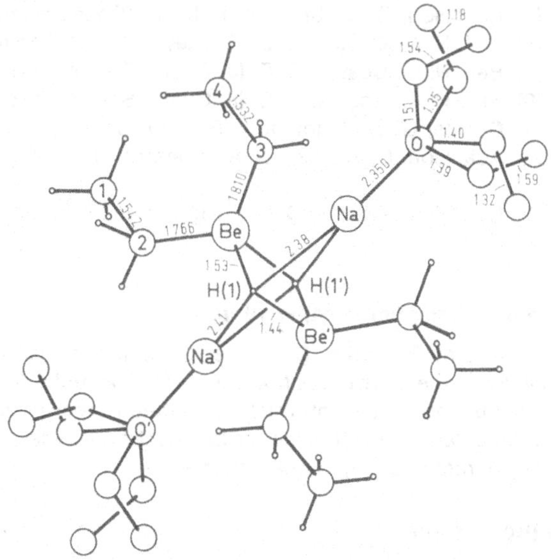

Fig. 11. Perspective view of $Na_2[Be(C_2H_5)_2H]_2·2O(C_2H_5)_2$ [10].

Some nonbonded distances (in Å) and bond angles (in °):

Na···Na'	3.620	O–Na–H(1)	150.3	C(3)–Be–Be'	118.1
Na···Be	3.245	O'–Na'–H(1)	127.3	H(1)–Be–H(1')	83.4
Na···Be'	3.051	H(1)–Na–H(1')	81.9	Na–H(1)–Na'	98.1
Na'···Be	3.675	C(1)–C(2)–Be	116.0	Na–H(1)–Be	110.1
Na'···Be'	3.183	C(4)–C(3)–Be	114.6	Na–H(1)–Be'	103.2
Be···Be'	2.219	C(2)–Be–C(3)	118.0	Na'–H(1)–Be	136.3
Na'–Na–O	167.2	C(2)–Be–Be'	123.9	Na'–H(1)–Be'	108.9
				Be–H(1)–Be'	96.6

References on p. 84

The molecule consists of $[Be_2(C_2H_5)_4H_2]^{2-}$ ions with pairs of Na^+ ions, each coordinated to one disordered ether molecule. Each Be is surrounded by a pseudotetrahedral arrangement of two H and two C atoms while the bridging H atoms are surrounded by an approximately tetrahedral arrangement of two Na and two Be atoms. There are no significant intermolecular interactions [10].

The ether is lost rapidly at room temperature [10]. Therefore, the ether-free compound No. 5, p. 81, is usually obtained during the normal procedure of drying in a vacuum [2, 4]. Reactions with $BeCl_2$ are as described for the Li analog, No. 12 [4, 8].

$Na[Be(C_4H_9\text{-}t)_2H]\cdot 2O(C_2H_5)_2$ (Table **18**, No. **16**) loses ether on exposure to the N_2 atmosphere of a glove-box and liquifies within 10 min. The same behavior is observed by filtration at normal atmosphere and subsequent suction. No definite product could be isolated. Heating of the liquid product to 60°C in a vacuum gives $CH_2{=}C(CH_3)_2$, $Be(C_4H_9\text{-}t)_2$, and a nonstoichiometric sodium berylliumhydride [9].

References:

[1] Wittig, G., Hornberger, P. (Liebigs Ann. Chem. **577** [1952] 11/25).
[2] Coates, G. E., Cox, G. F. (Chem. Ind. [London] **1962** 269/71).
[3] Bell, N. A., Coates, G. E. (Chem. Commun. **1965** 582/3).
[4] Bell, N. A., Coates, G. E. (J. Chem. Soc. **1965** 692/9).
[5] Adamson, G. W., Shearer, H. M. M. (Chem. Commun. **1965** 240).
[6] Bell, N. A., Coates, G. E., Emsley, J. W. (J. Chem. Soc. A **1966** 49/52).
[7] Bell, N. A., Coates, G. E. (J. Chem. Soc. A **1966** 1069/73).
[8] Bell, N. A., Coates, G. E. (J. Chem. Soc. A **1968** 628/31).
[9] Coates, G. E., Pendlebury, R. E. (J. Chem. Soc. A **1970** 156/60).
[10] Adamson, G. W., Bell, N. A., Shearer, H. M. M. (Acta Cryst. B **37** [1981] 68/70).

[11] Bell, N. A., Coates, G. E. (Proc. Chem. Soc. **1964** 59).

1.2.6.2 Halo(diorganyl)beryllates

General Remarks. Diethylberyllium forms discrete adducts with alkali or ammonium halides in the mole ratios MX:Be = 1:1, 1:2, and 1:3. The structures of these compounds are as yet unknown, and information on their properties is limited to a few examples, which were investigated as electrolytes for the electrochemical preparation of Be metal. All compounds are characterized by elemental analysis.

$K[Be(C_2H_5)_2F]$

The crystalline 1:1 complex of KF and $Be(C_2H_5)_2$ is obtained from the corresponding 1:2 complex in C_6H_6 at 70°C. The second equivalent of $Be(C_2H_5)_2$ is dissolved in C_6H_6 and decanted. The product is dried under a high vacuum [1]; see also [3, 5].

The complex is soluble in ether [1]. It is slightly soluble in C_6H_6 and may dissociate into KF and $Be(C_2H_5)_2$ under certain conditions [3].

$K[Be_2(C_2H_5)_4F]$

KF reacts with excess ether-containing $Be(C_2H_5)_2$ at room temperature to 65°C. The ether can be removed in vacuum. The solid product is washed with C_7H_{16}, and the colorless crystals are dried in a high vacuum [1].

The specific conductivity of the molten salt at 80°C is $1.9\times10^{-2}\,\Omega^{-1}\cdot cm^{-1}$ [5]. The complex is insoluble in C_7H_{16} [3, 5] and soluble in ether and THF [5]. With an excess of $Be(C_2H_5)_2$, two liquid phases are formed [5]. It dissociates in warm C_6H_6 into an equilibrium mixture with the 1:1 complex and $Be(C_2H_5)_2$ [1, 3, 5]. Distillation at 100 to 130°C under high vacuum gives $Be(C_2H_5)_2$ [3, 5] in 67% yield [3]. Electrolysis of the molten salt at 75°C/1.8 to 2.0 V gives a deposit of Be on the cathode, containing ~20 to 30% Be_2C. Electrolysis with an excess of $Be(C_2H_5)_2$ gives a similar result [4]. The process has been patented [7]. The mechanism of the electrolytic deposition has been studied. K^+ is discharged first, which then deposits Be in a cementation reaction [6].

$Rb[Be(C_2H_5)_2F]$

This 1:1 complex is obtained by treatment of the 1:2 complex with C_6H_6 at 80°C for 6 h. A white material separates, which is treated once more with C_6H_6 at 80°C for 6 h. The residue is washed with $C_{17}H_{16}$ and dried in a high vacuum (yield 100%). It is insoluble in C_6H_6, ether, and THF [5].

$Rb[Be_2(C_2H_5)_4F]$

RbF reacts with excess $Be(C_2H_5)_2$ or its etherate within 2 h at room temperature. The volatiles are removed in a vacuum, and the residue is washed with C_7H_{16} [2]; see also [3].

The off-white solid melts at 61 to 63°C [2]. The melt has a specific conductivity of $1.4\times10^{-2}\,\Omega^{-1}\cdot cm^{-1}$ at 80°C. The compound is soluble in $C_6H_5CH_3$ at 80°C and in THF and ether at 20°C. With an excess of $Be(C_2H_5)_2$, two phases are formed. Distillation in a high vacuum at 120°C affords $Be(C_2H_5)_2$. Treatment with C_6H_6 at 80°C gives a precipitate of the 1:1 complex, and the solution contains the 1:3 complex [5]. Electrolysis at a voltage of 1.8 V yields a Be metal coating of the cathode containing 20 to 30% Be_2C [4]. The complex salt is slowly decomposed by air [2].

$Rb[Be_3(C_2H_5)_6F]$

The C_6H_6 extracts obtained from the treatment of the 1:2 complex with C_6H_6 at 80°C (see the preparation of the 1:1 complex) leave a white crystalline residue after evaporation of the solvent in a vacuum. It is soluble in C_6H_6, THF, and ether [5].

$Cs[Be_2(C_2H_5)_4F]$

CsF already reacts with $Be(C_2H_5)_2$, or its etherate, below room temperature (−50°C [3]). For completion, the suspension is warmed to 50°C. The ether is evaporated at room temperature; after 10 h stirring at 0°C, crystals are formed which are treated with C_7H_{16} at 0°C [2].

The crystals melt at 29 to 31°C [2]. The melt shows a specific conductivity of $10\times10^{-3}\,\Omega^{-1}\cdot cm^{-1}$ at 80°C. It is soluble in $C_6H_5CH_3$ at 80°C, in ether, and in THF. With $Be(C_2H_5)_2$, two phases are formed. Distillation at 120°C in a high vacuum gives no $Be(C_2H_5)_2$ [5]. No 1:1 complex is obtained by treatment with C_6H_6 at 80°C [2, 3, 5]. The solid phase obtained contains a complex of the composition $CsF\cdot1.5\,Be(C_2H_5)_2$, and the composition in the solution phase is $CsF\cdot2.5\,Be(C_2H_5)_2$ [5]. Electrolysis of a 1:3 mixture with $Be(C_2H_5)_2$ at 1 to 2 V gives a deposit of Be and some BeC_2 on the cathode [4].

$[N(CH_3)_4]\,[Be(C_2H_5)_2F]$

The complex is obtained by heating the 1:2 complex for 4 h at 110°C. $Be(C_2H_5)_2$ is distilled in a high vacuum, and the residue is washed with C_7H_{16}. No crystals were obtained [5].

References on p. 87

$[N(CH_3)_4]$ $[Be_2(C_2H_5)_4F]$

Anhydrous $[N(CH_3)_4]F$ reacts with an excess of $Be(C_2H_5)_2$ containing ~7% ether at 40°C. A crystalline material remains after evaporation of all the volatiles in a high vacuum at 20°C, which is washed with C_7H_{16} and dried in a high vacuum (yield ~100%).

The complex melts between 80 to 90°C. The melt has a specific conductivity of $4.7\times10^{-3}\ \Omega^{-1}\cdot cm^{-1}$. It is soluble in $C_6H_5CH_3$ at 80°C, in THF, and in ether, and insoluble in C_6H_6. It dissolves in $Be(C_2H_5)_2$ until a ratio of MX:Be ≧ 1:5.2 is reached; then 2 phases are formed. Distillation at 120°C in a high vacuum gives pure $Be(C_2H_5)_2$. The complex is not decomposed by C_6H_6 at 80°C. The 1:1 complex remains after 4 h at 110°C [5].

$[N(CH_3)_4][Be(C_2H_5)_2Cl]$

The 1:1 complex is formed as a gelatinous white precipitate from samples of the 1:2 complex upon standing for 3 days at room temperature. It is washed with C_7H_{16} and dried in a high vacuum [5].

$[N(CH_3)_4][Be_2(C_2H_5)_4Cl]$

The complex is prepared from stoichiometric amounts of $[N(CH_3)_4]Cl$ and $Be(C_2H_5)_2$ containing ~7% ether at 50 to 70°C. The ether is removed in a high vacuum leaving a colorless liquid that is washed with C_7H_{16} and dried in a high vacuum (yield ~100%).

The specific conductivity is $17\times10^{-3}\ \Omega^{-1}\cdot cm^{-1}$ at 80°C. It is soluble in ether and THF and insoluble in C_7H_{16}. With $Be(C_2H_5)_2$, two phases are formed. Thermal decomposition leads to ill-defined products. With C_6H_6 at 80°C, two liquid phases are formed which do not contain stoichiometrically defined compounds. Standing at room temperature for 3 days gives a precipitate of the 1:1 complex [5].

$[N(C_2H_5)_4][Be_2(C_2H_5)_4Cl]$

Anhydrous $[N(C_2H_5)_4]Cl$ reacts with $Be(C_2H_5)_2$ containing 15% ether at room temperature within 0.5 h. A viscous liquid is obtained after evaporation of the ether in a high vacuum and treatment with C_7H_{16}. No crystals are formed upon cooling [3].

The specific conductivity is $10\times10^{-3}\ \Omega^{-1}\cdot cm^{-1}$. The complex is soluble in ether and THF. It dissolves in $Be(C_2H_5)_2$ until a ratio of MX:Be≧1:5.2 is reached. At higher concentrations of $Be(C_2H_5)_2$, two phases are formed. Distillation at 100°C in a high vacuum gives $Be(C_2H_5)_2$; at 140°C, $Be(C_2H_5)_2$ and $N(C_2H_5)_3$ are formed. In C_6H_6 at 80°C two phases are formed. The C_6H_6 layer contains $Be(C_2H_5)_2$:$[N(C_2H_5)_2]Cl$ in a ratio of 4:1 [5]. Electrolysis at 1.7 V affords a coating of metallic Be mixed with a resin [4].

$[N(CH_3)_3CH_2C_6H_5][Be_2(C_2H_5)_4F]$

$Be(C_2H_5)_2$ in C_7H_{16} is added to a stoichiometric amount of $[N(CH_3)_3CH_2C_6H_5]F$ at −10°C. The temperature is raised with stirring to 50 °C within 1 h, and the mixture is stirred for another 2 h. Ether and C_7H_{16} are removed in a high vacuum, the residue is washed with C_7H_{16} and dried in a high vacuum (yield ~100%).

The viscous oil has a specific conductivity of $2.2\times10^{-3}\ \Omega^{-1}\cdot cm^{-1}$ at 80°C. It is soluble in $C_6H_5CH_3$ at 80°C, in ether, in THF, and in $Be(C_2H_5)_2$, and insoluble in C_6H_6. Distillation in a high vacuum at 80 to 140°C gives $Be(C_2H_5)_2$, $C_6H_5CH_2F$, and $N(CH_3)_3$. Treatment with C_6H_6 at 80°C gives 2 phases, but the C_6H_6 layer contains no $Be(C_2H_5)_2$ [5].

$[N(CH_3)_3CH_2C_6H_5][Be_2(C_2H_5)_4Cl]$

For the preparation, see the previous compound (yield ~100%). The viscous liquid has a specific conductivity of $6.3\times10^{-3}\ \Omega^{-1}\cdot cm^{-1}$. It is soluble in $C_6H_5CH_3$ at 80°C, in ether, and in THF. Two phases are formed with $Be(C_2H_5)_2$. Distillation at 80°C in a high vacuum gives $Be(C_2H_5)_2$ and $N(CH_3)_3$. The product obtained at 100 and 140°C could not be identified. With C_6H_6 at 80°C, two phases are formed which do not contain stoichiometrically defined compounds [5].

References:

[1] Strohmeier, W., Gernert, F. (Z. Naturforsch. **16b** [1961] 760).
[2] Strohmeier, W., Gernert, F. (Z. Naturforsch. **17b** [1962] 128).
[3] Strohmeier, W., Gernert, F. (Chem. Ber. **95** [1962] 1420/7).
[4] Strohmeier, W., Gernert, F. (Z. Naturforsch. **20b** [1965] 829/31).
[5] Strohmeier, W., Haecker, W., Popp, G. (Chem. Ber. **100** [1967] 405/11).
[6] Strohmeier, W., Popp, G. (Z. Naturforsch. **23b** [1968] 38/41).
[7] Hans, G. (Ger. 1162576 [1964] 1/2; C.A. **60** [1964] 11624).

1.2.6.3 Pseudohalo(diorganyl)beryllates

General Remarks. Crystalline ionic adducts are formed from $Be(C_2H_5)_2$ with alkali or quarternary ammonium cyanides, thiocyanates, and azides. Some of the compounds are characterized spectroscopically, and structures are proposed.

$Na[Be_4(C_2H_5)_8CN]$

NaCN is readly dissolved in $Be(C_2H_5)_2$ containing 40% ether. Part of the ether can be removed in a vacuum. Treatment with C_7H_{16} results in the formation of two liquid phases. The upper layer contains excess $Be(C_2H_5)_2$. Analysis of the lower layer gives ratios of CN^-:Be from 1:4.2 to 1:4.8 and 16 to 24% ether. The higher Be content can be attributed to the presence of $Be(C_2H_5)_2$ etherate in the complex [2]. It decomposes during the electrolysis at 1.5 V and cannot be used for the deposition of Be [3].

$K[Be_4(C_2H_5)_8CN]$

In an initial report, KCN could not be reacted with $Be(C_2H_5)_2$ [1]. The product was later obtained by reaction of KCN with $Be(C_2H_5)_2$ containing 40% ether at 65°C for 10 h. The ether is removed at 70°C under vacuum. Treatment of the residual solution with C_7H_{16} gives the compound, which is washed with C_7H_{16} [2].

The white crystals melt at 52 to 53°C [2]. The specific conductivity of the melt is $6.9\times10^{-3}\ \Omega^{-1}\cdot cm^{-1}$ at 80°C. It is soluble in $C_6H_5CH_3$ at 80°C, in C_6H_6, ether, and THF. It dissolves in $Be(C_2H_5)_2$ until a ratio of CN^-:Be $\geqq 6.7$ is reached. At a higher $Be(C_2H_5)_2$ content, two phases are formed [4]. Distillation in a high vacuum at 120°C gives $Be(C_2H_5)_2$ [2, 4].

$[N(CH_3)_4][Be_4(C_2H_5)_8CN]$

The complex is obtained as a colorless viscous liquid from a reaction of $[N(CH_3)_4]CN$ with $Be(C_2H_5)_2$.

References on pp. 90/1

The ^1H NMR spectrum at room temperature shows only one set of C_2H_5 resonances indicating the equivalence of these groups. The CH_2 signal is shifted to higher field (19 Hz at 60 MHz) compared to $Be(C_2H_5)_2$, and the CH_3 signal is only slightly shifted (4 Hz). The coupling constant is J(H, H) = 5.9 Hz. The IR and Raman vibrations are given in Table 19. Both are also shown in a figure in the original. The spectra are assigned on the basis of a sandwich-like structure with CN^- bridging of two $(Be(C_2H_5)_2)_2$ units (Formula I) [6].

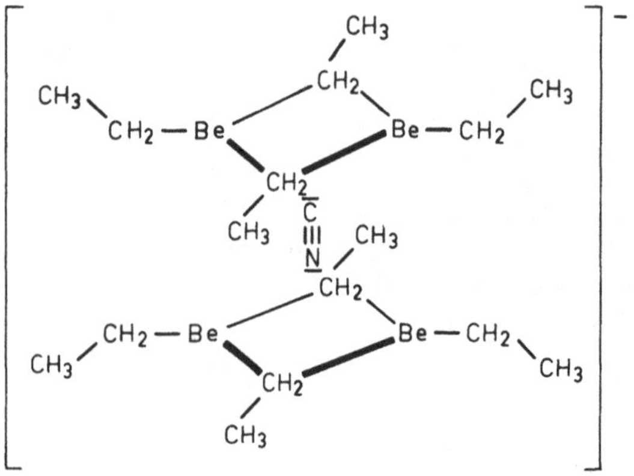

I

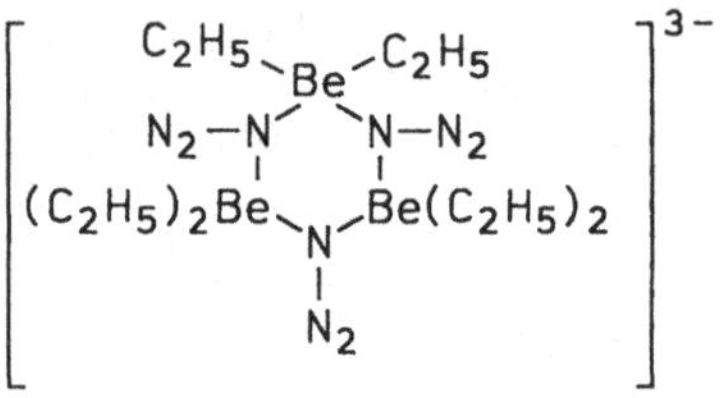

II

Table 19
IR and Raman Vibrations (in cm^{-1}) of Neat $[N(CH_3)_4][Be_4(C_2H_5)_8CN]$ [6].

IR	Raman	assignment
3035 (w)	3035 (w, dp)	$\nu(CH_3[N(CH_3)_4]^+)$
	2978 (w, p)	$\nu(CH_2, CH_3)$
2935 (s)		
	2922 (w, p)	
2860 (s)	2850 (m, p*)	
2840 (s)		
2780 (s)	2780 (w, p)	
2710 (vw)		overtones and combination bands
2620 (vw)		
2475 (vw)		
2195 (s)	2200 (s, p*)	$\nu(CN)$
1483 (s)		$\delta(CH_3)$
1465 (sh)	1460 (s, dp)	
1408 (s)	1415 (m, dp)	$\delta(CH_2)$
1372 (m)		$\delta(CH_2)$
1281 (vw)		
1225 (w)		
1185 (s)	1185 (m, p*)	$\tau(CH_2)$
990 (vs)	995 (m, p*)	$\nu(CC)$
945 (s)	952 (m, dp)	$\nu_{as}(NC_4)$
910 (sh)	905 (w, p)	$\nu(BeC)$
790 (s)	795 (m, p*)	$\varrho(CH_3)$
	752 (s, p)	$\nu_s(NC)$
740 (s)		$\nu(Be_2(CN)Be_2)$
680 (s)		$\varrho(CH_2)$
575 (s)		ν(ring) (B_{1u}, B_{3u})
	505 (s, p)	ν(ring) (A_g)
	475 (s, p*)	$\delta, \gamma(CC)(B_{1g})$
448 (w)		
375 (w)	375 (w, dp)	$\delta(NC_4)$
	250 (w, dp)	$\delta(BeCC)(B_{3g})$
220 (w)		$\delta(BeCC)$

p* = partially polarized

[$N(CH_3)_4$][$Be_2(C_2H_5)_4SCN$]

For the preparation of the complex, a 4-fold excess of $Be(C_2H_5)_2$ (containing ether) is condensed onto [$N(CH_3)_4$]SCN. Excess reagent and ether are removed in a high vacuum at 85°C.

The colorless viscous substance shows resonances in the ^1H NMR spectrum (solvent not given) at $\delta=0.32(q,CH_2)$, $1.42(t,CH_3C;\ J(H,H)=6.0$ Hz), and $3.78(CH_3N)$ ppm. The C_2H_5 resonances appear downfield from those of free $Be(C_2H_5)_2$. The IR and Raman data are summarized in Table 20. The spectra and the polarization behavior are given as a figure in the original. The ^1H NMR and the vibrational spectra give evidence that the S atom functions as the donor site [5].

Table 20
IR and Raman Vibrations (in cm^{-1}) of Neat [$N(CH_3)_4$][$Be_2(C_2H_5)_4SCN$] [5].

IR	Raman	assignment
3020 (w)	3029 (m)	$\nu(CH_3)[N(CH_3)_4]^+$
	2980 (m)	
2930 (sh)	2924 (m)	
2890 (s)		
2840 (s)	2828 (w)	$\nu(CH_2, CH_3)$
	2790 (m)	
2775 (vw)	2780 (m)	
2705 (vw)		
2620 (vw)		
2580 (vw)		
2500 (vw)		
2480 (vw)		
2420 (vw)		
2310 (vw)		
2100 (vs)	2113 (s)	$\nu(CN)$
1520 (sh)		
1500 (sh)		
1493 (sh)		
1485 (s)		
1475 (sh)		$\delta_{as}(CH_3)$
1460 (sh)	1453 (s)	
1410 (m)	1420 (m)	$\delta(CH_2)$
1390 (sh)		

IR	Raman	assignment
1370 (w)	1370 (w)	$\delta_s(CH_3)$
1285 (w)	1288 (w)	γ, $\tau(CH_2)$
1230 (w)		
1200 (sh)		
1185 (m)	1180 (m)	
1105 (vw)		
1055 (sh)		
990 (m)	977 (m)	$\nu(CC)$
945 (m)	950 (m)	$\nu_{as}(NC_4)$
920 (sh)		
865 (m)	870 (m)	ν_s, $\nu_{as}(Be_2S)$
760 (s)	752 (s)	$\nu_s(NC_4)$, $\varrho(CH_3)$
740 (s)	740 (vw)	$\nu(CS)$
665 (m)		$\varrho(CH_2)$
575 (sh)		
530 (sh)		
495 (sh)	497 (vw)	
450 (sh)	460 (vw)	$\delta(NC_4)$, $\delta(BeCC)$
415 (w, br)		$\gamma(SCN)$
355 (w, br)	370 (vw)	$\delta(SCN)$
228 (vw)	236 (vw)	$\delta(BeCC)$, $\delta(Be_2S)$
	169 (m)	

The complex is very sensitive towards moisture and O_2. It decomposes at 90°C to give $Be(C_2H_5)_2$, $N(CH_3)_3$, and CH_3SCN [5].

Cs[$Be(C_2H_5)_2N_3$]

CsN_3 and a ~4-fold excess of $Be(C_2H_5)_2$ (containing some ether) are reacted in $C_6H_5CH_3$ for 12 h at room temperature. The mixture is filtered, the residue washed with C_5H_{12}, and dried in a

References on pp. 90/1

high vacuum. More product can be obtained from the toluene phase of the filtrate. Toluene is completely removed only at temperatures above 50°C in a high vacuum. Without a solvent the reaction is not complete.

The white powder is extremely sensitive towards moisture and ignites in air. A consistent melting point could not be determined. The IR and Raman vibrations are summarized in Table 21. The number of observed N_3 vibrations suggests a cyclic trimeric structure for the anion (Formula II, p. 88) [7].

Table 21
IR and Raman Vibrations (in cm^{-1}) of $Cs[Be(C_2H_5)_2N_3]$ [7].

$\tilde{\nu}$	IR (mull)	Raman	assignment	$\tilde{\nu}$	IR (mull)	Raman	assignment
3460	sh		$\nu_{as}(N_3)+\nu_s(N_3)$	995	sh		$\nu(CC)$
3440	sh			985	ms	s	
3390	mw			950	w	m	
2945		m	$\nu(CH_3, CH_2)$	943	w	m	$\nu(BeC)$
2853		vs		912	m	w	
2795		m		830	s	s	$\varrho(CH_2), \nu(BeC)$
2780	m			775	w		
2174	m	m	$\nu_{as}(N_3)$	705		sh	
2140		m		685	s	vs	$\delta(N_3)$
2120	vs	s		617	ms	mw	
2070	m			600	w	w	
1462		s	δCH_3	590	w	w	
1410	vw	m	$\delta(CH_2)$	558		w	
1365		m	$\delta(CH_3)$	500	m		$\nu(BeN)$
1328	m	vs	$\nu_s(N_3)$	465		m	$\nu(BeN)$
1310	m	vs		445		s	$\delta(BeCC)$
1295	s	sh		430	w		δ(ring)
1289	sh	sh		400		vw	
1280	w			375		mw	
1233	vw	vw		360		vw	
1220	w			330	m		
1170	m	m	$\tau(CH_2)$	300		s	
1100	w			285		vs	
1050	w			205		m	
				195		vs	

References:

[1] Strohmeier, W., Gernert, F. (Z. Naturforsch. **16b** [1961] 760).
[2] Strohmeier, W., Gernert, F. (Chem. Ber. **95** [1962] 1420/7).
[3] Strohmeier, W., Gernert, F. (Z. Naturforsch. **20b** [1965] 829/31).
[4] Strohmeier, W., Haecker, W., Popp, G. (Chem. Ber. **100** [1967] 405/11).

[5] Atam, N., Müller, H., Dehnicke, K. (J. Organometal. Chem. **37** [1972] 15/23).
[6] Dehnicke, K., Atam, N. (Chimia [Switz.] **28** [1974] 663/4).
[7] Atam, N., Dehnicke, K. (Z. Anorg. Allgem. Chem. **427** [1976/77] 193/9).

1.2.6.4 Alkoxy(diorganyl)beryllates

General Remarks. This section describes some compounds that, formally, are addition products of alkali butoxides with $Be(C_2H_5)_2$. The structures of the compounds are still unknown, and only proposals are made based upon available physical and chemical properties. With $RbOC_4H_9$-t, a product of the composition $Rb_2Be_2(C_2H_5)_3(OC_4H_9\text{-t})_3$ is obtained, which is also included in this section since a structure with BeR_2 groups is proposed.

$Li[Be(C_2H_5)_2OC_4H_9\text{-t}]$

The complex is probably formed as the initial product from the reaction of equimolar quantities of $LiOC_4H_9$-t and $Be(C_2H_5)_2$ etherate in ether. After evaporation of the ether a liquid residue remains. Distillation at 36 to 39°C in a high vacuum (bath temperature 100 to 140°C) yields 36% of $Be(C_2H_5)OC_4H_9$-t which probably exists as an equilibrium of tetramers and dimers. An attempted purification of the initial adduct by crystallization from C_6H_{14} gave colorless crystals of unknown structure and composition (C_2H_5 : Be = 1 : 1.35), m.p. 36°C.

It is discussed that the title compound should have the dimeric structure shown in Formula I, p. 92 (M = Li).

$Na[Be(C_2H_5)_2OC_4H_9\text{-t}]$

$NaOC_4H_9$-t in ether is added to an equimolar amount of $Be(C_2H_5)_2$ etherate in ether at −78°C. The solution is warmed to room temperature and stirred for 15 min. Evaporation and crystallization from C_6H_5/C_6H_{14} (1:1) gives colorless prisms which soften at ~85°C and melt at 94 to 95°C.

^1H NMR ($C_6D_5CD_3$): δ = −0.35 (q, CH_2; J(H,H) = 8 Hz), 1.25 (s, CH_3C), and 1.58 ppm (t, C_4H_9). Lowering the temperature to −80°C causes the signals of the CH_3 and CH_2 groups to collapse into broad single lines at −0.49 and 1.65 ppm, respectively. The process that exchanges the bridging for terminal C_2H_5 groups is rapid even at this low temperature.

The molecular mass in C_6H_6 (by cryoscopy) corresponds to a dimer. The structure shown in Formula I, p. 92 (M = Na) is proposed. The complex reacts with t-C_4H_9OH in ether to afford $Na[Be(OC_4H_9\text{-t})_3]$.

$K[Be(C_2H_5)_2OC_4H_9\text{-t}]$

$Be(C_2H_5)_2$ etherate in ether is added to an equimolar amount of KOC_4H_9-t suspended in ether. The mixture is stirred for 30 min, filtered, and the ether is evaporated. Crystallization from C_6H_{14}/ether (10:1) or C_6H_6/C_6H_{14} (1:1) gives colorless waxy prisms in 29% yield.

M.p. 100 to 102°C (dec.); ^1H NMR (C_6H_6 or $C_6D_5CD_3$): δ = −0.28 (q, CH_2; J(H,H) = 8 Hz), 1.30 (s, C_4H_9), and 1.51 ppm (t, CH_3C).

The molecular mass (by cryoscopy) corresponds to a dimer. The structure given in Formula I, p. 92 (M = K) is assumed. The compound reacts with refluxing $C_6H_5CH_3$ to give a red product with the empirical formula $K_5Be_3(C_2H_5)_2(OC_4H_9\text{-t})_4(CH_2C_6H_5)_4$ but of unknown structure. With 1,3,5-$(CH_3)_3C_6H_3$, an orange-red material of the composition $K_3Be_2(C_2H_5)_2(OC_4H_9\text{-t})_2$-

References on p. 92

$(C_6H_2(CH_3)_3)_2$ is generated. Ethane is evolved in both reactions. Carboxylation of the products with CO_2, followed by hydrolysis, gives $C_6H_5CH_2COOH$ and 2,4,6-$(CH_3)_3C_6H_2COOH$, respectively.

$Rb_2[Be_2(C_2H_5)_3(OC_4H_9\text{-t})_3]$

Reaction of $Be(C_2H_5)_2$ etherate with $RbOC_4H_9$-t as described for the previous compound for 1 h gives, by crystallization from C_6H_{14}/ether (1:1) or from C_6H_6/C_6H_{14} (1:1), tan prisms in 15% yield.

The compound shrinks at ~125°C, softens at ~135°C, and melts at 180 to 185°C with decomposition. ^1H NMR (C_6H_6): $\delta = -0.17$, 0.14 (q, CH_2, ratio 2:1; J = 8 Hz), 1.36 (s, C_4H_9), and 1.62 ppm (m, CH_3C). Formula II has been tentatively assigned to this compound; Rb may be bound mainly ionically.

It reacts with 1,3,5-$(CH_3)_3C_6H_3$ at reflux temperature to give a red crystalline product of the empirical formula $Rb_5Be_3(C_2H_5)_3(OC_4H_9\text{-t})_2(C_6H_2(CH_3)_3)_4$. Ethane gas is evolved during the reaction. Carboxylation with dry ice followed by hydrolysis gives 2,4,6-$(CH_3)_3C_6H_2COOH$.

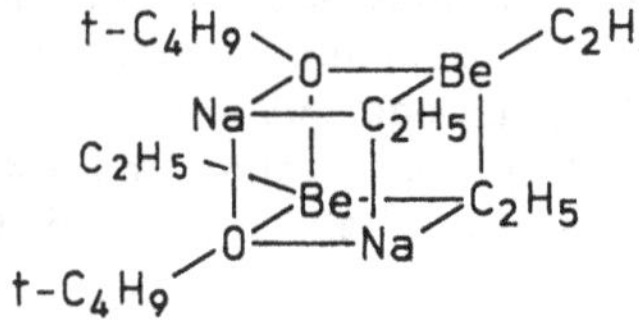

I

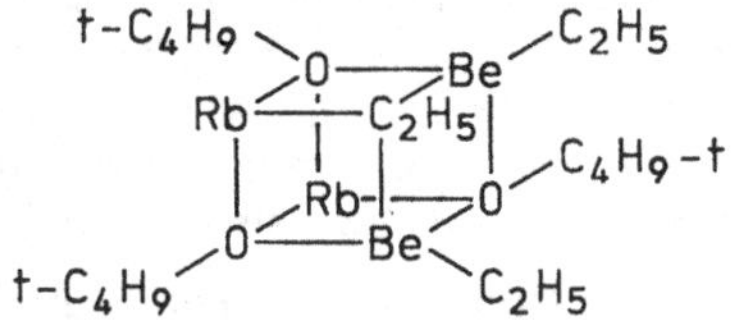

II

Reference:

Andersen, R. A., Coates, G. E. (J. Chem. Soc. Dalton Trans. **1974** 1729/36).

1.3 Monoorganoberyllium Compounds and Their Adducts

1.3.1 The Radical $BeC_2H_5 \cdot NC_5H_5$

The radical BeC_2H_5, stabilized by pyridine, is obtained in the electrolysis of an equimolar amount of $Be(C_2H_5)_2$ and pyridine with Cu electrodes at 60°C. A viscous black liquid is formed at the cathode, from which a highly viscous black material is isolated by washing with petroleum ether and drying in a high vacuum. The elemental analysis is in agreement with the formula.

The radical dissolves in C_6H_6. These solutions give an ESR signal with g = 2.004. But solutions in excess of pyridine (1:170) exhibit a hyperfine splitting of the signal into ~72 lines. An assignment was not possible.

Reference:

Strohmeier, W., Popp, G. (Z. Naturforsch. **22b** [1967] 891).

1.3.2 Organoberyllium Hydrides and Their Adducts

General Remarks. Organoberyllium hydrides, Be(R)H (R = alkyl or aryl), are highly electron-deficient compounds and, in the absence of suitable donor ligands, subject to oligomerization and polymerization processes. In most cases, the individual species are therefore intractable materials which are not or only partially characterized. Some were only prepared in solution and isolated as adducts, mainly with N-donors. Ether solutions are assumed to contain the etherates. When isolated, the ether adducts usually contain less than the stoichiometric amount of $O(C_2H_5)_2$. With other O-, N-, or P-containing ^2D ligands, 1:1 adducts are formed which were found to be dimeric in C_6H_6 in all examples investigated by a cryoscopic measurement. Formula I is suggested as a structure for these adducts without differentiating between cis- and trans-isomers. With ^4D donors, mainly $NR_2CH_2CH_2NR_2$ (R = CH_3, C_2H_5), 2:1 adducts were isolated which probably have a chelate structure, at least in solution (Formula II). A polymeric constitution with the diamine as chain propagator in the solid state cannot be totally excluded, but no convincing evidence exists.

The HBe ^1H NMR signal was only detected for $(Be(CH_3)H \cdot O(C_2H_5)_2)_2$. In all other cases where the ^1H NMR spectrum was measured, no signal was found for this proton. But the existence of bridging H was often verified by IR absorptions in the region ~1400 to 1200 cm^{-1}.

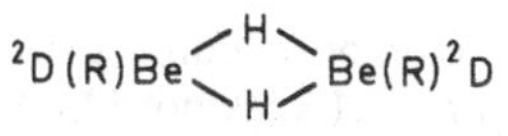

I

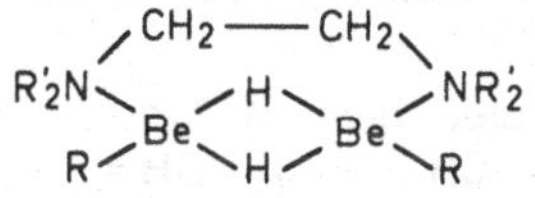

II

Owing to their virtual simplicity, the hypothetical Be(R)H monomers have been the target of numerous theoretical investigations on all levels of sophistication. The results of these studies are included in this chapter for the known compounds; for the hypothetical hydrides, see Chapter 5, pp. 204/15.

During the synthetic work directed towards the preparation of organoberyllium hydrides, products that analyze for $Be_3R_4H_2$ (R = CH_3, C_2H_5) and $Be_2(C_2H_5)_3H$ were also obtained. There is little evidence about their constitution. The materials can also be regarded as mixtures of several species. These compounds are therefore treated in the Section 4.1, pp. 199/200, dealing with compounds of unknown structure.

$Be(CH_3)H$

The compound was for the first time obtained from the reaction of $Al(CH_3)_2H$ with an excess of $Be(CH_3)_2$ in the absence of a solvent or in i-C_5H_{12}. A solid product appeared which yielded the calculated volume of H_2 and CH_4 gas upon hydrolysis. The compound was not further studied [1]. $Be(CH_3)H$ was again synthesized later through the pyrolysis of $Be(CH_3)C_4H_9$-t under N_2 for 2 h and then at 120°C/10^{-3} Torr for 10 h. The residual white C_6H_6 insoluble solid had an IR absorption at 1800 cm^{-1}. The product was characterized via its $N(CH_3)_3$ adduct (see below) [14].

Monomeric $Be(CH_3)H$ was the subject of several ab initio calculations with several levels of sophistication. The calculated total energies derived from optimized geometries range from −54.16140 to −54.98422 a.u. for a C_{3v} symmetry [18 to 20, 22, 24, 27, 28, 31]. The bond lengths are 1.713(1.690) (BeC), 1.291(1.341) (BeH), and 1.090(1.085) (CH) Å, while the BeCH angle is found to be 111.7°(111.8°) by a 3-21G (STO-3G) basis set [24]; see also [22, 27]. The charge distribution is interpreted in terms of two-center two-electron bonds with an ionic character of ca. 15% for the BeH bond (as Be^+H^-) and 13% for the BeC bond (as Be^+C^-) [18]. This yields a +0.296 charge on Be [32]. A later calculation postulates a transfer of 0.15 electrons from BeH

References on p. 104

to CH_3 and a dipole moment of 0.34 D [24]. See also [33] for the gross atomic populations. Calculations as to the degree of hyperconjugation (back donation of electrons from C to Be) result in a value of 7.1 kcal/mol for the π-bonding strength [18]. An ab initio calculation imposing a C_{2v} symmetry on the molecule (planar environment around C) clearly shows that the C_{3v} model with tetrahedral environment around C is energetically favored [20].

The equilibrium geometry and ground state electronic properties were also calculated by a simple FSGO model. The predicted equilibrium geometries of 1.693 (BeC), 1.392 (BeH), 1.132 (CH) Å, and 104.8° (HCH) are in reasonable agreement with the ab initio results. The BeC bond energy is estimated from the reaction $BeH^+(g) + CH_3^-(g) \rightarrow Be(CH_3)H(g)$ as 1263 kJ/mol. The fraction of electronic charge transferred from C to Be is estimated to be 0.48. The calculated dipole moment of 0.11 D is quite small. The calculated total energy amounts to −46.209 a.u.. The wave functions obtained have also been used to predict the electron momentum distributions and Compton profiles [26]. The vibrational frequencies (in cm^{-1}) are calculated for the monomer by the ab initio method with the STO-3G basis set: $\nu_{as}(CH_3)$ 3710(e), $\nu_s(CH_3)$ 3542(a_1), ν(BeH) 2644(a_1), $\delta(CH_3)$ 1829(e), $\delta_s(CH_3)$ 1623(a_1), ν(BeC) 966(a_1), δ(HBeC) 942(e), and $\delta(CH_3)$ 540(e) [33].

The geometrical parameters obtained from an MNDO approximation for an optimized C_{3v} symmetry with a HOMO of A_1 symmetry are 1.661 (BeC), 1.279 (BeH), 1.116 (CH) Å, and 111.6° (BeCH) [29].

The calculated heat of hydrogenation of the monomer according to the reaction $Be(CH_3)H + H_2 \rightarrow CH_4 + BeH_2$ is $\Delta H = -11.3$ kcal/mol [24] or −8.1 kcal/mol [28]. For the isodesmic reaction $CH_5^+ + 2Be(CH_3)H \rightarrow (BeH)_2CH_3^+ + CH_4$, ΔH values of −85.0 to −117.0 kcal/mol were obtained by ab initio calculations, depending on the basis sets employed (STO-3G or 5-21G) [27]. The isodesmic reaction $C_5H_6 + Be(CH_3)H \rightarrow CH_4 + Be(C_5H_5)H$ ($\Delta H = -54.8$ kcal/mol) indicates the relative stability of the two hydrides [25].

Calculations of the energy changes associated with the dimerization of $Be(CH_3)H$ were also performed. Dimerization through CH_3 bridges between two Be ($HBe(CH_3)_2BeH$) [18, 30] and through H bridges ($CH_3BeH_2BeCH_3$) [30] was considered. An ab initio calculation (STO-3G) for the dimerization to a CH_3-bridged $(Be(CH_3)H)_2$ with a planar $HBeC_2BeH$ framework shows that the dimer is 3.8 kcal/mol less stable than the two monomers. The dimer geometry was not optimized [18].

Calculations with the PRDDO method gave a total energy for the CH_3-bridged dimer of C_{2h} symmetry of −109.4727 a.u. and for the H-bridged dimer of −109.5011 a.u.. Optimized bond distances and bond angles for the CH_3-bridged dimer are: 1.811 (BeC), 1.852 (Be-Be) Å, 100.8° ($H^1CH^{1'}$), and 105.3° (H^1CH^2) [30]. For the H-bridged dimer, the optimized parameters (bond lengths in Å) calculated by a 3-21G (STO-3G) basis set are: 1.717 (1.697) (BeC), 1.492 (1.463) (BeH), 1.089 (1.083) (CH), 87.5° (BeHBe), and 111.3° (111.3°) (BeCH). The dimerization energy is 68.4 (63.6) kJ/mol [33]. Energies of dimerization and association reactions of $Be(CH_3)H$ obtained by the PRDDO method and the ab initio STO-4G method with standard geometries are as follows (ΔH in kcal/mol) [30]:

reaction	ΔH (PRDDO)	ΔH (STO-4G)
$Be(CH_3)H + BeH_2 \rightarrow CH_3BeH_2BeH$	−40.4	−18.3
$2Be(CH_3)H \rightarrow CH_3BeH_2BeCH_3$	−25.0	−13.1
$Be(CH_3)H + Be(CH_3)_2 \rightarrow CH_3Be(CH_3)_2BeH$	0.3	4.9
$2\,Be(CH_3)H \rightarrow HBe(CH_3)_2BeH$	− 7.2	0.7
$Be(CH_3)H + Be(CH_3)F \rightarrow HBe(CH_3)_2BeF$	0.4	7.2

The H-bridged dimer is favored over the CH_3-bridged dimer [30].

Vertical ionization of the monomer yields a degenerate cation whose structure on optimization belongs to a different point group [29]. Deprotonation at CH_3 to give $[Be(CH_2)H]^-$ is favored over $[BeCH_3]^-$ formation, since the carbanion is stabilized by the BeH group [31].

$Be(CH_3)H \cdot nO(C_2H_5)_2$ (n = 1, n < 1)

A suspension of LiH in ether is treated with $BeBr_2$ and $Be(CH_3)_2$ in the mole ratio 2:1:1. Almost all the solid is dissolved after 2 h of boiling with stirring. LiBr can be precipitated with C_6H_6. Evaporation of the solvents from the filtrate and exposure to a vacuum for a further 2 h at room temperature yielded a sticky, colorless glass for which the CH_3:H:Be:$O(C_2H_5)_2$ ratio was 1.006:1.00:1.018:0.433. Heating the glassy material in a vacuum at 60 to 70°C/10^{-3} Torr gave a material which analyzed for CH_3:H:$O(C_2H_5)_2$ = 1.00:1.02:0.156 and showed IR absorptions at 1802 and 1567 cm^{-1}. An oligomeric or polymeric structure with single H bridges is suggested for this product with a low ether content. Attempts to remove more ether by dissolving the glass in $C_6H_5CH_3$ and distilling at atmospheric pressure resulted in decomposition. $Be(CH_3)D \cdot n\,O(C_2H_5)_2$ (n < 1) is prepared analogously using LiD. It is obtained as a colorless glass which analyzes for CH_3:D:Be:$O(C_2H_5)_2$ = 1.00:1.00:1.10:0.262 [7].

When ether is evaporated from a solution of $Be(CH_3)H$ until the ether:Be ratio is 1:1, the resulting liquid is dimeric in C_6H_6 (by cryoscopy), Formula I, p. 93 (R = CH_3, ^{2}D = $O(C_2H_5)_2$). The characteristic IR absorptions due to the BeH_2Be bridge are in the 1350 to 1290 and 1170 to 1060 cm^{-1} regions. A spectrum of a solution of compound I in ether contains no strong IR absorptions in the 1500 to 1000 cm^{-1} region [7]. The ^{1}H NMR spectrum in c-$C_6D_{11}CD_3$ shows four resonances at δ = −0.89 (CH_3Be), 1.19 (CH_3C), 1.58 (HBe), and 3.66 ppm (CH_2). The HBe signal is broad. The spectrum in $C_6D_5CD_3$ is very similar. Changing the temperature of the solution from −90 to 100°C does not produce any extra peak in the spectra. Consequently, no information can be obtained about possible cis-trans equilibria [8].

When a suspension of NaH (in excess) in ethereal $Be(CH_3)_2$ is boiled for 48 h and then $BeCl_2$ is added, a solution is obtained which contains Be:CH_3:H in the ratio 3:4:2. The main solute species may be solvated $Be(CH_3)_2$ and solvated $Be(CH_3)H$ [6].

$(Be(CH_3)H \cdot O(C_2H_5)_2)_2$ slowly reduces CH_3I at 84°C quantitatively to CH_4. C_6H_5CHO and $C_6H_5CH{=}NC_6H_5$ are rapidly added with formation of $(Be(CH_3)OCH_2C_6H_5)_4$ and $(Be(CH_3)N(C_6H_5)CH_2C_6H_5)_2$, respectively. $(C_6H_5)_2CO$, in contrast, gives the monomeric $Be(CH_3)OCH(C_6H_5)_2 \cdot O(C_2H_5)_2$. Addition of $CH_2{=}CH_2$ in a sealed tube at 84°C gives a product mixture which is believed to consist mainly of $Be(CH_3)_2$ and $Be(C_2H_5)_2$ [22]. The reactions with t-C_4H_9CN [16] and C_6H_5CN [15, 16] in C_6H_6 or $C_6H_5CH_3$ in the absence of a donor ligand give colorless oils of $Be(CH_3)N{=}CHR$ (R = t-C_4H_9, C_6H_5) that cannot be crystallized; their hydrolysis yields RCHO (R = t-C_4H_9, C_6H_5) [15]. By slow addition of C_6H_5CN in ether to $Be(CH_3)H$ etherate, tetrameric $Be(CH_3)N{=}CHC_6H_5$ is obtained as an amorphous solid [16]. In the presence of $N(CH_3)_3$, the nitriles RC_6H_4CN (R = H, 2-CH_3, 3-CH_3) give the corresponding dimeric complexes $(Be(CH_3)N{=}CHC_6H_4R \cdot N(CH_3)_3)_2$ [15].

The ether can be replaced by $CH_3OCH_2CH_2OCH_3$ [6], $N(CH_3)_3$ [5 to 9, 15, 16], pyridine [2, 7, 14], $N(CH_3)_2CH_2CH_2N(CH_3)_2$ [5, 6, 13], $N(C_2H_5)_2CH_2CH_2N(C_2H_5)_2$ [13], 2,2'-bipyridine [7, 15], and 4-dimethylaminopyridine [16] (see below). $Be(CH_3)D \cdot n\,O(C_2H_5)_2$ is also easily converted into the $N(CH_3)_3$ adduct [6].

$Be(CH_3)H \cdot 0.5\,CH_3OCH_2CH_2OCH_3$

The adduct is obtained by addition of $CH_3OCH_2CH_2OCH_3$ to ethereal "$Be_3(CH_3)_4H_2$" (obtained from $Na[Be(CH_3)_2H]$ and $BeCl_2$). After sublimation of $Be(CH_3)_2 \cdot CH_3OCH_2CH_2OCH_3$, a viscous oil remains which is insoluble in C_6H_6 and which analyzes for the title compound. Its

References on p. 104

constitution is probably similar to that of the $NR_2CH_2CH_2NR_2$ (R = CH_3, C_2H_5) complexes below (Formula II, p. 93) [6], mentioned in [5].

$Be(CH_3)H \cdot N(CH_3)_3$

The adduct is best prepared from the components NaH, $Be(CH_3)_2$, $BeCl_2$, and $N(CH_3)_3$ in the appropriate mole ratio. The reaction of NaH and $Be(CH_3)_2$ in boiling ether leads to a solution of $Na_2[Be(CH_3)_2H]_2$ which, on addition of $BeCl_2$, affords a precipitate of NaCl and a solution of a complex hydride, "$Be_3(CH_3)_4H_2$". This solution is treated with $N(CH_3)_3$ to give the final product together with $Be(CH_3)_2 \cdot N(CH_3)_3$. The two compounds can be separated by vacuum condensation [5, 6]. A second method uses $Be(CH_3)_2$ and $Sn(C_2H_5)_3H$ as the starting materials. Their reaction in ether at 65 to 70°C affords crude $Be(CH_3)H$ as a viscous oil after evaporation of most of the solvent. Treatment with $N(CH_3)_3$ and sublimation at 40 to 45°C/~10^{-3} Torr gives the adduct as colorless prisms [9]. The adduct is also obtained from $Be(CH_3)H$ (produced by pyrolysis of $Be(CH_3)C_4H_9$-t) and excess $N(CH_3)_3$ and isolated by sublimation at 70°C/10^{-3} Torr [14].

Melting points are given at 70 to 72°C [9], 73 to 73.5°C [5, 6], and 99 to 107°C [14]. The melting behavior varies with the amount of material in the melting point tube and with the rate of heating. It may be associated with a cis-trans isomerization. Slow overnight sublimation, slightly above ambient temperature, gives several mm long crystals which melt at 109 to 112°C [14]. The vapor pressure is 1.8 Torr at 64°C and log p = 7.483–2439 T^{-1} (p in Torr, T in K) for the liquid phase between 73 and 115°C. From the vapor pressure equation, it follows that $\Delta H_v = 11.2$ kcal/mol, the extrapolated boiling point is 257°C, and Trouton's constant is $\Delta H_v/T_v = 21.1$ cal·mol^{-1}·K^{-1}. There seems to be no change in the degree of association during the process of vaporization, and the vapor consists mainly of dimers at least up to 110°C. Some decomposition of the vapor is observed at 145 to 175°C, and the vapor appears to be monomeric at least at 175°C. The molecular mass in C_6H_6 (by cryoscopy) corresponds to dimers [6].

The ^1H NMR spectrum at 25°C in c-C_6H_{12} shows absorptions at $\delta = -2.21$ (CH_3Be), 2.27 (cis-CH_3N), and 2.36 ppm (trans-CH_3N), and in $CD_3C_6D_5$ at -0.69 (CH_3Be), 2.055 (cis-CH_3N) and 2.17 ppm (trans-CH_3N). At -50°C in $CD_3C_6D_5$, the CH_3N resonances are observed at 1.87 and 1.98 ppm. The splitting of the resonance for the CH_3N protons can be interpreted as arising from the presence of cis- and trans-isomers (Formula III). Between -60 and 10°C in $CD_3C_6D_5$, the relative intensity of the two CH_3N peaks changes from 0.7 to 2.0. Above this temperature, the two peaks begin to coalesce, until at 40°C a single broad peak appears. At 100°C, both the CH_3N and CH_3Be peaks are narrow lines. The CH_3Be peak divides at -60°C. ΔH for the cis-trans isomerization is measured as 3.1 ± 2 kcal/mol and ΔS as 13 ± 2 cal·mol^{-1}·K^{-1}. A resonance for the bridging hydrogens was not detected [8].

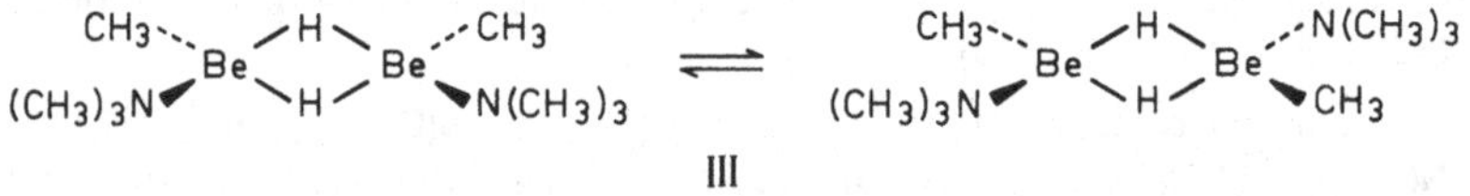

III

The IR spectrum in c-C_6H_{12} contains a strong absorption at 1344 cm^{-1} for $\nu(BeH_2Be)$ as well as an absorption at 1007 cm^{-1} for ν(NC). The saturated vapor phase has a similar IR absorption at 1342 cm^{-1} for $\nu(BeH_2Be)$; but at 65°C and particularly at 80°C, a new sharp band appears at 2141 cm^{-1} which is assigned to terminal BeH bonds of the monomers [6].

The compound adds RC_6H_4CN (R = H, 2-CH_3, 3-CH_3) to give the corresponding $(Be(CH_3)N{=}CHC_6H_4R)_2$ [15, 16].

The deuterated complex $Be(CH_3)D \cdot N(CH_3)_3$ is obtained like its hydrido analog using NaD instead of NaH. Sublimation yields a colorless crystalline complex, m.p. 75 to 76°C. The IR absorption of the BeD_2Be bridges occurs at 1020 cm^{-1} in c-C_6H_{12} [6].

$Be(CH_3)H \cdot N(CD_3)_3$ is prepared from $Be(CH_3)H$ and $N(CD_3)_3$ in C_6H_6/ether and isolated by sublimation at 50 to 60°C/10^{-3} Torr, m.p. 74 to 75°C. The ^1H NMR spectrum in $C_6D_5CD_3$ contains no resonance that can be attributed to the bridging hydrogens [8].

$Be(CH_3)H \cdot 0.5\,N(CH_3)_2CH_2CH_2N(CH_3)_2$

An excess of the ligand is condensed onto an ethereal solution of "$Be_3(CH_3)_4H_2$" (prepared from $Na[Be(CH_3)_2H]$ and $BeCl_2$). Warming to room temperature gives a white amorphous precipitate [6], mentioned in [5]. The complex is also obtained by addition of the diamine to a solution of $Be(CH_3)H$ etherate in C_6H_6 [13].

The nonvolatile material [6,13] can be crystallized from C_6H_5Cl as needles [13], but is insoluble in C_6H_6, ether, CS_2, and CCl_4 [6]. A polymeric structure with bridging H and diamine groups was suggested [6], but a comparison with other diamine complexes (see below) makes a chelate structure (Formula II, p. 93, $R = R' = CH_3$), at least in solution, more feasable [13]. The compound is not pyrophoric and does not fume in air [6].

$Be(CH_3)H \cdot 0.5\,N(C_2H_5)_2CH_2CH_2N(C_2H_5)_2$

Ethereal $Be(CH_3)H$ and the ligand react in C_6H_6 in an exothermic reaction. After evaporation of the solvent, the residue can be crystallized from $C_6H_5CH_3/C_6H_{14}$ (3:1) in a yield of 10%, m.p. 125°C. The diamine probably acts as a chelating ligand in C_6H_6, as the compound is a monomer in this solvent (by cryoscopy), and the structure is given as in Formula II, p. 93 ($R = CH_3$, $R' = C_2H_5$) [13].

$Be(CH_3)H \cdot NC_5H_5$

An equimolar amount of pyridine reacts with $Be(CH_3)H$ in ether at −20°C [7,16]. Colorless crystals are deposited and recrystallized from ether at −20°C; m.p. 81 to 82°C with decomposition [7].

The adduct is a dimer in C_6H_6 (by cryoscopy) [7,16]. A solution in ether shows no sign of decomposition during 3 h at room temperature [7], but addition of an excess of pyridine gives orange $Be(CH_3)NC_5H_6 \cdot 2NC_5H_5$ (see Section 4.2, p. 201) [7,16].

$Be(CH_3)H \cdot NC_5H_4N(CH_3)_2$-4

The adduct appears as a white precipitate when an equimolar amount of 4-dimethylaminopyridine is added to $Be(CH_3)H$ in ether at −78°C.

The solid is shown to be crystalline by its X-ray powder photograph (not given) and darkens at 140 to 150°C and melts to a red liquid at 152°C. A $\nu(BeH_2Be)$ absorption at 1333 cm^{-1} in the IR spectrum suggests a dimeric structure [16].

$Be(CH_3)H \cdot n\,P(CH_3)_3$ ($n < 1$)

A crystalline complex is formed from $Be(CH_3)H$ and $P(CH_3)_3$ in ether at −50°C. The material loses $P(CH_3)_3$ very easily, and an analysis of such a product gave a ratio of $CH_3 : H : P = 1 : 1 : 0.64$ [9].

$Be(C_2H_5)H$ and $Be(C_2H_5)H \cdot nO(C_2H_5)_2$ ($n<1$)

Donor-free $Be(C_2H_5)H$ still remains to be prepared in a pure state. $Be(C_2H_5)_2$ and $Sn(C_2H_5)_3H$ give in C_6H_{14} at 60 to 75°C a solid material that analyzes for $Be(C_2H_5)_{0.61}H_{1.41}$, $Be(C_2H_5)_{0.29}H_{1.71}$, or $Be(C_2H_5)_{0.26}H_{1.76}$, depending on the reaction and workup conditions [9]. A glassy residue, evidently consisting mainly of $Be(C_2H_5)H$, is obtained after 8 h heating of $Be_3(C_2H_5)_4H_2$ at 70 to 80°C [6].

$Sn(C_2H_5)_3H$ reacts with $Be(C_2H_5)_2$ in ether at 40°C for 2 h and then at 75°C for 2 h, after which most of the ether is evaporated. All volatile material is evaporated at 10^{-3} Torr, and a viscous glassy residue is obtained which turns to a white solid upon prolonged pumping. This product, dissolved in ether, gives equal amounts of H_2 and C_2H_6 upon hydrolysis [6]. Ether solutions of $Be(C_2H_5)H$ are available from $Na[Be(C_2H_5)_2H]$ and $BeCl_2$ [5], or from LiH, $Be(C_2H_5)_2$, and $BeBr_2$ [7, 13]. The latter reaction gives a colorless glass after heating at 40 to 45°C/10^{-3} Torr. The ratio of $C_2H_5:H:Be:O(C_2H_5)_2$ was determined as 1.00:0.98:0.98:0.15 [7]. The etherate is also accessible from the reaction between $Be(C_2H_5)_2$, $BeCl_2$, and two equivalents of $Na[B(C_2H_5)_3H]$ in ether, but no details are given [9, 10].

Ab initio calculations with standard geometry give total energies of −93.69655 a.u. (5-21G) for the eclipsed conformation [19] and −93.70152 (5-21G) [19, 22, 23] for the staggered conformation. The theoretical rotational barrier of 3.1 kcal/mol is close to that in C_2H_6 (3.3 kcal/mol) [19].

The glassy ether-containing $Be(C_2H_5)H$ reacts with C_2H_4 in a sealed tube to give $Be(C_2H_5)_2$. The addition of decene-1 is very fast at 84°C (<0.17 h); addition of pentene-1 at 50°C takes 4.2 h for completion; whereas pentene-2 reacts slowly, and no reaction is observed with 2-methylbutene-2 [7]. The reaction with $Na[Al(C_2H_5)_3H]$ leads to mixed sodium alkylhydridoberyllates [10].

Adducts with $N(CH_3)_3$ [6, 9] and $N(CH_3)_2CH_2CH_2N(CH_3)_2$ [13] have been isolated from the ether solutions. With excess $4\text{-}(CH_3)_2NC_5H_4N$, an adduct is initially formed which reacts further to an orange-brown oil which is probably a hydropyridine derivative in analogy to the reaction of $Be(CH_3)H$ [16]. An excess of pyridine reacts with $Be(C_2H_5)H$ in ether to give $Be(C_2H_5)NC_5H_6 2NC_5H_5$ (see Section 4.2, p. 201). An intermediate adduct was not detected [7].

$Be(C_2H_5)H \cdot N(CH_3)_3$

$N(CH_3)_3$ is added to a solution of $Be(C_2H_5)H$ in ether and cooled to liquid N_2 temperature. Ether and excess amine are evaporated [6], also mentioned in [9]. The compound is also formed from $Be_2(C_2H_5)_3H$ and $N(CH_3)_3$ in C_6H_{14} at −10°C [12].

Sublimation at 40 to 45°C/10^{-2} Torr gives colorless prisms, m.p. 90 to 91°C [6]; recrystallization from C_6H_{14} gives colorless needles with a m.p. of 91°C [12]. The material is dissolved in C_6H_6 as a dimer (by cryoscopy). The IR spectrum in c-C_6H_{12} shows a $\nu(BeH_2Be)$ absorption at 1333 cm^{-1}. The compound fumes in air and reacts vigorously with H_2O [6]. Ethylene is added at 84°C in a sealed tube to give a liquid product that yields only C_2H_6 on hydrolysis [7]. Reaction with $Hg(C_2H_5)_2$ at 60°C in a sealed tube gives C_2H_6, Hg, and $Be(C_2H_5)_2 \cdot N(CH_3)_3$ [9].

$Be(C_2H_5)H \cdot 0.5\,N(CH_3)_2CH_2CH_2N(CH_3)_2$

The adduct is formed from ether/C_6H_6 solutions of the components. Colorless needles crystallize from $C_6H_5CH_3$, m.p. 151°C. The diamine is probably a chelating ligand, since the compound is dimeric in C_6H_6 (by cryoscopy), Formula II, p. 93 ($R=C_2H_5$, $R'=CH_3$) is proposed [13].

$Be(C_2H_5)H \cdot NC_5H_4N(CH_3)_2$-4

When $Be(C_2H_5)H$ in ether is added to a ca. 3-fold excess of 4-$(CH_3)_2NC_5H_5N$ in C_6H_6, an orange-yellow solution and a white precipitate form. The precipitate is probably the 1:1 adduct in analogy to the corresponding reaction of $Be(CH_3)H$; it was not isolated. The solution was boiled, and an orange-brown oil was isolated which could not be purified. It is probably a hydridopyridine derivative as was found for the reaction of $Be(CH_3)H$ with excess of pyridine [16].

$Be(C_3H_7$-i)H

The compound is formed upon pyrolysis of $Be(C_3H_7\text{-i})_2$ at 200°C. A colorless, viscous, nonvolatile oil remains which is rapidly hydrolyzed by H_2O to BeO, C_3H_8, and H_2. A polymeric structure is proposed. Further heating of $Be(C_3H_7$-i)H to 220 to 250°C results in the formation of Be, H_2, C_3H_6, C_3H_8, and an orange-colored residue. The reaction with $NH(CH_3)_2$ yields C_3H_8 and H_2 in the ratio 2.5:1, and $(Be(C_3H_7\text{-i})N(CH_3)_2)_n$ [3].

$Be(C_4H_9$-s)H·n$O(C_2H_5)_2$ (n<1)

$Be(C_4H_9\text{-s})_2$ etherate is heated to reflux at 10^{-3} Torr. Butene is evolved and the product gradually becomes viscous within ~24 h. The viscous product dissolves in C_6H_6 and analyzes for Be:C_4H_9:H:$O(C_2H_5)_2$=1.0:1.0:0.95:0.10. A 2:1 adduct is formed with $N(CH_3)_2CH_2CH_2N(CH_3)_2$ [13].

$Be(C_4H_9$-s)H·0.5$N(CH_3)_2CH_2CH_2N(CH_3)_2$

The diamine is added to the previously described hydride in C_6H_6. After evaporation of the volatile material, the residue is recrystallized from C_6H_6/C_6H_{14} as prisms, m.p. 163 to 165°C. It dissolves in C_6H_6 as a monomeric 2:1 adduct (by cryoscopy), Formula II, p. 93 (R=C_4H_9-s, R'=CH_3) [13].

$Be(C_4H_9$-i)H

The compound is obtained as a glassy viscous liquid in the pyrolysis of $Be(C_4H_9\text{-i})_2$ at 85°C. A conversion of 95% is achieved. It can also be prepared by the thermal decomposition of $Be(C_4H_9\text{-i})_2$ etherate at 150°C for 1 h. The product is treated with C_5H_{12} to remove undecomposed $Be(C_4H_9\text{-i})_2$. The compound is also a constituent of the pyrolysis residue generated from $Be(C_4H_9\text{-t})_2$ at 110°C, but it was not separated from $Be(C_4H_9$-t)H.

The ^{1}H NMR spectrum in C_6H_6 shows only the resonances of i-C_4H_9: δ=0.51, 0.62 (CH_2), 1.04, 1.14 (CH_3), and 2.1 ppm (sept, CH). The IR spectra in Nujol and c-C_6H_{12} show absorptions in the 1400 to 1200 cm^{-1} region for ν(BeH_2Be). Bands at 1820 and 1580 cm^{-1} are strong and broad in Nujol, but weak in c-C_6H_{12}.

The hydride dissolves only slowly in aromatic solvents. It is extensively associated in C_6H_6. From the small freezing point suppressions, it is concluded that the degree of association in C_6H_6 is of the order of 10 to 30. An equimolar amount of pentene-1 is added. At 33.5°C, $t_{1/2}$ for the reaction is 10 to 15 min in C_6H_6, as measured by ^{1}H NMR. When $Be(C_4H_9$-i)H etherate or the $N(CH_3)_3$ adduct is used, the reaction is slowed, and $t_{1/2}$=7 to 9 h and 20 to 25 h, respectively. The title compound does not detectably react with $CH_3C(CH_3){=}CHCH_3$ after 7 days at 33.5°C or 4 days at 85°C, but at 110°C the $t_{1/2}$ is 50 h. The complex forms adducts with equimolar amounts of $O(C_2H_5)_2$ and $N(CH_3)_3$, which were not isolated but used for the reaction with pentene-1. The adducts with THF and $NR_2CH_2CH_2NR_2$ (R=CH_3 or C_2H_5) were isolated (see p. 100) [11].

$Be(C_4H_9\text{-}i)H \cdot O(C_2H_5)_2$ and **$Be(C_4H_9\text{-}i)H \cdot N(CH_3)_3$**

The adducts are prepared from equimolar amounts of $Be(C_4H_9\text{-}i)H$ and the ligands in C_6H_6. They were neither isolated nor further characterized, but the addition of pentene-1 was studied by ^1H NMR. See also $Be(C_4H_9\text{-}i)H$ [11].

$Be(C_4H_9\text{-}i)H \cdot OC_4H_8$ (OC_4H_8 = tetrahydrofuran)

$Be(C_4H_9\text{-}i)H$ in C_6H_6 reacts with an excess of THF. The solvents are evaporated, and the liquid residue is dissolved in C_6H_{14} and filtered. Evaporation of C_6H_{14} gives the adduct as a liquid that dissolves in C_6H_6 as a dimer and probably has the structure given in Formula I, p. 93 (R = i-C_4H_9, D = OC_4H_8) [11].

$Be(C_4H_9\text{-}i)H \cdot 0.5\,N(CH_3)_2CH_2CH_2N(CH_3)_2$

An excess of the diamine reacts with $Be(C_4H_9\text{-}i)H$ in C_6H_6 with evolution of a little heat. The white precipitate, obtained after concentration and addition of C_6H_{14}, is recrystallized from C_6H_6/C_6H_{14} as colorless needles, m.p. 96 to 98°C, which sublime at 80°C/10^{-3} Torr.

IR absorptions in c-C_6H_{12} are observed at 1379 (vs) and 1333 (vs) cm^{-1}. One absorption is likely to be due to $\nu(BeH_2Be)$. The adduct is readily soluble in C_6H_6, in which its molecular mass (by cryoscopy) corresponds to Formula II, p. 93 (R = C_4H_9-i, R′ = CH_3) [11].

$Be(C_4H_9\text{-}i)H \cdot 0.5\,N(C_2H_5)_2CH_2CH_2N(C_2H_5)_2$

This adduct is prepared like the previous one and recrystallized from a small volume of C_5H_{12} as needles, m.p. 62 to 63°C.

The IR spectra in c-C_6H_{12} or in Nujol show five strong absorptions between 1400 and 1300 cm^{-1}. The $\nu(BeH_2Be)$ absorption could not be identified. The compound is a monomeric 2:1 adduct in C_6H_6 (by cryoscopy) and probably has the structure given in Formula II, p. 93 (R = C_4H_9-i, R′ = C_2H_5) [11].

$Be(C_4H_9\text{-}t)H$ and **$Be(C_4H_9\text{-}t)H \cdot O(C_2H_5)_2$**

The compound is prepared from $Be(C_4H_9\text{-}t)_2 \cdot 0.6\,O(C_2H_5)_2$, $BeCl_2 \cdot 2\,O(C_2H_5)_2$, and LiH in the mole ratio 1:1:2 in ether at 60°C for 24 h. LiCl can be filtered after ether has been replaced by C_6H_6. The compound is not isolated, but the solution is used for the reactions described below [13]. Pyrolysis of $Be(C_4H_9\text{-}t)_2$ at 60 to 110°C gives $Be(C_4H_9\text{-}t)H$ only as a minor product in a mixture with $Be(C_4H_9\text{-}i)H$ [11]. It is stated in a former publication that pyrolysis of $Be(C_4H_9\text{-}t)_2$ above 100°C gives mainly BeH_2 and i-C_4H_8 [4].

The Be–H bond of $Be(C_4H_9\text{-}t)H$ etherate adds to the C≡N bond of RCN (R = t-C_4H_9, C_6H_5) to give $Be(C_4H_9\text{-}t)N{=}CHR$ (R = t-C_4H_9, C_6H_5) as oils that could not be crystallized, but addition of pyridine gives $(Be(C_4H_9\text{-}t)N{=}CHR \cdot NC_5H_5)_2$ (R = t-C_4H_9, C_6H_5) [15]. With $N(CH_3)_2CH_2CH_2N(CH_3)_2$ a 2:1 complex is formed (see below) [13].

$Be(C_4H_9\text{-}t)H \cdot 0.5\,N(CH_3)_2CH_2CH_2N(CH_3)_2$

The adduct is obtained from the components in C_6H_6 after evaporation of the volatiles. It crystallizes from C_6H_6/C_6H_{14} as prisms, m.p. 195 to 196°C. It is a monomeric 2:1 adduct in C_6H_6 (by cryoscopy), and the structure given in Formula II, p. 93 (R = t-C_4H_9, R′ = CH_3) is proposed [13].

$Be(C_5H_{11})H$

The compound is prepared from $Be(C_5H_{11})_2$ etherate, $BeCl_2$, and LiH in the mole ratio 1:1:2 at 50°C bath temperature for 30 h. The ether is replaced by C_6H_6 and LiCl is filtered. Analysis of the solution gives a Be:H ratio of 1:1. The compound was isolated as its adduct with $N(CH_3)_2CH_2CH_2N(CH_3)_2$. An adduct with $N(CH_3)_3$ is also mentioned (see below) [13].

$Be(C_5H_{11})H \cdot N(CH_3)_3$

The adduct was prepared by addition of pentene-1 to $(BeH_2 \cdot N(CH_3)_3)_2$ in C_6H_6 in an NMR tube at 33.5°C. The reaction is complete in ~48 h, as shown by the disappearance of the vinyl protons in the NMR spectrum. A sample of the adduct is also obtained by condensing the amine onto $Be(C_5H_{11})H$, then removing unreacted amine under reduced pressure. Physical data are not given [13].

$Be(C_5H_{11})H \cdot 0.5\,N(CH_3)_2CH_2CH_2N(CH_3)_2$

The diamine is added to $Be(C_5H_{11})H$ in C_6H_6 which results in a perceptible warming. After 0.5 h the volatile matter is evaporated leaving a viscous liquid which solidifies under continuous vacuum after 2 days. It forms needles from C_6H_{14}, m.p. 57 to 58°C, and dissolves in C_6H_6 as a monomeric 2:1 adduct (by cryoscopy), Formula II, p. 93 ($R = C_5H_{11}$, $R' = CH_3$) [13].

$Be(CH_2C_4H_9\text{-}t)H$

The compound was only prepared in solution from $Na[B(C_2H_5)_3H]$, $BeBr_2$, and $Be(CH_2C_4H_9\text{-}t)_2$ etherate in ether/C_6H_6. After removal of the precipitated NaBr, $Be(CH_2C_4H_9\text{-}t)H$ is isolated as an adduct with $N(CH_3)_2CH_2CH_2N(CH_3)_2$ (see below) [17].

$Be(CH_2C_4H_9\text{-}t)H \cdot 0.5\,N(CH_3)_2CH_2CH_2N(CH_3)_2$

$N(CH_3)_2CH_2CH_2N(CH_3)_2$ is added in stoichiometric amounts to a solution of $Be(CH_3C_4H_9\text{-}t)H$ in ether/C_6H_{14}. Evaporation of the solvents gives a white solid that crystallizes as colorless plates from C_6H_{14}, m.p. 123 to 125°C. The ^1H NMR spectrum in C_6H_6 shows at 30°C a broad resonance at $\delta = 0.36$ ppm for CH_2Be which sharpens at 60°C. The compound is a monomeric 2:1 adduct in C_6H_6 (by cryoscopy) and is described by Formula II, p. 93 ($R = CH_2C_4H_9\text{-}t$, $R' = CH_3$) [17].

$Be(C_6H_5)H$

A 2:1:1 mixture of LiH, $BeBr_2$ [7] or $BeCl_2$ [13], and $Be(C_6H_5)_2$ in ether is stirred for 30 h at 40 to 50°C bath temperature [13] or boiled for 108 h [7]. The ether is evaporated, replaced by C_6H_6, and filtered [7, 13]. Removal of C_6H_6 by evaporation gives a sticky glass [7]. The compound is also obtained by reaction of $Na[B(C_2H_5)_3H]$, $BeCl_2$, and $Be(C_6H_5)_2$ (ratio 2:1:1) as a white precipitate. The solvent and some $B(C_2H_5)_3$ is removed from the supernatant liquid, and the residue is taken up in C_6H_6 [9]. Reaction of $Be(C_6H_5)_2$ with $Sn(C_6H_5)_3H$ in ether/C_6H_6 at 70 to 75°C gives a precipitate and a solution, containing Be:H in an approximate 1:1 ratio. A pure compound could not be isolated since the product is contaminated with phenyltin compounds [9].

Total energies were calculated with the PRDDO approximation using standard geometries (1.68 (BeC terminal) and 1.81 Å (BeC bridging)) for the monomer and H- or C_6H_5-bridged dimers:

References on p. 104

molecule (configuration of C_6H_5)	total energy (a. u.)
$Be(C_6H_5)H$	−245.0283
$C_6H_5BeH_2BeC_6H_5$ (planar)	−490.1147
$C_6H_5BeH_2BeC_6H_5$ (perpendicular)	−490.0934
$HBe(C_6H_5)_2BeH$ (perpendicular)	−490.1323
$HBe(C_6H_5)_2BeH$ (planar)	−489.9631

The calculated energies of dimerization (in kcal/mol) are:

reaction (configuration of C_6H_5)	ΔH(PRDDO)
$2Be(C_6H_5)H \rightarrow C_6H_5BeH_2BeC_6H_5$ (planar)	−36.4
$2Be(C_6H_5)H \rightarrow C_6H_5BeH_2BeC_6H_5$ (perpendicular)	−23.1
$2Be(C_6H_5)H \rightarrow HBe(C_6H_5)_2BeH$ (perpendicular)	−47.5

For the bridging hydrogen dimers, partial geometry optimization results in a change of ΔH of only −1.3 kcal/mol. The planar dimer is 13.4 kcal/mol more stable than the conformation with the BeH_2Be plane perpendicular to the C_6H_5 rings. This is related to π-donation effects. In the coplanar configuration, the system favors delocalization to the extent of 0.14 e (e = electrons) into the vacant Be π orbitals; whereas in the perpendicular conformer, delocalization is negligible. The electron population on Be in the monomer is 0.11 e. The increase of π-donation in the planar dimer is probably associated with direct Be–Be interaction, as indicated by the increase in the Be–Be overlap population from 0.361 (perpendicular conformer) to 0.374 (planar conformer). For the C_6H_5-bridged dimer, the planar conformation is calculated to be 106 kcal/mol less stable than the perpendicular configuration, due to steric interactions between the C_6H_5 hydrogens and the hydrogens bound to Be. The C_6H_5 groups strongly favor dimerization, due to interactions between the π electrons in the C_6H_5 rings and the Be 2s and $2p_x$ orbitals (with Be on the z axis and bridging C on the x axis). The Be–C overlap population increases from 0.40 in $HBe(CH_3)_2BeH$ to 0.48 in $HBe(C_6H_5)_2BeH$. Similarly, the 2s and $2p_x$ populations on Be increase by 0.04 and 0.09e, respectively, upon going from the CH_3-bridged to the C_6H_5-bridged dimer [30].

The heat of formation of $Be(C_6H_5)H$ is calculated from CH_3 stabilization energies as $\Delta H_f^\circ = 54.6$ kcal/mol. The heats of formation of the singlet cation $[Be(C_6H_4)H]^+$ are estimated from theoretical energies as 293, 298, and 301 kcal/mol for the ortho, meta, and para positions, respectively [21].

The solutions of the compound in C_6H_6 form adducts with $N(CH_3)_3$ [9], $N(C_2H_5)_2CH_2CH_2N(C_2H_5)_2$ [13], and $P(CH_3)_3$ [9]. With $N(CH_3)_2CH_2CH_2N(CH_3)_2$, an adduct is formed from the solutions of $Be(C_6H_5)H$ in ether [7], ether/C_6H_6 [9], or C_6H_6 [13].

$Be(C_6H_5)H \cdot N(CH_3)_3$

The adduct is prepared by addition of $N(CH_3)_3$ to a solution of $Be(C_6H_5)H$ in C_6H_6. A crystalline solid separates overnight which crystallizes from C_6H_6 as colorless needles. It does not melt when heated in a sealed tube under N_2, but is charred in the range 180 to 230°C. It is a dimer in C_6H_6 (by cyroscopy). The IR spectrum in Nujol has the characteristic ν(BeH_2Be) absorptions at 1342 and 1316 cm^{-1}. Formula I, p. 93 (R = C_6H_5, $^2D = N(CH_3)_3$) is suggested [9].

$Be(C_6H_5)H \cdot 0.5\,N(CH_3)_2CH_2CH_2N(CH_3)_2$

The adduct is formed from solutions of $Be(C_6H_5)H$ in ether [7], ether/C_6H_6 [9], or C_6H_6 [13] by addition of the diamine. The precipitated complex (pale yellow in [9]) is washed with ether and dried under reduced pressure [7,9], or is recrystallized from CH_2Cl_2 [13].

Melting points are given at 245°C [7] and 250 to 252°C with decomposition [13]. The adduct is insoluble in C_6H_6 [7] and not appreciably soluble in C_6H_5Cl at 80 to 90°C [13]. Spectral and molecular mass data are not given, but the compound is assumed to have the structure given in Formula II, p. 93 ($R=C_6H_5$, $R'=CH_3$) [13].

$Be(C_6H_5)H \cdot 0.5\,N(C_2H_5)_2CH_2CH_2N(C_2H_5)_2$

The ligand is added to a solution of $Be(C_6H_5)H$ in C_6H_6. The precipitated complex is crystallized from C_6H_5Cl/C_6H_{14}, m.p. 255 to 260°C with decomposition. It is insoluble in C_6H_6. Formula II, p. 93 ($R=C_6H_5$, $R'=C_2H_5$), is assumed to be valid [13].

$Be(C_6H_5)H \cdot P(CH_3)_3$

$P(CH_3)_3$ is added to a solution of $Be(C_6H_5)H$ in C_6H_6. A crystalline product separates after much of the solvent has been evaporated, and a Be:P ratio of only 1:0.8 is found. Since the complex evidently loses phosphine fairly readily, it is dissolved in $C_6H_5CH_3/P(CH_3)_3$ and cooled to −78°C. The resulting crystalline complex is washed with C_6H_{14} containing $P(CH_3)_3$ and dried at −15°C under reduced pressure. The analytical data are then in agreement with a 1:1 adduct. Cryoscopic measurements in C_6H_6 again show that it loses some $P(CH_3)_3$ (found 271, calculated 326). The adduct chars at 190 to 220°C when heated in a sealed capillary under N_2. The IR spectrum in Nujol contains absorptions at 1356 and 1312 cm^{-1}, characteristic of $\nu(BeH_2Be)$ [9].

$Be(C_6H_4CH_3\text{-}2)H$

$Be(C_4H_9\text{-t})C_6H_4CH_3\text{-}2$ is heated in boiling xylene for 24 h to give a white precipitate of the compound [14]. Reaction of LiH, $BeCl_2$, and $Be(C_6H_4CH_3\text{-}2)_2$ (mole ratio 2:1:1) in ether also gives the compound, but it was not isolated [13].

It is insoluble in boiling C_6H_6 [14]. Addition of $N(CH_3)_2CH_2CH_2N(CH_3)_2$ gives a 2:1 adduct (see below) [13, 14].

$Be(C_6H_4CH_3\text{-}2)H \cdot 0.5\,N(CH_3)_2CH_2CH_2N(CH_3)_2$

The adduct is prepared from the components [13, 14]. It decomposes above 265°C [13].

$Be(C_6H_4CH_3\text{-}3)H$

$Be(C_6H_4CH_3\text{-}3)_2$ and $Be(C_4H_9\text{-t})_2$ are stirred in C_6H_6 for 24 h. The solid residue, left after volatile material has been evaporated, is heated at 140°C/10^{-3} Torr for 24 h to give the hydride. It assumes a glassy appearance at ~200°C and decomposes at ~260°C. It is also obtained from LiH, $BeCl_2$, and $Be(C_6H_4CH_3\text{-}3)_2$ (mole ratio 2:1:1) in boiling ether for 30 h. The ether is evaporated and replaced by C_6H_6. LiCl is removed. The hydride was not isolated. A 1:1 adduct is formed with $N(CH_3)_3$ (see below) [14].

$Be(C_6H_4CH_3\text{-}3)H \cdot N(CH_3)_3$

The adduct is obtained by addition of an excess of amine to $Be(C_6H_4CH_3\text{-}3)H$ with subsequent removal of the excess of $N(CH_3)_3$. It is also obtained by addition of $N(CH_3)_3$ to the

References on p. 104

C_6H_6 solution of $Be(C_6H_4CH_3\text{-}3)H$ at $-78°C$. The mixture is warmed to room temperature, and all the volatile material is evaporated.

The adduct melts at 165 to 167°C or, after recrystallization from C_6H_6/C_6H_{14} (needles), at 168 to 169°C. It is a dimer in C_6H_6 (by cryoscopy). The ^1H NMR spectrum shows no BeH signal, $\delta = 2.01$ (s, CH_3N) and 2.59 ppm (s, CH_3C). The IR spectrum was measured, but not given [14].

References:

[1] Barbaras, G. D., Dillard, C., Finholt, A. E., Wartik, T., Wilzbach, K. F., Schlesinger, H. I. (J. Am. Chem. Soc. **73** [1951] 4585/90).
[2] Coates, G. E., Glockling, F., Huck, N. D. (J. Chem. Soc. **1952** 4512/5).
[3] Coates, G. E., Glockling, F. (J. Chem. Soc. **1954** 22/7).
[4] Coates, G. E., Glockling, F. (J. Chem. Soc. **1954** 2526/9).
[5] Bell, N. A., Coates, G. E. (Proc. Chem. Soc. **1964** 59).
[6] Bell, N. A., Coates, G. E. (J. Chem. Soc. **1965** 692/9).
[7] Bell, N. A., Coates, G. E. (J. Chem. Soc. A **1966** 1069/73).
[8] Bell, N. A., Coates, G. E., Emsky, J. W. (J. Chem. Soc. A **1966** 1360/2).
[9] Coates, G. E., Tranah, M. (J. Chem. Soc. A **1967** 615/7).
[10] Bell, N. A., Coates, G. E. (J. Chem. Soc. A **1968** 628/31).

[11] Coates, G. E., Roberts, P. D. (J. Chem. Soc. A. **1969** 1008/12).
[12] Coates, G. E., Francis, B. R. (J. Chem. Soc. A **1971** 1308/10).
[13] Blindheim, U., Coates, G. E., Srivastava, R. C. (J. Chem. Soc. Dalton Trans. **1972** 2302/5).
[14] Coates, G. E., Smith, D. L., Srivastava, R. C. (J. Chem. Soc. Dalton Trans. **1973** 618/22).
[15] Coates, G. E., Smith, D. L. (J. Chem. Soc. Dalton Trans. **1974** 1737/40).
[16] Bell, N. A., Coates, G. E., Fishwick, A. H. (J. Organometal. Chem. **198** [1980] 113/20).
[17] Coates, G. E., Francis, B. R. (J. Chem. Soc. A **1971** 1305/8).
[18] Baird, N. C., Barr, R. F., Datta, R. K. (J. Organometal. Chem. **59** [1973] 65/81).
[19] Dill, J. D., Schleyer, P. von R., Pople, J. A. (J. Am. Chem. Soc. **98** [1976] 1663/8).
[20] Collins, J. B., Dill, J. D., Jemmis, E. D., Apeloig, Y., Schleyer, P. von R., Seeger, R., Pople, J. A. (J. Am. Chem. Soc. **98** [1976] 5419/27).

[21] Dill, J. D., Schleyer, P. von R., Pople, J. A. (J. Am. Chem. Soc. **99** [1977] 1/8).
[22] Apeloig, Y., Schleyer, P. von R., Pople, J. A. (J. Am. Chem. Soc. **99** [1977] 1291/5).
[23] Apeloig, Y., Schleyer, P. von R., Pople, J. A. (J. Am. Chem. Soc. **99** [1977] 5901/9).
[24] Dill, J. D., Schleyer, P. von R., Binkley, J. S., Pople, J. A. (J. Am. Chem. Soc. **99** [1977] 6159/73).
[25] Jemmis, E. D., Aleksandratos, S., Schleyer, P. von R., Streitwieser, A., Schaefer III, H. F. (J. Am. Chem. Soc. **100** [1978] 5695/700).
[26] Ray, N. K., Mehandru, S. P., Bhargave, S. (Intern. J. Quantum Chem. **13** [1978] 529/36).
[27] Jemmis, E. D., Chandrasekhar, J., Schleyer, P. von R. (J. Am. Chem. Soc. **101** [1979] 527/33).
[28] Pross, A., Radom, L. (Tetrahedron **36** [1980] 673/6).
[29] Glidewell, C. (J. Organometal. Chem. **217** [1981] 273/80).
[30] Marynick, D. S. (J. Am. Chem. Soc. **103** [1981] 1328/33).

[31] Pross, A., DeFrees, D. J., Levi, B. A., Pollack, S. K., Radom, L., Hehre, W. J. (J. Org. Chem. **46** [1981] 1693/9).
[32] Armstrong, D. R., Perkins, P. G. (Coord. Chem. Rev. **38** [1981] 139/275).
[33] Hashimoto, K., Osamura, Y., Iwata, S. (Nippon Kagaku Kaishi **1986** 1377/83).

1.3.3 Organoberyllium Hydridoborates

$Be(CH_3)BH_4$

Methylberyllium tetrahydridoborate can be prepared by the reaction of $Be(CH_3)_2$ with gaseous B_2H_6 in a vacuum system at 95°C for 90 min [1] or 1 h [2]. A glassy material is formed in the initial reaction period which is then transformed into a nonvolatile, mobile liquid and finally into a rather unstable sublimable solid on further treatment with B_2H_6 [1]. The entire reaction vessel is frozen with liquid N_2 and then warmed to −80°C; $B(CH_3)_3$ and excess B_2H_6 are pumped away. This process is repeated 5 times to give a white volatile material [2]. An analysis of the solid gave a ratio of Be:CH_3:B:H=1:0.88:1.03:4.21. The deviations from the expected stoichiometry are ascribed to a BeB_2H_8 impurity [1]. In another method of preparation, BeB_2H_8 and $Zn(CH_3)_2$ are reacted for 1.5 h at room temperature in the mole ratio 2.2:1.3 (85% yield). No reaction occurs with $Sn(CH_3)_4$, and the compound cannot be isolated from the reaction mixture obtained with $Al_2(CH_3)_6$ [5]. The compound is purified by passing it slowly, with continuous pumping, through a −45°C [2] or −30°C [5] trap and collecting it at −80°C [2,5]. It was found to decompose slowly at room temperature and should be stored at −80°C [2].

The isotopically enriched species $Be(CH_3)^{10}BH_4$ and $Be(CH_3)^{10}BD_4$ are prepared from $Be(CH_3)_2$ and $^{10}B_2H_6$ (96% ^{10}B, 4% ^{11}B) or $^{10}B_2D_6$, respectively, by the same method described for $Be(CH_3)BH_4$ [2].

The physical data given below indicate the existence of monomers and dimers. Vapor density measurements at 1.2 Torr (unsaturated) of $Be(CH_3)^{10}BD_4$ and 5.6 Torr (saturated) of $Be(CH_3)BH_4$ showed that the compound is predominantly dimeric in the gas phase at 24°C [4], whereas the unsaturated gas phase IR spectra (see below) indicate the existence of monomers and dimers. $Be(CH_3)^{10}BH_4$ is a dimer in C_6H_6 solution, as determined by cryoscopy [2]. The dimeric nature in solution is confirmed by ^1H NMR spectra. The spectrum of $C_6H_6/C_6D_5CD_3$ (20:80) solutions at ambient temperature consists of a singlet at $\delta = -0.11$ (CH_3) and a quartet at 0.80 ppm (BH_4; $J(^{11}B,^1H) = 86.0$ Hz) with the expected intensity ratios. These values are similar to those obtained for $Be(CH_3)_2$ and $Be(BH_4)_2$, respectively. It is therefore concluded that $Be(CH_3)_2Be$ and BeH_2BH_2 bridging units are the characteristic structural features. The spectrum is slightly temperature-dependent in the mixed solvent (−0.2 Hz/°C). The chemical shift values for c-C_5H_{10} solutions are not temperature-dependent and are assumed to represent the "true" hydrogen environment: $\delta = 0.01$ (CH_3) and 0.28 ppm (BH_4) [2].

The ^9Be NMR spectrum in C_6F_6 shows a single resonance at $\delta = -12.2$ ppm. In C_5H_{12}/CF_3Br, the line appears at −11.7 ppm; the line widths at half-height are temperature-dependent showing 6, 7, and 9 Hz at 37.7°C, ambient temperature, and −32.5°C, respectively [6].

The gas phase IR spectrum was recorded between KBr windows coated with a thin polyethylene film to prevent reaction with the KBr windows. The data are listed in Table 22 together with those of the isotopically labelled compounds $Be(CH_3)^{10}BH_4$ and $Be(CH_3)^{10}BD_4$. From intensity differences of the bands at 2170, 1265, 1223, and 1108 cm^{-1}, at different pressures in the gas cell, it was concluded that both monomers and dimers are present in the gas phase at low pressure. The assignments are made by (1) comparison of the band positions with those of $Be(CH_3)_2$ and $Al(BH_4)_3$, (2) comparison of observed rotational contours and theoretical contours which are predicted on the basis of the structures shown in Formulas I and II, p. 106, and (3) correlation of deuterium and boron-10 shifts [2].

References on p. 111

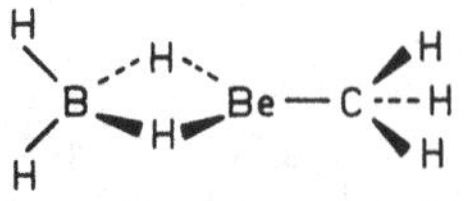

I

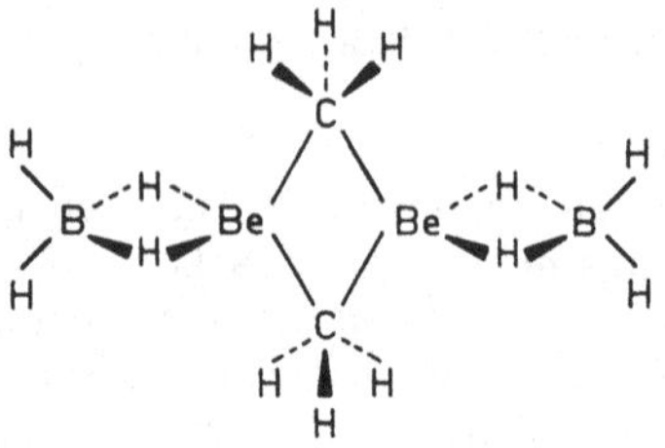

II

Table 22
Gas Phase IR Vibrations (in cm^{-1}) of $Be(CH_3)BH_4$, $Be(CH_3)^{10}BH_4$, and $Be(CH_3)^{10}BD_4$ Monomers and Dimers [2].

$Be(CH_3)BH_4$	$Be(CH_3)^{10}BH_4$	$Be(CH_3)^{10}BD_4$	assignment
monomers			
2170 (w)	2172 (w)	1642 (w)	BeH_2Be expansion
1223 (m)	1224 (m)	1225 (m)	$\delta_s(CH_3)$
1108 (m)	1108 (m)	obscured	$\nu_{as}(BeC)$
dimers			
3948 (w)	3959 (w)		
3018 (w)	3018 (w)	3018 (w)	CH_4
2947 (m)	2947 (m)	2947 (m)	$\nu_{as}(CH)$
2856 (w)	2858 (w)	2855 (w)	$\nu_s(CH)$
2629 (w)	2641 (w)		
		2515 (w)	
		2054 (w)	
2532 (s)	2543 (s)	1926 (s)	$\nu_{as}(BH)$
2469 (s)	2476 (s)	1831 (s)	$\nu_s(BH)$
		1803 (m)	
2087 (s)	2095 (s)	1582 (s)	BeH_2B expansion
		1508 (m)	
2010 (s)	2018 (s)	1478 (s)	$\nu_{as}(BeH_2B)$
1768 (w)	1782 (w)		
1637 (m)	1648 (m)	obscured	
1494 (vs)	1498 (vs)	1111 (vs)	$\nu_s(BeH_2B)$
1315 (w)	1315 (w)	1315 (w)	CH_4
1265 (m)	1265 (m)	1265 (m)	$\delta_s(CH_3)$
1138 (m)	1146 (m)	880 (m)	$\delta(BH_2)$
1020 (w)	1035 (w)	997 (w)	$\nu_{as}(BeB)$
898 (w)	898 (w)	667 (w)	$\delta(BH_2)$
810 (sh)	810 (sh)	823 (w)	
754 (vs)	763 (vs)	742 (vs)	$\varrho(CH_3)$
587 (w)	593 (w)	584 (w)	$\nu_{as}(BeC_2Be)$

Table 23
IR and Raman Vibrations (in cm^{-1}) of Solid $Be(CH_3)BH_4$ and $Be(CH_3)^{10}BD_4$ Dimers at 20 K [4].

$(Be(CH_3)BH_4)_2$ IR	$(Be(CH_3)BH_4)_2$ Raman	$(Be(CH_3)^{10}BD_4)_2$ IR	$(Be(CH_3)^{10}BD_4)_2$ Raman	assignment
2952 (w)	2959 (m)	2962 (vw)	2960 (m)	$\nu_{as}(CH_3)$
2926 (w)	2933 (m)	2925 (vw)	2934 (m)	
2890 (m)	2896 (s)	2891 (m)	2896 (s)	$\nu_s(CH_3)$
2493 (s)	2502 (m)	1897 (s)	1909 (s)	$\nu_{as}(BH_{2t})$
			2497 (w)	} $\nu(BHD_t)$
			1880 (m)	
2437 (s)	2444 (s)	1793 (s)	1810 (s)	$\nu_s(BH_{2t})$
			1782 (m)	
		1747 (s)	1748 (m)	
2097 (s)	2103 (s)	1498 (s)	1503 (s)	$\nu_s(BH_{2b})$
2056 (s)	2063 (m)	1607 (m)	1614 (m)	
2017 (s)	2023 (m)	1580 (m) }	1586 (m)	$\nu_{as}(BH_{2b})$
		1554 (m)	1559 (m)	
1607 (s)	1615 (vw)	1095 (s) }	1117 (w)	$\nu(BeH_{2b})$
1468 (br, vs)	1480 (br, w)	1032 (m)	1032 (vw)	
1397 (m)	1430 (vw)	1429 (br, w)	1430 (br, w)	$\delta_{as}(CH_3)$
1353 (m)	1408 (w)	1388 (w, sh)		
1257 (s)		1258 (s)	1270 (vvw)	$\delta_s(CH_3)$
1143 (w)	1143 (m)		819 (m)	$\tau(BH_{2t})$
1138 (s)	1134 (m)	873 (s)	843 (m)	$\delta(BH_{2t})$
1093 (m)				
1029 (m)	1026 (w)	788 (s)	789 (w)	$\varrho(BH_{2t})$
924 (m)	937 (m)	942 (m)	945 (m)	$\nu_{as}(BeC)$
874 (m)				$\varkappa(BH_2)$
831 (m)	796 (w)	814 (m)	712 (w)	$\varrho, \varkappa(CH_3)$
	627 (vw)		628 (m)	
740 (br, s)		730 (br, s)		$\nu(BeB)$
594 (s)	526 (s)	588 (m, sh)	524 (s)	$\nu_s(BeC)$
	357 (m)		332 (m)	$\nu(BeBe)$
		310 (w)		$\delta(BBeC_2Be)$
277 (s)	273 (m)	268 (m)	261 (m)	$\delta(CBeC_2Be)$
152 (s)	194 (w)	132 (w)	162 (w)	$\delta(BeH_2B)$
120 (s)		105 (s)		$\delta(BeC_2Be)$

The IR and Raman spectra for the annealed crystalline phase, obtained upon sublimation of the compound onto a polished aluminium block, were measured at 20 K. The results are presented in Table 23 together with the spectra of $Be(CH_3)^{10}BH_4$. The IR spectra are reflection

References on p. 111

spectra. They show very little difference to those in the vapor phase described above, implying that the solid is also made up of dimeric units with CH_3 and H bridging between two Be atoms and between Be and B, respectively (Formula II, p. 106). The spectra are assigned with the assumption of a C_{2h} symmetry, and the BeH_2BH_2 and CH_3 group vibrations are compared with similar molecules. Efforts to obtain matrix spectra of monomeric $Be(CH_3)BH_4$ have not been fulfilled [4].

The monomer and dimer were studied by MNDO [7] and PRDDO [8] calculations. The structures shown in Formulas III to VI are considered. Calculated ionization potentials and dipole moments by MNDO for these isomers are as follows [7]:

Formula	point group	I.P. (eV)	μ(D)
III	C_{3v}	11.46	0.49
IV	C_{2v}	11.57	1.35
V	C_{2h}	11.48	
VI	C_{2h}	11.04	

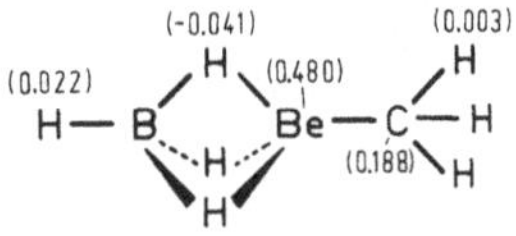

III

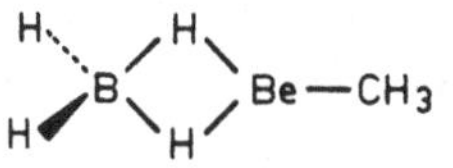

IV

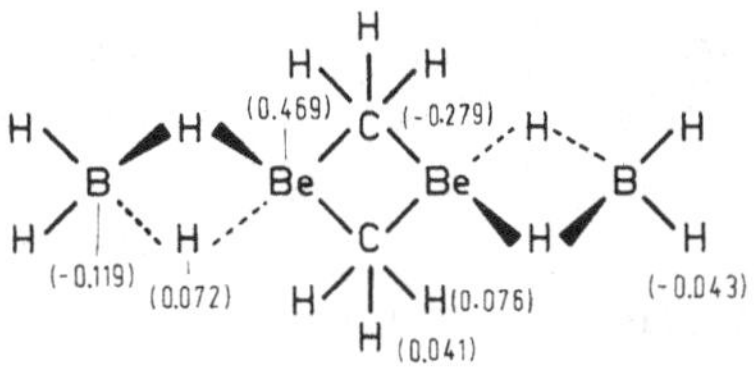

V

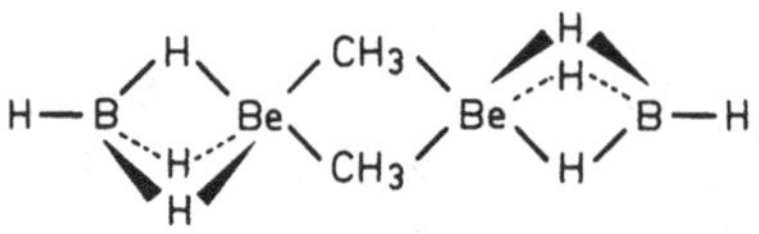

VI

The MNDO calculation favors the triple-bridged monomer III by 3.6 kcal/mol over the double-bridged monomer IV [7], whereas the double-bridged form IV is calculated to be 0.4 kcal/mol less stable than III by the PRDDO method [8]. No experimental information about the structure of the monomer is known. The heat of dimerization is calculated as 21.8 kcal/mol by the MNDO method, and the triply bridging form of the dimer VI is 21 kcal/mol less stable than the structure shown in Formula V. The dimer of Formula V is the most stable species [7]. This also results from the PRDDO calculations. Dimerization to the double-bridged form V is predicted to be exothermic by −12.8 kcal/mol, and is favored by ~16 kcal/mol over $Be(CH_3)_2$ dimerization [8]. The net atomic charges are included in Formulas III to VI [7]. The calculated results for the dimer are in agreement with the experimental observations.

The compound reacts with excess B_2H_6 to give $Be(BH_4)_2$ and methylboranes [1, 3]. $Be(CH_3)B_3H_8$ appears to be the product of the reactions of $Be(CH_3)BH_4$ with $Be(B_3H_8)_2$ and with CsB_3H_8 [5].

$Be(CH_3)B_3H_8$

Reaction of 0.30 mmol $Be(B_3H_8)_2$ and 0.23 mmol $Zn(CH_3)_2$ for 1 h at room temperature produces the compound as a nonvolatile solid in moderate yield. It can be isolated by condensation into a $-22°C$ trap in a vacuum system. The compound appears to be also a product of the reaction of $Be(CH_3)BH_4$ with $Be(B_3H_8)_2$ or CsB_3H_8, but these preparations are less convenient and the products have not been well described [5].

No determination of the molecular mass is described, but the existence of dimers can be concluded from the ^1H, ^{11}B NMR, and IR spectra.

In the ^1H NMR spectrum in CD_2Cl_2, the B_3H_8 portion of the spectrum at room temperature is qualitatively the same as for $Be(B_3H_8)_2$ in the static form except that the Be–H–B hydrogens appear to be equivalent. The assignments are tentative owing to the complexity of the signals and their broad features. Even at 7°C the sharpening is not sufficient for this part of the spectrum: δ(CD_2Cl_2 at 7°C) = ~ −0.7 (H-5,6), 0.89 (H-1,3; J(B,H) = 131 Hz), 1.06 (H-7,8; J(B,H) ~ 60 Hz), and 2.17 ppm (H-4). See Formula VII for the numbering. The CH_3Be protons give rise to a 1:2:1 triplet at 7°C which collapses to a broad peak at 40°C:

t in °C	δ(CH_3) in ppm
7	−0.39, −0.14, 0.13
22	−0.34, −0.12, 0.11
40	−0.15
60	−0.10

These results can be explained by the existence of two stereoisomers with two equivalent and nonequivalent CH_3 groups, respectively, as shown in Formulas VII and VIII. Rapid interconversion of these isomers at elevated temperatures leads to an equilibration of the CH_3 positions [5].

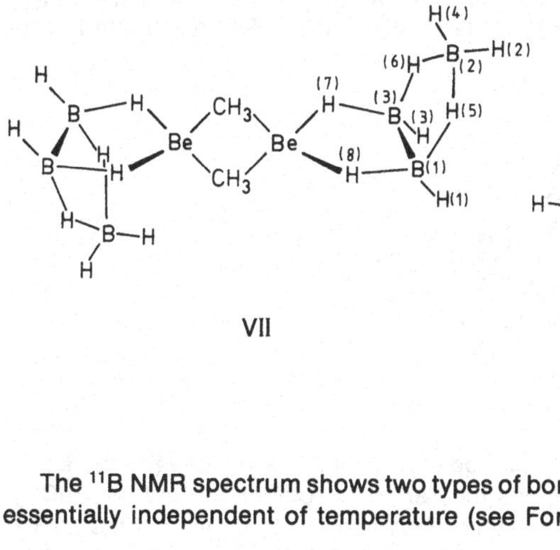

VII

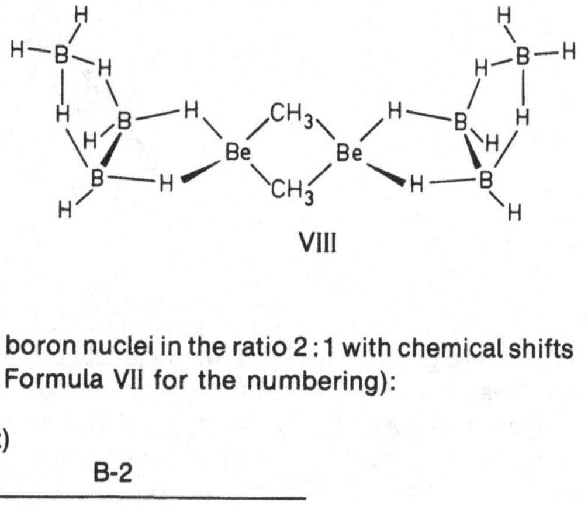

VIII

The ^{11}B NMR spectrum shows two types of boron nuclei in the ratio 2:1 with chemical shifts essentially independent of temperature (see Formula VII for the numbering):

solvent	t in °C	δ in ppm (J(B,H) in Hz) B-1, B-3	B-2
CD_2Cl_2	−50	−42.4 (J = 135, 59)	−12.1
CD_2Cl_2	23	−42.4 (d of t; J(B,H-2,4) = 123, J(B,H-7,8) = 60)	−12.0 (J = 117)
C_6D_6	23	−42.2 (J = 127, 56)	−11.5 (J = 116)
C_6D_6	37	−42.3 (q; J = 78)	−11.8 (t; J = 114)

References on p. 111

Upon ^1H decoupling, the signals for B-1, B-3, and for B-2 collapse to singlets. At 70°C, only broad singlets are observed suggesting rapid exchange between all H positions [5].

The ^9Be NMR spectrum shows a single resonance at $\delta = 7.6$ ppm in C_6F_6 with a line width at half-height of 18 Hz. ^1H decoupling reduces the line width to 11 Hz. The signal appears at 6.6 ppm in CH_2Cl_2/CH_3Br with temperature-dependent line widths at half-height of 17, 19, and 26 Hz at 37.7°C, ambient temperature, and −32.5°C, respectively [6].

The IR spectrum of a solid film at −196°C on Plexiglas coated NaCl has the following absorptions in cm^{-1} (±10): 2940(w), 2550(vs), 2490(s), 2220(vs), 2150(s,sh), 1425(m,sh), 1390(m), 1340(m), 1260(s), 1155(m), 1040(s), 1015(m), 985(m), 890(s), 835(s), 780(m), 730(m), and 690(m). Assignments were not made [5].

$Be(CH_3)B_5H_{10}$

The compound is prepared by reaction of $Be(Br)B_5H_{10}$ with $Al_2(CH_3)_6$ for 1 h at room temperature. It is isolated by a high vacuum trap-to-trap distillation at −30 to −50°C. The volatile liquid is very air- and moisture-sensitive.

The ^1H NMR spectrum in C_6D_6 or $C_6D_5CD_3$ shows resonances at $\delta = -2.13$ (H-9), −1.27 (H-8, 10), −0.20 (H-1; J(B,H)=141 Hz), 0.30, (s, CH_3), 0.82 (d, H-7, 11; J(B,H)=75 Hz), 2.77 (H-3,6; J(B,H)=136 Hz), and 3.54 ppm (H-4,5; J(B,H)=164 Hz). The ^{11}B NMR spectrum in C_6D_6 or $C_6D_5CD_3$ contains three resonances at $\delta = -54.6$ (B-1; J(B,H)=146 Hz), −2.0 (B-3,6; J(B,H)=138,83 Hz), and 9.6 ppm (B-4,5; J(B,H)=161 Hz). The pseudotriplet at −2.0 ppm can be resolved into a doublet of doublets by line narrowing (see Formula IX for the numbering).

The IR spectrum of the gas phase shows absorptions at 3000(w), 2640(s), 2600(s), 2250(s), 1880(w,br), 1580(w), 1520(m), 1440(m), 1280(m), 1220(s), 1070(m), 990(m), 940(m), and 880(w) cm^{-1} [9].

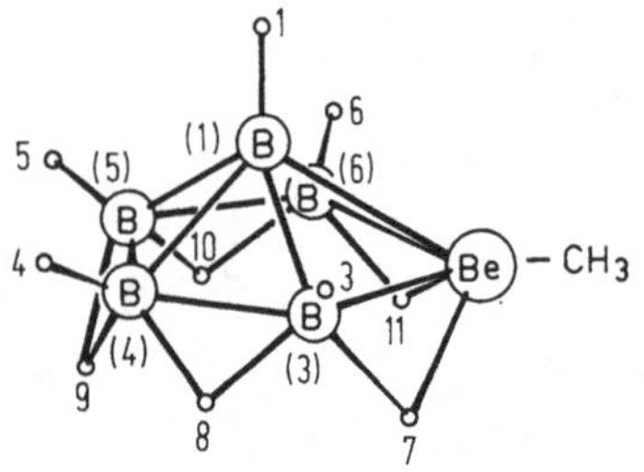

IX

Calculations were carried out by the PRDDO method assumming C_s symmetry (Formula IX). A total energy of −183.266 a.u. is calculated, and the barrier to rotation around the $BeCH_3$ bond was found to be only 0.9 kcal/mol. The distances between the hydrogens of the BeHB bridges and those of the CH_3 group increase from 2.495 to 2.701 Å, during a rotation of 60°, from the staggered to the eclipsed conformation. The bonding is analyzed in terms of charge stability, static reactivity indices, degrees of bonding, overlap populations, and fractional bonds obtained from localized molecular orbitals. The gap between the HOMO and LUMO is 0.545 a.u.. This large value is indicative of some resistance to oxidation and reduction. The BeC bond is a strong single bond (degree of bonding 0.963) [10].

References:

[1] Burg, A. B., Schlesinger, H. I. (J. Am. Chem. Soc. **62** [1940] 3425/9).
[2] Cook, T. H., Morgan, G. L. (J. Am. Chem. Soc. **92** [1970] 6487/92).
[3] Cook, T. H., Morgan, G. L. (J. Am. Chem. Soc. **92** [1970] 6493/8).
[4] Allamandola, L. J., Nibler, J. W. (J. Am. Chem. Soc. **98** [1976] 2096/100).
[5] Gaines, D. F., Walsh, J. L., Morris, J. H., Hillenbrand, D. F. (Inorg. Chem. **17** [1978] 1516/22).
[6] Gaines, D. F., Coleson, K. M., Hillenbrand, D. F. (J. Magn. Resonance **44** [1981] 84/8).
[7] Dewar, M. J. S., Rzepa, H. S. (J. Am. Chem. Soc. **100** [1978] 777/84).
[8] Marynick, D. S. (J. Am. Chem. Soc. **103** [1981] 1328/33).
[9] Gaines, D. F., Walsh, J. L. (Inorg. Chem. **17** [1978] 1238/41).
[10] Bicerano, J., Lipscomb, W. N. (Inorg. Chem. **18** [1979] 1565/71).

1.3.4 Organoberyllium Halides, Pseudohalides, and Their Adducts

$Be(CH_3)F$

The preparation of $Be(CH_3)F$ is not described. But the molecular and electronic structure of the compound was investigated by an ab initio calculation [32] (see Chapter 5, p. 204).

$Be(CH_3)Cl$

$Be(CH_3)Cl$ was obtained for the first time from the reaction of $Be(CH_3)_2$ and HCl in the absence of a solvent. No reaction occurs below −5°C. The process is complete, however, after 1 h at 120 to 130°C. One equivalent of CH_4 is evolved. The remaining white solid is less volatile than $Be(CH_3)_2$, and evidently highly polymerized. The vapor pressure is $\log p = 6.52 - 2614 \cdot T^{-1}$ (p in Torr, T in K). The pressure at 128°C corresponds to 1 Torr. No elemental analysis was performed [5]. The compound can also be prepared by a redistribution reaction between $Be(CH_3)_2$ and $BeCl_2$ in ether [18] or $S(CH_3)_2$ [22]. Addition of 1,4-dioxane (Be:dioxane ratio = 2:1) to an equimolar solution of $Be(CH_3)_2$ and $BeCl_2$ in ether gives a white precipitate of the empirical formula $Be(CH_3)Cl$ (Be:Cl = 1:1.16) which is a monodioxanate, $\mathbf{Be(CH_3)Cl \cdot O_2C_4H_8}$ [18].

The ^1H NMR signal of the compound is identified from studies of $Be(CH_3)_2/BeCl_2$ mixtures in ether (ratios 1:0, 1:1, 2:1, 3:1). At −85°C, the signal at $\delta = -1.35$ ppm can be attributed to $Be(CH_3)Cl$. It is shifted relative to the $Be(CH_3)_2$ signal (−1.44 ppm). At higher temperatures the signals coalesce [18]. A 4:1 mixture of $Be(CH_3)_2$ and $BeCl_2$ in $S(CH_3)_2$ shows only one ^1H NMR signal at ambient temperature, which is split in two resonances at −45°C: $\delta = -0.89$ and −1.17 ppm. The high field signal is assigned to $Be(CH_3)Cl$ [26]. The ^9Be NMR spectrum of the solution of $Be(CH_3)Cl$ in $S(CH_3)_2$ exhibits a signal at $\delta = -4.2$ ppm. The species present in solution is believed to be $\mathbf{Be(CH_3)Cl \cdot 2S(CH_3)_2}$ [23, 34].

The ebullioscopically determined association data for 1:1 mixtures of $Be(CH_3)_2$ and $BeCl_2$ in ether (36°C) show that $Be(CH_3)Cl$ is monomeric in 0.031 to 0.123 molar solutions [18].

$Be(CH_3)Br$

To a solution of equimolar amounts of $Be(CH_3)_2$ and $BeBr_2$ in ether, 1,4-dioxane is added in excess or in an equimolar amount. A white precipitate is formed. The analytically determined ratio of Be:Br is 1:1 [15, 18]. An excess of $Be(CH_3)_2 \cdot O_2C_4H_8$ can be easily washed from the precipitate [15]. It is assumed that the precipitate is a monodioxanate, $\mathbf{Be(CH_3)Br \cdot O_2C_4H_8}$ [18].

References on pp. 118/9

The ^1H NMR spectrum of $Be(CH_3)Br$ in ether was determined from 1:0, 1:1, 2:1, and 1:2 mixtures of $Be(CH_3)_2$ and $BeBr_2$. At −75°C, the signal at $\delta = -0.125$ ppm is assigned to $Be(CH_3)Br$. At higher temperatures the signal coalesces with that of $Be(CH_3)_2$ [15, 18].

The ebullioscopically determined association data in ether (36°C) for 1:1 mixtures of $Be(CH_3)_2$ and $BeBr_2$ show that $Be(CH_3)Br$ is monomeric in 0.056 to 0.255 molar solutions [15, 18].

The solutions of $Be(CH_3)Br$, obtained from $Be(CH_3)_2$ and $BeBr_2$ in ether, react with LiH to give $(Be(CH_3)H \cdot O(C_2H_5)_2)_2$ [13, 27]. Reaction of $Be(CH_3)Br$ with $(CH_3)_2C{=}O$ in ether gives $(Be(OC_4H_9\text{-}t)Br \cdot O(C_2H_5)_2)_2$ [27].

$Be(CH_3)I$

The compound was claimed to be prepared for the first time from Be and CH_3I in boiling ether with some $HgCl_2$ as a catalyst. The "$Be(CH_3)I$" obtained was insoluble in ether [2]. However, the experiment could not be repeated under these conditions [3], and also no reaction of Be with CH_3I was observed in a previous study [1]. In a revised procedure, Be powder, $HgCl_2$, and excess CH_3I were heated in ether in a sealed tube to 80 to 90°C for 15 h [3]. The compound is also assumed to be formed as a liquid by reaction of $Be(CH_3)_2$ and I_2, standing for several days in a cork stoppered flask. The solution in ether does not fume. After driving off the ether, a liquid residue results and further heating gives off dense white fumes, probably BeO. Heating the ether solution yields $Be(CH_3)_2$. The compound decomposes in H_2O and does not react with CO_2 [4, 11]. A slowly developing positive color test is observed with Michler's ketone, and reaction with C_6H_5CNO [3] or $C_6H_5C{=}NH$ [4] gives $C_6H_5NHC(O)CH_3$.

$Be(CH_3)CN$

$Be(CH_3)_2$ in ether and an equimolar amount of HCN in C_6H_6 are simultaneously added dropwise to ether with stirring; CH_4 gas is evolved, and a white insoluble material separates upon filtration. The filtrate is assumed to contain $Be(CH_3)CN$ which is coordinated to ether and which is not highly polymeric. Evaporation of the colorless filtrate gives a solid product. The product loses ether when pumped at 70°C. The white nonvolatile and insoluble residue is probably impure polymeric $Be(CH_3)CN$. Analysis indicated a purity of 97 to 98%, and the $Be:CH_3$ ratio was 1:1.

The IR spectrum in Nujol shows ν(CN) at 2222 cm^{-1}. Addition of 2,2′-bipyridine to the solution of $Be(CH_3)CN$ in ether gives an orange-yellow solution. Probably an adduct is formed, but it is not specified. Excess HCN gives $Be(CN)_2$ [12].

$Be(CH_3)CN \cdot N(CH_3)_3$

Reaction of HCN with $Be(CH_3)_2 \cdot N(CH_3)_3$ in C_6H_6 results in the deposition of a little insoluble matter. Decantation and removal of C_6H_6 from the solution by pumping at room temperature and then at 80°C gives a colorless, amorphous product which is nonvolatile at 200°C/0.01 Torr. No melting or decomposition is noted below 300°C. The compound is evidently polymeric. The IR spectrum in Nujol shows ν(CN) at 2200 cm^{-1}. The compound is air-sensitive [12].

$Be(C_2H_5)Cl$

A solution of equimolar quantities of $Be(C_2H_5)_2$ and $BeCl_2$ in ether is assumed to mainly contain $Be(C_2H_5)Cl$ [4, 17, 25]. The pure compound can be prepared from the ether-free anhydrous components. $BeCl_2$ is stirred with a 3- to 4-fold excess of $Be(C_2H_5)_2$ for 12 h. Then $c\text{-}C_6H_{12}$ is added, and the slurry is stirred for another 24 h. The compound is isolated by filtration, washed with c-C_6H_{12}, and dried under high vacuum [31]. $Be(C_2H_5)Cl$ is assumed to be formed by the reaction $TiCl_3 + Be(C_2H_5)_2 \rightarrow Ti(C_2H_5)Cl + Be(C_2H_5)Cl$ during the polymerization of propene with this catalyst [9].

The compound melts at 175 to 179°C. The IR and Raman vibrations, and their assignments, are given in Table 24. It is assumed that the molecule is a dimer of C_{2h} symmetry with Cl bridges. Some intermolecular Cl–Be association is also postulated [31].

Table 24

IR and Raman Vibrations (in cm^{-1}) of $Be(C_2H_5)Cl$ [31].

$\tilde{\nu}$	intensity IR(mull)	intensity Raman	assignment	$\tilde{\nu}$	intensity IR(mull)	intensity Raman	assignment
2957	—	mw		670	s	—	$\varrho(CH_2)$
2947	—	mw		585	vs	—	
2935	—	mw	$\nu(C_2H_5$	550	—	sh	$\nu(BeCl_2Be)$
2880	—	mw		505	—	w	
2860	—	mw		475	w	—	
1450	—	w, br	$\delta(CH_3)$	435	—	vs	$\delta(BeCC)$
1395	w	w, br	$\delta(CH_2)$	400	vs	—	
1385	w	—	$\delta(CH_2)$	305	m	—	
1220	m	w, br	$\tau(CH_2)$	285	sh	vw	δ(ring) and
1010	m	—	$\nu(CC)$	235	sh	—	$\delta(BeCC)$
995	m	—		250	—	vw	
965	—	w	$\nu(BeC)$	200	vs	—	
925	m	w	$\nu(BeC)$				

The compound is extremely sensitive to air and moisture [31]. With $N(CH_3)_2CH_2CH_2N(CH_3)_2$ [25] and 2,2′-bipyridine [17], 1:1 adducts precipitate from solutions of $Be(C_2H_5)Cl$ in ether.

$Be(C_2H_5)Cl \cdot N(CH_3)_2CH_2CH_2N(CH_3)_2$

Addition of $N(CH_3)_2CH_2CH_2N(CH_3)_2$ to an ethereal solution of $Be(C_2H_5)Cl$, followed by evaporation of the solvent and crystallization from C_6H_6/C_6H_{14}, gives colorless needles. The ratio of C_2H_5 : Be : Cl was found to be 0.67 : 1.00 : 1.32. The product may be a 2:1 mixture of the title compound (Formula I, $R = C_2H_5$, $X = Cl$) and $BeCl_2 \cdot N(CH_3)_2CH_2CH_2N(CH_3)_2$ [25].

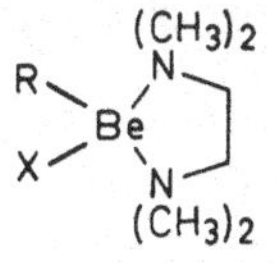

I

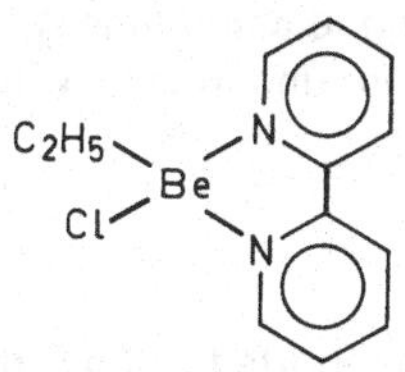

II

$Be(C_2H_5)Cl \cdot N_2C_{10}H_8$ ($N_2C_{10}H_8$ = 2,2′-bipyridine)

The adduct (Formula II) forms as an orange-yellow precipitate from ethereal solutions of $Be(C_2H_5)Cl$ by addition of 2,2′-bipyridine. It is washed with ether and C_6H_6 and dried under reduced pressure.

It decomposes at 167°C and turns white in air. A Soxhlet extraction with boiling C_6H_6 gives $BeCl_2 \cdot N_2C_{10}H_8$ [17].

$Be(C_2H_5)Br$ and **$Be(C_2H_5)I$**

The compounds are reported in the literature, but no physical data are given. They are assumed to be obtained by the reaction of powdered Be metal with C_2H_5X (X = Br, I) in the presence of $HgCl_2$ (or $BeBr_2$ or Br_2 for X = Br) as a catalyst in a sealed tube for 15 h at 80 to 90°C [3]. However, in a previous investigation, no $Be(C_2H_5)I$ could be obtained by reaction of Be with C_2H_5I under various conditions [1]. The gray mass formed by this method could not be analyzed [2]. $Be(C_2H_5)I$ is also obtained as an insoluble precipitate from Be metal and C_2H_5I at 130°C and 60 h reaction time [10]. Reaction of $Be(C_2H_5)_2$ with I_2 is reported to give the compound, but no details are given [4]. The compounds are suggested to be formed from equimolar mixtures of $Be(C_2H_5)_2$ and $BeBr_2$ [4, 18, 25] or BeI_2 [4] in ether. They were isolated as adducts with 1,4-dioxane (the stoichiometry is not given) [18] or with $N(CH_3)_2CH_2CH_2N(CH_3)_2$ [5]. $^7Be(C_2H_5)Br$ is obtained in the redistribution reaction of $^7Be(C_2H_5)_2$ and $BeBr_2$ [18].

Ethereal solutions do not fume in air, but the residues give white fumes on further heating after driving off the ether. Hydrolysis with H_2O gives C_2H_6. A color test with Michler's ketone is positive only after 10 to 15 min [3]. Reaction of $Be(C_2H_5)I$ with cyclohexanone gives c-$C_6H_{10}(C_2H_5$-1)OH-1 after hydrolysis [10]. $Be(C_2H_5)Br$ in ether reacts with LiH to give $Be(C_2H_5)H$ [13].

$Be(C_2H_5)Br \cdot N(CH_3)_2CH_2CH_2N(CH_3)_2$

For the preparation of the adduct, an excess of the diamine is added to equimolar amounts of $Be(C_2H_5)_2$ and $BeBr_2$ in ether. Evaporation of the ether and crystallization from C_6H_{14} gives the compound as long needles, m.p. 107 to 109°C.

The ^1H NMR spectrum in C_6H_6 at 30°C shows signals at $\delta = -0.13$ (q, CH_2), 1.65 (t, CH_3), and 2.09 to 2.42 ppm (m, CH_2N, CH_3N). At 95°C, $\delta = -0.14$ (q, CH_2), 1.63 (t, CH_3), 2.19 (s, CH_3N), and 2.32 ppm (CH_2N). The adduct is very soluble in C_6H_6, in which it is a monomer (by cryoscopy); see Formula I, p. 113 (R = C_2H_5, X = Br). It is sparingly soluble in C_6H_{14} [25].

$Be(C_4H_9)Br$

Powdered Be is reported to react with C_4H_9Br in the absence of ether and catalyst to give, after a sufficiently long reaction time (12 to 60 h at 130°C), $Be(C_4H_9)Br$ in satisfactory yield as a precipitate. The compound reacts with cyclohexanone to give c-$C_6H_{10}(C_4H_9$-1)OH-1 after hydrolysis [10].

$Be(C_4H_9)I$

The compound is reported to be formed when Be metal reacts with C_4H_9I in ether in the presence of $HgCl_2$ as a catalyst in a sealed tube at 80 to 90°C for 15 h [3]. A reaction of the same components at 130°C for 12 h in the absence of a solvent and a catalyst gives the compound as a precipitate in 65 to 70% yield [10].

The ethereal solutions do not fume in air. However, driving off the ether and further heating gives white fumes of BeO. On heating, $Be(C_4H_9)I$ is converted into $Be(C_4H_9)_2$. It is decomposed by H_2O. The color test with Michler's ketone is developed only after standing 10 to 15 min before hydrolysis [3]. Reaction with cyclohexanone gives c-$C_6H_{10}(C_4H_9$-1)OH-1 after hydrolysis [10].

$Be(C_4H_9$-t)Cl

$Be(C_4H_9$-t$)_2$ reacts with a suspension of $BeCl_2$ in C_6H_6 within 24 h. The compound obtained after removal of the solvent under reduced pressure melts at 123 to 124°C with decomposition. The sublimate (at 120 to 130°C/10^{-3} Torr) softens at ~125°C and does not melt up to 320°C. It is insoluble in C_6H_6. Reaction with quinuclidine (1-azabicyclo[2.2.2]octane) gives a 1:1 adduct (see below) [28].

$Be(C_4H_9$-t)Cl·$O(C_2H_5)_2$

The etherate is one of the products of the reaction of $BeCl_2$ with $Be(C_4H_9$-t$)_2$·$O(C_2H_5)_2$ maintained at room temperature, which also gives overnight ether-free $Be(C_4H_9$-t$)_2$ [7, 16, 30]. It is also formed by the reaction of $BeCl_2$·$2O(C_2H_5)_2$ and $Mg(C_4H_9$-t)Cl in ether. No details are given [3].

The compound is described as a white solid, m.p. ~50°C, which sublimes at <50°C in a vacuum [7]. Sublimation of the residue remaining after evaporation of $Be(C_4H_9$-t$)_2$ at 60°C/0.01 Torr gives colorless needles, m.p. 79 to 80°C [16]. Colorless platelets with a m.p. of 86 to 88°C are obtained after sublimation at 60°C/0.15 Torr [33]. The ^{1}H NMR spectrum in C_6H_6 shows signals at $\delta=0.94$(t, CH_3CO), 1.25(s, CH_3CBe), and 3.75 ppm (q, CH_2; $^3J(H,H)=7$ Hz). The compound is a dimer in C_6H_6 (by cryoscopy), and Formula III (X = Cl) is proposed. In ether, the degree of associaton is 1.07 from 0.002 to 0.020 M, and Formula IV is proposed [16].

Reaction with (t-$C_4H_9)_2C{=}O$ or with t-C_4H_9CHO in ether gives $(Be(OCH(C_4H_9$-t$)_2)Cl)_2$ and $(Be(OCH_2C_4H_9$-t)Cl·$O(C_2H_5)_2)_n$ (n = 1, 2), respectively [30].

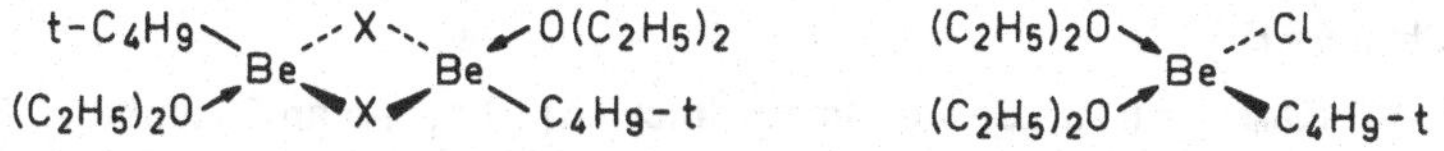

III IV

$Be(C_4H_9$-t)Cl·NC_7H_{13} (NC_7H_{13} = quinuclidine)

The amine in excess is added to a suspension of unsublimed $Be(C_4H_9$-t)Cl (solvent not specified), resulting in perceptible heat evolution. After 0.5 h of stirring the solid product is separated and washed with warm C_6H_6. It decomposes above 200°C [28].

$Be(C_4H_9$-t)Cl·NCC_4H_9-t

The nitrile is added to $Be(C_4H_9$-t)Cl in ether. The adduct precipitates during addition. After stirring for 15 min, the ether is evaporated, and the residue is crystallized from C_6H_6/C_6H_{14} (2:1). It shrinks at ~50°C, softens at ~70°C, turns brown at ~150°C, and becomes white at ~180°C. It is a monomer in C_6H_6 (by cryoscopy), and the IR spectrum contains an absorption at 2300 cm^{-1} [30].

$Be(C_4H_9$-t)Br·$O(C_2H_5)_2$

The compound is obtained by reaction of $BeBr_2$ with $Be(C_4H_9$-t$)_2$ in ether at room temperature. It is purified by sublimation at 70°C/0.01 Torr. The colorless needles melt at 55 to 56°C. It is a dimer in C_6H_6 (by cryoscopy), and the structure given in Formula III, p. 115 (X = Br) is suggested [16].

Be(C_5H_{11})I and **Be(C_8H_{17})I**

Beryllium powder reacts with $C_5H_{11}I$ or $C_8H_{17}I$ at 130°C within 12 to 60 h in the absence of ether or a catalyst. The title compounds are obtained as precipitates. They react with cyclohexanone to give (after hydrolysis) c-C_6H_{10}(R-1)OH-1 (R = C_5H_{11} and C_8H_{17}) [10].

Be($CH_2C_4H_9$-t)Cl and **Be($CH_2C_4H_9$-t)Br**

The compounds are formed by reaction of Be($CH_2C_4H_9$-t)$_2$ with BeX_2 (X = Cl or Br) in ether. The bromide was isolated as an adduct with $N(CH_3)_2CH_2CH_2N(CH_3)_2$ (see below). Reaction between Be($CH_2C_4H_9$-t)Br in ether and $N(CH_3)_3$ results in quantitative disproportionation at, or below, room temperature producing $BeBr_2 \cdot N(CH_3)_3$ and Be($CH_2C_4H_9$-t)$_2 \cdot N(CH_3)_3$. The intermediate formation of **Be($CH_2C_4H_9$-t)Br·$N(CH_3)_3$** is assumed. The analogous reaction with Be($CH_2C_4H_9$-t)Cl resulted in partial disproportionation, but no experimental details are given for the chloride. Be($CH_2C_4H_9$-t)Br seems to be inert towards LiH, but Be($CH_2C_4H_9$-t)H can be prepared by reaction with Na[B(C_2H_5)$_3$H] and is isolated as an adduct with $N(CH_3)_2CH_2CH_2N(CH_3)_2$ [25].

Be($CH_2C_4H_9$-t)Br·$N(CH_3)_2CH_2CH_2N(CH_3)_2$

An excess of $N(CH_3)_2CH_2CH_2N(CH_3)_2$ is added to a solution of Be($CH_2C_4H_9$-t)Br in ether. The white solid, remaining after volatile material has been evaporated, is crystallized from C_6H_6/C_6H_{14} (1:2) as colorless plates, m.p. 104 to 106°C.

^1H NMR (C_6H_6) at 30°C: $\delta = -0.02$ (s, CH_2Be), 1.54 (s, CH_3C), 1.88 and 2.35 (s, CH_3N), 2.17 ppm (m, CH_2N). At 90°C, the CH_2N and CH_3N signals collapse to two singlets ($\delta = 2.24$ (s, CH_3N) and 2.04 ppm (s, CH_2N)) indicating that dissociation takes place at higher temperatures. The compound is a monomer in C_6H_6 (by cryoscopy); see Formula I, p. 113 (R = t-$C_4H_9CH_2$, X = Br) [25].

Be($C_{10}H_{17}$)Cl ($C_{10}H_{17}$ = ((1S,2S)-6,6-dimethylbicyclo[3.1.1]heptan-2-yl)methyl, Formula V)

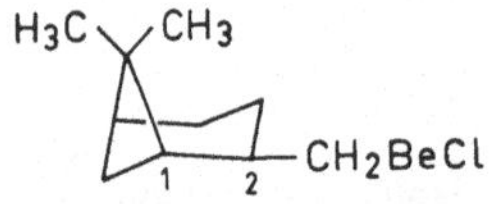

V

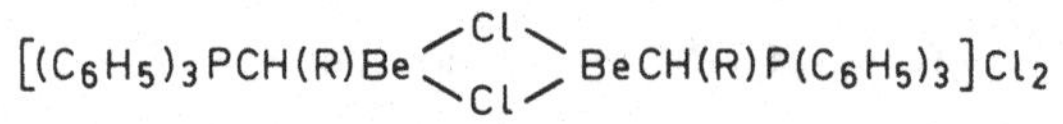

VI

The compound is prepared from cis-$C_{10}H_{17}MgCl$ and anhydrous $BeCl_2$ in ether at 0°C. After 4 h at reflux, the upper layer is transferred to a container, and the solvent is evaporated. The crude product is stirred at 0.005 Torr for 4 h to give 75% yield of pure material.

Reaction of the product with $C_6H_5C(O)C_3H_7$-i, $C_6H_5C(O)C_4H_9$-t, and t-$C_4H_9C(O)C{\equiv}CC_4H_9$ yields the corresponding alcohols upon hydrolysis with dilute H_2SO_4 in 35 to 40% conversion and 30 to 39% enantiomeric purity (R-configuration) [35].

Be(C_6H_5)Br

Evidence is presented in the literature that the compound is present in a solution of equimolar amounts of Be(C_6H_5)$_2$ and $BeBr_2$ in ether. Ebullioscopic measurements indicate the presence of monomeric Be(C_6H_5)Br in the concentration range 0.035 to 0.130 M. Addition of 1,4-dioxane gives a precipitate of **Be(C_6H_5)Br·$O_2C_4H_8$** [18]. Thus, an earlier claim that no

$^7Be(C_6H_5)Br$ is formed from $Be(C_6H_5)_2$ and $^7BeBr_2$ is invalidated [8]. Earlier attempts to prepare the compound from Be metal and C_6H_5Br failed [1]. The ether solution of $Be(C_6H_5)Br$ reacts with LiH to give $Be(C_6H_5)H$, which is isolated as an adduct with $N(CH_3)_2CH_2CH_2N(CH_3)_2$ [13].

$Be(C_6H_5)I$

The compound is said to be formed in the reaction of Be metal and C_6H_5I in ether in the presence of $HgCl_2$ or $BeCl_2$ as catalysts in a sealed tube at 150 to 175°C for 15 h. A positive color test with Michler's ketone is reported. No other details are available [3].

$Be(CCl_3)Cl$

The compound is claimed to be obtained beside HCl in the reaction of anhydrous $BeCl_2$ and $CHCl_3$ at 20°C. After hydrolysis, the $Be^{2+}:Cl^-$ ratio was found to be close to 1 [6]. The reaction could not be repeated with carefully purified components [24].

Be(C(O)R)X ($R = CH_3$, C_2H_5, C_3H_7, C_4H_9, C_5H_{11}, X = Cl, Br, I)

The acyl halide RC(O)X is added dropwise to an equimolar amount of finely divided Be metal at room temperature. Then the mixture is heated in a water bath for 1 h. RC(O)Br reacts slowly with Be in solvents like ether or dioxane. The reaction is rapid in anhydrous $CH_3CO_2C_2H_5$ [19 to 21, 26] in the presence of some $HgCl_2$ [20, 21, 26]. The products were not isolated in the pure state, but their structure as RC(O)BeX was established from their chemical reactions [14].

Hydrolysis with H_2O gives aldehydes. Thus CH_3CHO, C_2H_5CHO, and C_3H_7CHO are obtained from Be(C(O)R)X ($R = CH_3$, C_2H_5, C_3H_7, X = Cl, Br, I). When the amount of Be is reduced to a mole ratio of Be : RC(O)X = 1 : 2 during the preparation of RC(O)BeX, the acylberyllium halide reacts further with RC(O)X to give α-diketones for $R = CH_3$ to C_5H_{11} [14]. Be(C(O)R)Br reacts with ketones R'C(O)R" to give α-hydroxyketones RC(O)C(OH)R'R" after hydrolysis ($R = CH_3$ and $R' = CH_3$, $R'' = C_2H_5$, C_4H_9 [14], $R' = R'' = C_3H_7$; $R = C_2H_5$ and $R' = R'' = CH_3$, C_3H_7, $R' = CH_3$, $R'' = C_2H_5$, C_4H_9, R' = H, $R'' = C_6H_5$; $R = C_3H_7$ and $R' = R'' = CH_3$, C_3H_7, $R' = CH_3$, $R'' = C_2H_5$, C_4H_9 [19]) in 55 to 20% yields [14,19]. With aromatic ketones and $Be(C(O)CH_3)Br$ in $CH_3CO_2C_2H_5$, pinacols HOCRR"CR'R'''OH ($R'' = R''' = C_6H_5$, $R = R' = 2\text{-}FC_6H_4$, $4\text{-}FC_6H_4$, $4\text{-}BrC_6H_4$, $2,4\text{-}Cl_2C_6H_3$; $R'' = R''' = 4\text{-}CH_3C_6H_4$, $R = R' = 2\text{-}CH_3\text{-}4\text{-}BrC_6H_3$) are formed upon hydrolysis. The presence of excess $CH_3C(O)Br$ leads to pinacolones [21]. Be(C(O)R)Br ($R = CH_3$, C_2H_5, C_4H_9) and aromatic aldehydes R'CHO are reacted and hydrolyzed. The products are stilbenes R'CH=CHR' ($R' = C_6H_5$, $4\text{-}CH_3C_6H_4$, $3,4\text{-}(CH_3)_2C_6H_3$, $4\text{-}(i\text{-}C_3H_7)C_6H_4$, $4\text{-}CH_3OC_6H_4$, $4\text{-}ClC_6H_4$, $4\text{-}BrC_6H_4$, $C_6H_5CH{=}CH$, $4\text{-}ClC_6H_4CH{=}CH$), $BeBr_2$, and $Be_4O(O_2CR)_6$. The yields are highest (50 to 78%) with $Be(C(O)CH_3)Br$ [20]. A study of the reactions of $Be(C(O)CH_3)Br$ with aromatic aldehydes, containing various substituents (Cl, Br, CH_3, CH_3O, CH_3CO_2, etc.), showed that the introduction of two or more substituents into the ring does not affect the course of the reaction. The low yields observed in some cases are due to steric hindrance [29].

Aromatic nitro compounds $R'NO_2$ are converted into carboxamides R'NHC(O)R by reaction with Be(C(O)R)Br and subsequent hydrolysis ($R = CH_3$, $R' = C_6H_5$, $2\text{-}CH_3C_6H_4$, $4\text{-}CH_3C_6H_4$, $2,4,6\text{-}(CH_3)_3C_6H_2$; $R = C_2H_5$, $R' = C_6H_5$, $2\text{-}CH_3C_6H_4$, $4\text{-}CH_3C_6H_4$; $R = C_3H_7$, $R' = C_6H_5$, $2\text{-}CH_3C_6H_4$, $4\text{-}CH_3C_6H_4$; $R = C_4H_9$, $R' = C_6H_5$, $2\text{-}CH_3C_6H_4$; $R = C_5H_{11}$, $R' = C_6H_5$, $2\text{-}CH_3C_6H_4$, $2,4,6\text{-}(CH_3)_3C_6H_2$). Be appears as $Be_4O(O_2CR)_6$. The yields vary from 12 to 79%. Increased steric crowding in the nitro compound reduces the yield [26].

$[Be(C(R)HP(C_6H_5)_3)Cl]Cl$ ($R = H$, CH_3)

The compounds are prepared by addition of $BeCl_2$ to a solution of $(C_6H_5)_3P{=}CHR$ ($R = H$, CH_3) (mole ratio 2 : 1) in THF. The mixture is stirred for 10 to 12 h at room temperature. The

References on pp. 118/9

white precipitate is filtered, washed with ether, and dried in a vacuum. The yields are 86.5 (R = H) and 36.0% (R = CH_3), respectively.

Decomposition occurs at 224°C (R = H) and 220°C (R = CH_3). It follows from the molecular weight that the compounds are dimeric. The existence of a free anion is concluded from the equivalent conductance of the corresponding Mg compound (R = H). Therefore, the structure of Formula VI, p. 116 (R = H, CH_3) is suggested. The compounds dissolve only in aqueous mineral acids. They are unstable in 3.6 M D_2SO_4 at −34°C, giving $[(C_6H_5)_3PCH_3]Cl$ [36].

References:

[1] Gilman, H. (J. Am. Chem. Soc. **45** [1923] 2693/5).
[2] Durand, J. F. (Compt. Rend. **182** [1926] 1162/4).
[3] Gilman, H., Schulze, F. (J. Am. Chem. Soc. **49** [1927] 2904/7).
[4] Gilman, H., Schulze, F. (J. Chem. Soc. **1927** 2663/9).
[5] Coates, G. E., Glockling, F., Huck, N. D. (J. Chem. Soc. **1952** 4512/5).
[6] Silber, P. (Ann. Chim. [Paris] [12] **7** [1952] 182/233).
[7] Head, E. L., Holley, C. E., Rabideau, S. W. (J. Am. Chem. Soc. **79** [1957] 3687/9).
[8] Dessy, R. E. (J. Am. Chem. Soc. **82** [1960] 1580/2).
[9] Firsov, P., Sandomirskaya, N. D., Tsvetkova, V. I., Chirkov, N. M. (Vysokomol. Soedin. **3** [1962] 1352/8; Polym. Sci. [USSR] **3** [1962] 343/9).
[10] Zakharkin, L. I., Okhlobystin, O. Yu., Strunik, B. N. (Izv. Akad. Nauk SSSR Otd. Khim. Nauk **1961** 2254; Bull. Acad. Sci. USSR Div. Chem. Sci. **1961** 2114).

[11] Schulze, F. (Iowa State Coll. J. Sci. **8** [1933] 225/8; C. A. **1934** 2325).
[12] Coates, G. E., Mukherjee, R. N. (J. Chem. Soc. **1963** 229/33).
[13] Bell, N. A., Coates, G. E. (J. Chem. Soc. A **1966** 1069/73).
[14] Lapkin, I. I., Anvarova, G. Ya., Povarnitsyna, T. N. (Zh. Obshch. Khim. **36** [1966] 1952/4; J. Gen. Chem. [USSR] **36** [1966] 1945/6).
[15] Ashby, E. C., Sanders, R., Carter, J. (Chem. Commun. **1967** 997/8).
[16] Coates, G. E., Roberts, P. D. (J. Chem. Soc. A **1968** 2651/5).
[17] Bell, N. A. (J. Organometal. Chem. **13** [1968] 513/5).
[18] Sanders, J. R., Ashby, E. C., Carter, J. H. (J. Am. Chem. Soc. **90** [1968] 6385/90).
[19] Lapkin, I. I., Povarnitsyna, T. N. (Zh. Obshch. Khim. **38** [1968] 99/102; J. Gen. Chem. [USSR] **38** [1968] 96/9).
[20] Lapkin, I. I., Zinnatullina, G. Ya. (Zh. Obshch. Khim. **39** [1969] 1132/4; J. Gen. Chem. [USSR] **39** [1969] 1102/3).

[21] Lapkin, I. I., Zinnatullina, G. Ya. (Zh. Obshch. Khim. **39** [1969] 2708/10; J. Gen. Chem. [USSR] **39** [1969] 2647/8).
[22] Kovar, R. A., Morgan, G. L. (J. Am. Chem. Soc. **91** [1969] 7269/74).
[23] Kovar, R. A., Morgan, G. L. (J. Am. Chem. Soc. **92** [1970] 5067/72).
[24] Coates, G. E., Morgan, G. L. (Advan. Organometal. Chem. **9** [1970] 195/257).
[25] Coates, G. E., Francis, B. R. (J. Chem. Soc. A **1971** 1305/8).
[26] Lapkin, I. I., Tenenboim, N. F., Evstafeeva, N. E. (Zh. Obshch. Khim **41** [1971] 1554/7; J. Gen. Chem. [USSR] **41** [1971] 1558/9).
[27] Andersen, R. A., Bell, N. A., Coates, G. E. (J. Chem. Soc. Dalton Trans. **1972** 577/82).
[28] Coates, G. E., Smith, D. L., Srivastava, R. C. (J. Chem. Soc. Dalton Trans. **1973** 618/22).
[29] Lapkin, I. I., Evstafeeva, N. E., Sinani, S. V. (Zh. Obshch. Khim. **43** [1973] 1984/6; J. Gen. Chem. [USSR] **43** [1973] 1968/9).
[30] Andersen, R. A., Coates, G. E. (J. Chem. Soc. Dalton Trans. **1974** 1171/80).

[31] Atam, N., Dehnicke, K. (Z. Anorg. Allgem. Chem. **427** [1976/77] 193/9).
[32] Baskin, C. P., Bender, C. F., Lucchese, R. R., Bauschlicher, C. W., Schaefer III, H. F. (J. Mol. Struct. **32** [1976] 125/31).
[33] Baker, R. W., Brendel, G. J., Lowrance, B. R., Maugham, J. R., Marlett, E. M., Shepherd, L. H. (J. Organometal. Chem. **159** [1978] 123/30).
[34] Gaines, D. F., Coleson, K. M., Hillenbrand, D. F. (J. Magn. Resonance **44** [1981] 84/8).
[35] Giacomelli, G., Falorni, M., Lardicci, L. (Gazz. Chim. Ital. **115** [1985] 289/91).
[36] Yamamoto, Y. (Bull. Chem. Soc. Japan **56** [1983] 1772/4).

1.3.5 Organoberyllium Alcoholates and Their Adducts

1.3.5.1 Be(R)OR′ Type Compounds

General Remarks. The compounds of this type are summarized in Table 25. Most of the compounds described in this chapter are dimeric or tetrameric in solution. The degree of association depends on the steric bulk of the organic groups, but also the solvent and concentration may be important. The dimers are assumed to have the structure shown in Formula I [7]. For the tetramers, the structure shown in Formula II is suggested in analogy to $Be(CH_3)OSi(CH_3)_3$ (No. 13), for which an X-ray structure analysis was performed [15].

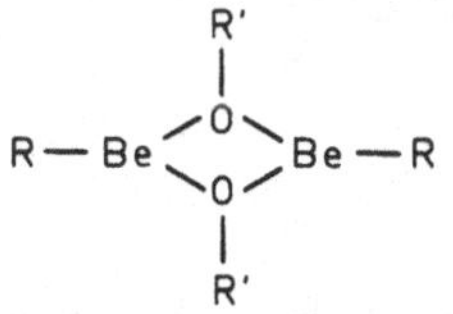

I

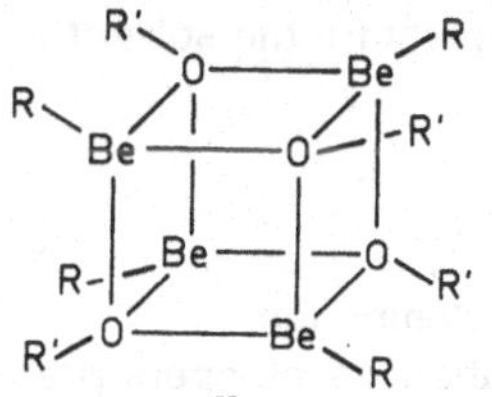

II

It is described that $Be(CH_3)_2$ reacts with C_6H_5OH, 2-,3-, and 4-$CH_3C_6H_4OH$, and 2-,3-, and 4-ClC_6H_4OH, and that $Be(C_2H_5)_2$ reacts with C_6H_5OH, 4-$CH_3C_6H_4OH$, and 4-ClC_6H_4OH in ether $>-80°C$ to give first the appropriate primary products Be(R)OR′, and $>-55°C$ finally $Be(OR')_2$. The Be(R)OR′ compounds were not isolated [3].

For the preparation of the compounds in Table 25, the following methods are used:

Method I: Reaction of BeR_2 with R′OH in a 1:1 mole ratio.
The appropriate alcohol is added to BeR_2 in ether at −196°C (Nos. 1 [1, 7], 2 to 5 [7], 10 [8]); at −80 to −40°C (Nos. 11 [8], 15 [7], 17 [8], 18 [7], 27 [13]); at ~0°C (No. 12 [8]); at room temperature (No. 7 [5]); or without a solvent at −196°C (No. 18 [2]). The mixture is warmed to room temperature, and the ether is evaporated (Nos. 1 to 5, 10, 15, 17) after filtration (Nos. 7, 18) or upon stirring for 15 min (No. 27). No. 11 deposits as the reaction proceeds. Nos. 8, 9 [7], 30 [11], 31 [6] are prepared similarly in C_6H_6 at room temperature. The solvent is evaporated, and the residue is boiled three times with C_6H_{14}, the latter being removed each time under reduced pressure (Nos. 8, 9), or the product precipitates (No. 31). No. 30 is prepared from $Be(C{\equiv}CC_6H_5)_2 \cdot {}^2D$ (${}^2D = N(CH_3)_3$ or $O(C_2H_5)_2$) and precipitates after addition of C_6H_{14}. Nos. 21 and 23 are prepared by mixing the reactants in C_6H_{14} at −60°C (No. 21) or at room temperature (No. 23). The mixtures are boiled. On cooling, crystals deposit (No. 21), or the solvent is evaporated (No. 23) [7]. No. 28 is prepared by mixing the reactants in C_5H_{12} at −40°C. Then the mixture is boiled for 5 min, and the solvent is evaporated. The product is treated with boiling C_6H_{14}, which is subsequently evaporated [9].

Method II: Reaction of BeR_2 with R'_2CO, R'CHO, or ethylene oxide in a 1:1 mole ratio.
The appropriate aldehyde (Nos. 4 [7], 19, 24 [13]), ketone (Nos. 5 [7], 6 [13], 15 [7], 16, 20, 22, 25, 26, 29, 32 [13]), or ethylene oxide (No. 3 [7]) is added to BeR_2 in ether at −196°C (Nos. 3 to 5), at ~ −80°C (Nos. 6, 15, 22, 24 to 26), in C_6H_{14} at ~ −80°C (Nos. 16, 19, 20, 25), or in C_6H_6 at room temperature (Nos. 29, 32). The mixtures are warmed to room temperature, and the solvent is evaporated (Nos. 3 to 6, 15) after stirring for 15 min (Nos. 16, 19, 24 to 26, 29), 30 min (Nos. 20, 32), or 1 h (No. 22).

Method III: Reaction of BeR_2 with $Be(OR')_2$.
A suspension of $Be(OCH_3)_2$ (in excess) in ethereal $Be(CH_3)_2$ is boiled for 1 h. Filtration and evaporation of the ether gives No. 1 [7]. Alternatively, $Be(OC_4H_9\text{-t})_2$ in toluene is added to $Be(CH_3)_2$ (mole ratio ~1:1). Workup as above gives No. 5 [7, 14].

Method IV: Reaction of BeR_2 with a peroxide.
Di-t-butylperoxide is added to a suspension of $Be(CH_3)_2$ (mole ratio 1:1) in toluene at −196°C. Warming to room temperature, filtration, and evaporation of the solvent yields No. 5 [7].

Method V: Reaction of Be(R)H with an aldehyde.
C_6H_5CHO in C_6H_6 is added to $Be(CH_3)H$ in C_6H_6 (mole ratio 1:1). After 1 h at room temperature the solvent is evaporated (No. 7) [5].

Table 25

Be(R)OR' Type Compounds.

Further information on numbers preceded by an asterisk is given at the end of the table. Concentrations given are calculated for the monomer. Other explanations, abbreviations, and units on p. X.

No.	compound method of preparation [Ref.]	properties and remarks [Ref.]
*1	$Be(CH_3)OCH_3$ I [1, 7], III [7]	clear crystals [1], m.p. 23 to 25° [1, 7] sublimes at 70°/10^{-3} Torr [7] ^{1}H NMR(C_6H_6): −0.61 (s, CH_3Be), 3.24 and 3.50 (d, CH_3O) [7] IR (contact film): 2941 (s), 2849 (s), 1538 (w, sh), 1464 (w), 1408 (w, sh), 1262 (w), 1208 (m), 1104 (s), 1039 (s), 958 (m), 910 (w), 579 (s, br), 752 (s, br), 667 (m) [7] dimeric in C_6H_6 [1] and ether [7] (rather dilute solutions, by ebullioscopy) tetrameric in C_6H_6 (0.49 and 0.99 wt%, by cryoscopy) [7]
2	$Be(CH_3)OC_2H_5$ I [7]	m.p. 28 to 30° (by evaporation of C_6H_{14}) ^{1}H NMR(C_6H_6): −0.71 (s, CH_3Be), 1.07 (t, CH_3C), 3.91 (q, CH_2O) IR (contact film): 2963 (s), 2915 (s), 2890 (s), 2825 (m), 1462 (w, sh), 1445 (w), 1389 (m), 1366 (w, sh), 1300 (w), 1263 (m), 1170 (m), 1122 (s), 1075 (s), 1050 (s), 1026 (m, sh), 917 (m), 901 (m, sh), 769 (s, br) dimeric in ether (0.004 to 0.015 M, by ebullioscopy) tetrameric in C_6H_6 (0.31 and 0.94 wt%, by cryoscopy) [7]

Table 25 (continued)

No.	compound method of preparation [Ref.]	properties and remarks [Ref.]
3	$Be(CH_3)OC_3H_7$ I, II [7]	plates, m.p. 38 to 40° (from C_6H_{14}) ^{1}H NMR(C_6H_6): −0.77 (s, CH_3Be) IR (Nujol): 1312 (w), 1290 (w), 1266 (m), 1211 (m), 1174 (w), 1117 (m, sh), 1098 (m), 1044 (s), 1015 (s), 964 (s), 922 (m), 890 (m), 880 (m), 798 (s), 754 (m, br), 676 (m), 667 (m) dimeric in ether (0.004 to 0.015 M, by ebullioscopy) tetrameric in C_6H_6 (0.64 and 1.28 wt%, by cryoscopy) addition of a further mole of ethylene oxide gives no reaction in C_6H_{14} up to 50° [7]
4	$Be(CH_3)OC_3H_7$-i I, II [7]	m.p. 134 to 136° (from C_6H_{14} at −30° or sublimation at 100°/10^{-3} Torr) ^{1}H NMR(C_6H_6 at 33°): −0.77 (s, CH_3Be), 2.19 (d, CH_3C; ^3J(H,H) = 6), 4.10 (sept, CH); in $C_6D_5CD_3$ at −45° the CH_3Be signal is split into an unequal d, and the i-C_3H_7 signals are unaffected IR (Nujol): 1335 (w), 1287 (w), 1258 (m), 1210 (m), 1173 (m), 1140 (m), 1117 (s), 1015 (s), 936 (s), 790 (s, br), 722 (m, sh), 576 (m, br), 538 (m, br) dimeric in ether (0.004 to 0.15 M, by ebullioscopy) tetrameric in C_6H_6 (0.30 and 0.60 wt%, by cryoscopy) [7]
5	$Be(CH_3)OC_4H_9$-t I, II [7], III [7,14], IV [7]	m.p. 93° (from ether) [7], shrinks at 90° and melts at 185 to195° with dec. (from C_6H_6/ether 5:2) [14] ^{1}H NMR(C_6H_6, 1.2 M): −0.46 (s, CH_3Be), 1.47 (s, CH_3C); the spectrum is unchanged at 33, 50, and 60° [7] ^{1}H NMR(c-$C_6D_{11}CD_3$ at 33°): −0.94, −0.71 (2d, CH_3Be), 1.37, 1.48 (2d, CH_3C); the spectrum is unchanged from 33 to 100° and from 0.9 to 1.8 M, no explanation is given [7] IR (Nujol): 1366 (s), 1256 (s), 1236 (m), 1200 (s), 1080 (m, sh), 1055 (m), 1027 (s), 1012 (s), 990 (s), 971 (s, sh), 847 (s), 819 (s), 806 (s), 676 (m), 671 (m, sh), 598 (w), 532 (w), 494 (w) [7] dimeric in ether (0.004 to 0.015 M, by ebullioscopy), tetrameric in ether (0.088 M, by isopiestic measurement) and C_6H_6 (0.43 and 0.86 wt%, by cryoscopy) [7] no reaction is observed with NaH [14] reaction with $(CH_3)_2CO$ gives $Be(OC_4H_9\text{-t})_2$ [7] with pyridine a 1:1 adduct is formed (see Table 26, No. 3) [7]
6	$Be(CH_3)OC(C_4H_9\text{-t})_2CH_3$ II [13]	colorless needles, m.p. 89 to 91° (from C_6H_{14}) dimeric in C_6H_6 (0.89 and 1.19 wt%, by cryoscopy) hydrolysis gives $(t\text{-}C_4H_9)_2C(CH_3)OH$ [13]

References on pp. 130/1

Table 25 (continued)

No.	compound method of preparation [Ref.]	properties and remarks [Ref.]
7	$Be(CH_3)OCH_2C_6H_5$ I, V [5]	viscous liquid, which cannot be distilled up to 80° in a vacuum ^{1}H NMR(C_6D_6): −0.84(s, CH_3), 5.83(s, CH_2), 7.05(m, C_6H_5) tetrameric in C_6H_6 (0.78 to 1.28 wt%, by cryoscopy) reaction with 2,2'-bipyridine gives $Be(CH_3)_2 \cdot N_2C_{10}H_8$ [5]
8	$Be(CH_3)OCH(C_6H_5)_2$ I [7]	star-like clusters, m.p. 97 to 108° (from C_6H_{14}) ^{1}H NMR(C_6D_6): −1.06(s, CH_3Be), 5.82(s, CH), 6.96 to 7.40 (m, C_6H_5) IR(Nujol): 1517(m), 1264(m), 1245(w), 1198(m), 1088(s), 1060(s), 1027(s), 923(w), 914(m), 887(m), 871(m), 844(w), 792(m), 741(m), 697(s), 633(w), 609(w), 507(w, br) dimeric in C_6H_6 (0.87 and 1.73 wt%, by cryoscopy) [7]
9	$Be(CH_3)OC(C_6H_5)_3$ I [7]	m.p. 183 to 184° (from C_6H_{14}) ^{1}H NMR(C_6D_6): −1.43(s, CH_3), 6.84 to 7.38(m, C_6H_5) IR(Nujol): 1600(w), 1261(m), 1216(m), 1205(m), 1158(w), 1083(m), 1037(m), 1012(s), 995(s), 934(m), 921(m), 905(m), 890(m), 800(m,br), 759(m), 752(m), 707(m), 697(s), 684(m), 650(w), 633(m), 532(w), 519(w), 500(w) dimeric in C_6H_6 (0.51 and 1.98 wt%, by cryoscopy) [7]
10	$Be(CH_3)OC_2H_4OCH_3$ I [8]	large rhombic prisms, m.p. 103 to 104° (from C_6H_{14}) ^{1}H NMR (C_6H_6): −0.94 and −0.88(d, CH_3Be), 3.08 and 3.12(d, CH_3O), complex m for C_2H_4 tetrameric in C_6H_6 (0.42 to 0.85 wt%, by cryoscopy) reaction with 2,2'-bipyridine gives $Be(CH_3)_2 \cdot N_2C_{10}H_8$ [8]
11	$Be(CH_3)OC_2H_4N(CH_3)_2$ I [8]	white crystals, m.p. 208 to 212° with dec. (washed with ether) ^{1}H NMR(C_6H_6): 1.90(s, CH_3Be), 4.83(s, CH_3N) oligomeric in C_6H_6 (n ~ 7, 1.33 to 2.00 wt%, by cryoscopy) dissolves readily in C_6H_6, sparingly in ether, and not at all in C_6H_{14} [8]
*12	$Be(CH_3)OC_9H_6N$ (OC_9H_6N = quinolinyl-8-oxy) I [8]	pale yellow crystals, dec. at 184 to 201° (washed with ether) tetrameric in C_6H_6 (0.57 to 0.89 wt%, by cryoscopy) not very soluble in ether, C_6H_{14}, and $C_6H_5CH_3$ [8]
*13	$Be(CH_3)OSi(CH_3)_3$ see further information	colorless crystals, m.p. 131 to 133° (from $C_6H_5CH_3$/C_6H_{14} 1:4 at −40°) tetrameric in C_6H_6 (0.77 and 1.54 wt%, by cryoscopy) [15]
*14	$Be(C_2H_5)OC_4H_9$-t see further information	colorless viscous liquid IR (neat liquid): 2975(s), 2935(s), 2895(s), 2855(s), 2800(m, sh), 2715(w), 1470(m), 1457(m), 1410(m), 1388(s), 1365(s), 1252(s), 1228(m, sh), 1202(s), 1190(s), 1137(w), 1037(s), 1005(s), 950(w), 932(s), 912(w), 855(w), 823(m), 713(m), 615(m), 597(m), 635(m, ?) [14]

Table 25 (continued)

No.	compound method of preparation [Ref.]	properties and remarks [Ref.]
15	$Be(C_2H_5)OC(C_2H_5)_3$ I, II [7]	liquid IR(contact film): 2959(s), 2924(s, sh), 2874(m, sh), 1458(m), 1418(w), 1385(w), 1311(w), 1264(m), 1233(w), 1209(w), 1152(m), 1121(m), 1078(s), 1049(m), 1020(s), 963(m), 918(m), 859(m), 818(m, sh), 793(s), 760(s, br), 689(m, sh), 667(w), 565(w, br) trimeric in C_6H_6 (0.45 and 0.90 wt%, by cryoscopy) a cyclic Be_3O_3 skeleton is suggested [7]
16	$Be(C_2H_5)OCH(C_4H_9\text{-}t)_2$ II [13]	colorless plates, m.p. 147 to 148° (from C_6H_{14}) dimeric in C_6H_6 (1.20 and 1.60 wt%, by cryoscopy) hydrolysis yields $(t\text{-}C_4H_9)_2CHOH$ no reaction is observed with excess of $(t\text{-}C_4H_9)_2CO$ in boiling xylene within 3 h [13]
17	$Be(C_2H_5)OCH_2CH_2OCH_3$ I [8]	needles, m.p. 79 to 80° (from C_6H_{14}) ^1H NMR(C_6H_6): −0.16(q, CH_2Be), 2.08(t, CH_3C), 3.16(s, CH_3O), 3.22 to 3.65(m, C_2H_4) tetrameric in C_6H_6 (1.25 and 1.66 wt%, by cryoscopy) [8]
18	$Be(C_3H_7\text{-}i)OCH_3$ I [2, 7]	transparent [2] plates [2, 7], m.p. 133° [2], m.p. 133 to 135° (from C_6H_{14}) [7] ^1H NMR(C_6H_6): 1.50(d, CH_3C), 3.44(s, br, CH_3O) [7] IR(Nujol): 1263(w), 1222(w), 1199(w), 1152(w, sh), 1109(m), 1042(s), 936(m, sh), 846(s), 795(s, br), 781(s, sh, br), 727(m, sh), 513(w, br) [7] tetrameric in C_6H_6 (0.49 and 0.97 wt%, by cryoscopy) [7] unaffected by short exposure to air [2] vigorously hydrolyzed by H_2O [2]
19	$Be(C_4H_9\text{-}i)OCH_2C_4H_9\text{-}t$ II [13]	liquid, distillation at 80 to 85°/10^{-3} Torr IR(Nujol): 1478(m), 1462(m), 1400(w), 1382(w), 1360(m), 1310(m), 1185(m), 1078(s), 1018(m), 978(m), 925(w), 900(w), 880(w), 830(w), 805(w), 785(w), 635(w) trimeric in C_6H_6 (0.77 and 1.04 wt%, by cryoscopy) reaction with additional $t\text{-}C_4H_9CHO$ gives $Be(OCH_2C_4H_9\text{-}t)_2$ [13]
20	$Be(C_4H_9\text{-}i)OCH(C_4H_9\text{-}t)_2$ II [13]	colorless plates, m.p. 171 to 172° (from C_6H_{14}/C_6H_6 2:1) dimeric in C_6H_6 (1.07 and 1.47 wt%, by cryoscopy) no reaction with additional $(t\text{-}C_4H_9)_2CO$ observed [13]
21	$Be(C_4H_9\text{-}t)OCH_3$ I [7]	colorless crystals which soften at 155° and melt at 199° ^1H NMR(C_6H_6): 1.03(s, CH_3C), 3.41(s, CH_3O) IR(Nujol): 1266(m), 1210(m), 1182(w), 1099(s), 1033(s), 1004(s, sh), 934(w), 893(s), 800(s), 719(s), 690(s, sh), 505(w) sparingly soluble in C_6H_6, and tetrameric in this solvent (0.35 and 0.70 wt%, by cryoscopy) [7]

References on pp. 130/1

Table 25 (continued)

No.	compound method of preparation [Ref.]	properties and remarks [Ref.]
22	$Be(C_4H_9\text{-}t)OCH(CH_3)_2$ II [13]	colorless prisms, m.p. 95° (from C_6H_{14} at 0°) dimeric in C_6H_6 (0.76 and 1.01 wt%, by cryoscopy) [13]
23	$Be(C_4H_9\text{-}t)OC_4H_9\text{-}t$ I [7]	sublimes at 40 to 50°/10^{-3} Torr softens at 53°, appears to decompose at 108°, and melts to a brown liquid at 177 to 189° ^{1}H NMR (C_6H_6): 1.29 (s, C_4H_9Be), 1.42 (s, C_4H_9O) IR (Nujol): 1366 (m), 1252 (m), 1193 (m), 1033 (m), 1008 (m), 958 (s), 930 (m, sh), 917 (m), 843 (s), 813 (s), 800 (m, sh), 694 (w), 667 (w), 617 (w), 547 (m), 540 (w, sh) dimeric in C_6H_6 (0.49 to 1.36 wt%, by cryoscopy) reaction with 1 equivalent of pyridine gives a mixture of the 1:1 adduct (see Table 26, No. 8, p. 129) and $Be(C_4H_9\text{-}t)_2\cdot 2NC_5H_5$ [7]
24	$Be(C_4H_9\text{-}t)OCH_2C_4H_9\text{-}t$ II [13]	colorless needles, m.p. 213 to 215° (from C_6H_{14} at −10°) dimeric in C_6H_6 (0.68 and 0.91 wt%, by cryoscopy) [13]
25	$Be(C_4H_9\text{-}t)OCH(C_4H_9\text{-}t)_2$ II [13]	colorless prisms that shrink at ~60° and melt at 84 to 86° (from C_6H_{14}) dimeric in C_6H_6 (0.52 to 1.95 wt%, by cryoscopy) hydrolysis gives $(t\text{-}C_4H_9)_2CHOH$ reaction with another equivalent of $(t\text{-}C_4H_9)_2CO$ gives $Be(OCH(C_4H_9\text{-}t)_2)_2$ 1:1 adducts are obtained with quinuclidine and with 4-N,N-dimethylaminopyridine (see Table 26, Nos. 9 and 10, p. 129) no adduct is obtained with $N(CH_3)_2CH_2CH_2N(CH_3)_2$ [13]
26	$Be(C_4H_9\text{-}t)OCH(C_6H_5)_2$ II [13]	colorless prisms which soften at ~80° and melt at 101 to 103° with dec. (from C_6H_{14} at −20°) dimeric in C_6H_6 (0.58 and 0.85 wt%, by cryoscopy) stable in boiling $C_{10}H_{22}$ for 3 h hydrolysis gives $(C_6H_5)_2CHOH$ [13]
27	$Be(C_4H_9\text{-}t)OC(C_6H_5)_3$ I [13]	colorless prisms which soften at ~120° and melt at 141 to 143° (from C_6H_6) dimeric in C_6H_6 (0.60 and 0.85 wt%, by cryoscopy) [13]
*28	$Be(C_4H_9\text{-}t)OCH_2CH_2N(CH_3)_2$ I [9]	small colorless needles, m.p. 98 to 101° with dec. (from C_5H_{12}) dimeric in C_6H_6 (0.78 and 1.67 wt%, by cryoscopy) [9]
29	$Be(CH_2Si(CH_3)_3)O\text{-}C(C_6H_5)_2CH_2Si(CH_3)_3$ II [13]	colorless prisms, m.p. 118 to 120° (from C_6H_{14}) dimeric in C_6H_6 (1.38 and 1.83 wt%, by cryoscopy) [13]
*30	$Be(C{\equiv}CC_6H_5)OC_4H_9\text{-}t$ I [11]	IR: 2118 (w) [11]
31	$Be(C_6H_5)OCH_3$ I [6]	m.p. 177 to 178° tetrameric in C_6H_6 (0.93 and 1.85 wt%, by cryoscopy) [6]

Table 25 (continued)

No.	compound method of preparation [Ref.]	properties and remarks [Ref.]
32	$Be(C_6H_5)OC(C_6H_5)_3$ II [13]	colorless prisms (from $C_6H_5CH_3/C_6H_{14}$) which shrink at ~90° and melt at 162 to 165° with dec. dimeric in C_6H_6 (1.15 and 1.53 wt%, by cryoscopy) the solid is not hydrolyzed by 2N H_2SO_4 or 6N HNO_3; however, solutions in ether are readily hydrolyzed by 2N H_2SO_4 [13]

* Further information:

$Be(CH_3)OCH_3$ (Table **25**, No. **1**). Direct air oxidation of $Be(CH_3)_2$ at −78.5 or 25°C gives $Be(OCH_3)_2$. $Be(CH_3)OCH_3$ is discussed as an intermediate, but this was not proved [4].

The ^1H NMR spectrum shows no temperature dependence in $C_6D_5CD_3$. The signal for CH_3Be at −0.76 ppm remains unchanged at −90°C. Similarly, the doublet for CH_3O does not change at 33, 70, and 100°C. It is assumed that the CH_3O groups in the tetramer are nonequivalent [7]. The dimer is assumed to have Formula I, p. 119 (R = R′ = CH_3) [4, 9]. The compound disproportionates easily. Heating, particularly >120°C, gives a white solid of $Be(OCH_3)_2$ and a colorless liquid of $Be(CH_3)_2$. The mixture becomes homogeneous again on slow cooling due to reformation of $Be(CH_3)OCH_3$; but at 150°C, irreversible decomposition sets in with evolution of CH_4 [1]. $Be(CH_3)OCH_3$ reacts slowly with moist air and vigorously with H_2O with formation of CH_4 and $Be(OH)_2$ [1]. $Be(OCH_3)_2$ does not dissolve in a boiling solution of $Be(CH_3)OCH_3$ in C_6H_{14} [7].

With excess pyridine, the 1:2 adduct is formed (Table 26, No. 2, p. 128). With an equimolar amount of pyridine per Be, disproportionation to insoluble $Be(OCH_3)_2$ and $Be(CH_3)_2 \cdot 2NC_5H_5$ occurs. The 1:1 adduct (Table 26, No. 1) is an intermediate [7]. It is claimed in [8] that reaction with 2,2′-bipyridine gives $Be(CH_3)_2 \cdot N_2C_{10}H_8$ according to [7], but [7] does not contain such a statement.

$Be(CH_3)OC_9H_6N$ (Table **25**, No. **12**). It is proposed that the tetrameric compound has a structure which consists of two nonplanar dimeric units (Formula IV), parallel to each other, with coordination between each oxygen in one dimer and a Be atom in the other to give a tetramer containing three-coordinate oxygen and four-coordinate Be [8].

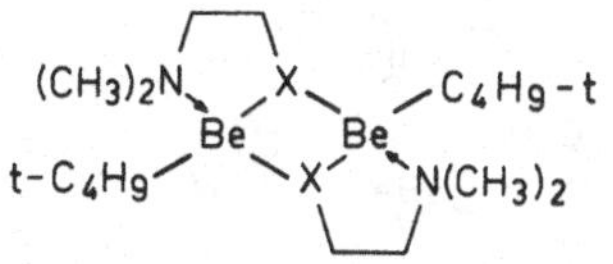

III

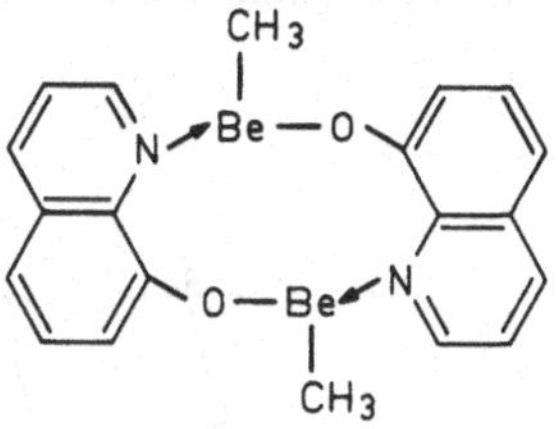

IV

$Be(CH_3)OSi(CH_3)_3$ (Table **25**, No. **13**) is prepared by heating a mixture of $Be(CH_3)_2$ and $(-Si(CH_3)_2O-)_4$ (mole ratio 5:1) in $C_6H_5CH_3$ under reflux for 40 h. The cooled solution is filtered, and the solvent is removed under reduced pressure [15]. The compound can also be prepared by reaction of $Be(CH_3)_2$ and $Si(CH_3)_3OH$, but no details are reported [10, 15].

References on pp. 130/1

Crystallization from petroleum ether gives a material, the structure of which was determined by single crystal X-ray diffraction. The compound crystallizes in the monoclinic space group C2/c-C^6_{2h} (No. 15). The structure, bond lengths, and bond angles are shown in **Fig. 12.** The data are based on ~900 reflections [10].

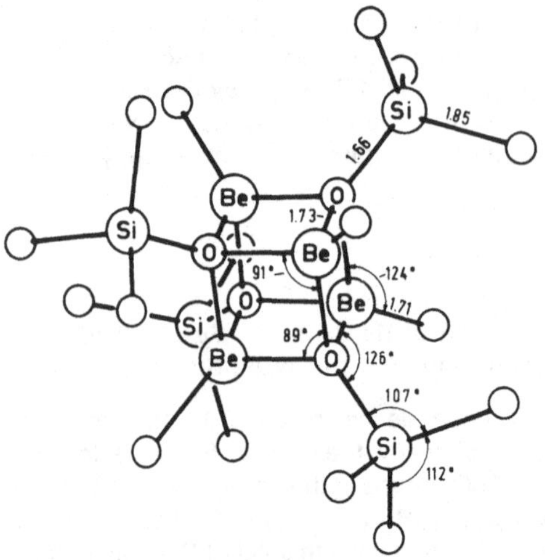

Fig. 12. Molecular structure of $Be(CH_3)OSi(CH_3)_3$ [10].

$Be(C_2H_5)OC_4H_9$-t (Table **25**, No. **14**) is formed in the reaction of $Be(C_2H_5)_2$ in ether with $LiOC_4H_9$-t (mole ratio 1:1) also in ether. The solution is stirred for 1 h, the ether is evaporated, and the residue is distilled at 36 to 39°C under a diffusion pump vacuum in a 36% yield.

The degree of association of the liquid alkoxide in C_6H_6 is between 2 and 3. The measured equilibrium constants of 0.12 and 0.16 mol/L for 1.06 and 1.48 wt% concentrations, respectively, indicate an equilibrium between tetramers and dimers. $Be(C_2H_5)OC_4H_9$-t reacts with an equimolar amount of $NaOC_4H_9$-t in ether to give $Na[BeC_2H_5(OC_4H_9\text{-t})_2]$ [14].

$Be(C_4H_9\text{-t})OCH_2CH_2N(CH_3)_2$ (Table **25**, No. **28**) ist proposed to have the structure shown in Formula III, p. 125 (X=O) [9].

$Be(C{\equiv}CC_6H_5)OC_4H_9$-t (Table **25**, No. **30**) is insoluble in organic solvents suggesting a polymeric constitution as shown in Formula V.

It is remarkably resistant to chemical attack, being insoluble and unaffected by H_2O, 2N H_2SO_4, methanolic H_2SO_4, ethanolic NaOH, and CH_3CO_2H. It dissolves in warm concentrated HNO_3 with decomposition [11].

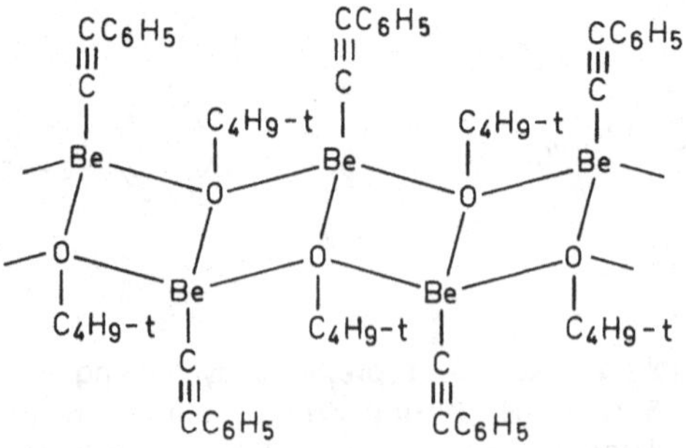

V

1.3.5.2 Be(R)OR′·n²D (n = 1,2) Type Compounds

General Remarks. The compounds Be(R)OR′ usually form 1:1 adducts with O- and N-donors. Only with pyridine was a 1:2 adduct also isolated (Table 26, No. 2, p. 128). When determined, the 1:1 adducts are found to exist as either monomers, dimers, or a mixture of monomers and dimers in solution. It is assumed that the compounds have the structures shown in Formulas I and II. The adducts are summarized in Table 26. They are prepared by the methods given below.

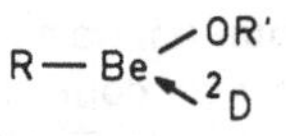

I

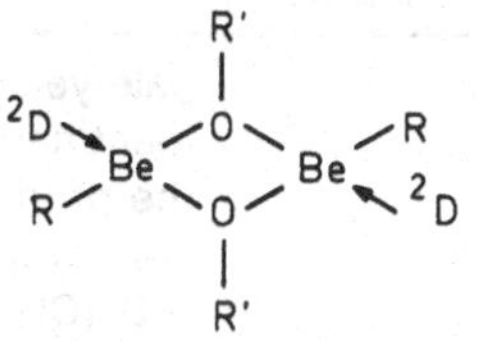

II

Method I: Reaction of the ligand with Be(R)OR′.
Pyridine in ether is added to Be(R)OR′ in ether (mole ratio 1:1). Crystals are formed (No. 8) when a small amount of solvent is evaporated (No. 3) or by evaporation of part of the solvent followed by addition of C_6H_{14} at −78°C (No. 1). Addition of an excess of pyridine to $Be(CH_3)OCH_3$ in C_6H_{14} results in slow deposition of No. 2 [7]. No. 10 is prepared in toluene, the mixture is stirred for 15 min, and toluene is evaporated [13].

Method II: Reaction of alcohol or ketone with $BeR_2 \cdot n^2D$ (n = 1,2).
The appropriate alcohol in ether is added to $BeR_2 \cdot n^2D$ in ether (mole ratio 1:1) at −196°C (No. 4 [5]), −50°C (No. 6 [7]), or at room temperature (No. 7 [7]). The compounds precipitate (No. 7) after warming to room temperature (Nos. 4, 6). No. 4 is extracted 5 times with ether. No. 13 is prepared by this method in $C_6H_5CH_3$/ether [7] or in C_6H_6/ether [6] at room temperature. The solvents are evaporated [6, 7]. No. 9 is obtained by addition of $(t\text{-}C_4H_9)_2CO$ in ether to $Be(CH_3)_2 \cdot NC_7H_{13}$ (NC_7H_{13} = quinuclidine) in ether at −78°C or to $Be(CH_3)_3 \cdot O(C_2H_5)_2$ and subsequent addition of NC_7H_{13}. The mixtures are warmed to room temperature and stirred for 15 or 30 min, respectively. The ether is evaporated [13]. For Nos. 11 and 12, the alcohol in C_6H_6 is added to $Be(C{\equiv}CC_6H_5)_2 \cdot OC_4H_8$ (OC_4H_8 = tetrahydrofuran) in C_6H_6. The product precipitates and is redissolved at 80°C and cooled again to give No. 11. Concentration and addition of C_6H_{14} gives No. 12 [11].

Method III: Reaction of a ketone with Be(R)H.
$(C_6H_5)_2CO$ in ether is added to a solution of $Be(CH_3)H$ in ether. No. 4 precipitates [5].

Method IV: Reaction of Be(R)OR′·²D with ²D′.
The ether adduct No. 4 is dissolved in warm ether and an excess of THF is added. The solvent volume is reduced by evaporation, and the mixture is heated to boiling. No. 5 separates upon cooling to −10°C [5].

References on pp. 130/1

Table 26

Be(R)OR'·²D and Be(R)OR'·2²D Compounds.

Abbreviations: NC_5H_5 = pyridine, NC_7H_{13} = quinuclidine (1-azabicyclo[2.2.2]octane), $NC_7H_{10}N$ = 4,4'-(dimethylamino)pyridine, OC_4H_8 = tetrahydrofuran. Concentrations given are calculated for the monomer. Other explanations, abbreviations, and units on p. X.

No.	compound method of preparation [Ref.]	properties and remarks [Ref.]
1	$Be(CH_3)OCH_3 \cdot NC_5H_5$ I [7]	pale yellow crystals (from C_6H_{14}/ether at −78°) identified only by elemental analysis the product does not dissolve completely in ether, but disproportionates to insoluble $Be(OCH_3)_2$ and soluble $Be(CH_3)_3 \cdot 2NC_5H_5$ [7]
2	$Be(CH_3)OCH_3 \cdot 2NC_5H_5$ I [7]	m.p. 112 to 114° (from C_6H_6/C_6H_{14}) IR(Nujol): 1608(m), 1572(w), 1259(m), 1212(m), 1179(m), 1151(m), 1095(s), 1069(s), 1040(s), 1014(s,sh), 876(m,sh), 794(s,br), 754(s), 700(s), 687(s,sh), 650(m,sh), 631(w,sh), 582(w) monomeric in C_6H_6 (0.60 and 1.21 wt%, by cryoscopy) [7]
3	$Be(CH_3)OC_4H_9\text{-t} \cdot NC_5H_5$ I [7]	m.p. 135 to 136° (from warm ether by addition of C_6H_{14}) ^1H NMR (C_6D_6): −0.26(s,CH_3Be), 1.18(s,CH_3C), 6.75 to 6.83(m,C_5H_5N) IR(Nujol): 1605(w), 1259(s), 1208(m), 1176(m), 1088(s), 1070(s), 1042(s), 1013(s), 977(m), 838(m), 798(s), 760(m), 732(m), 699(m), 649(w), 634(w) scarcely soluble in C_6H_{14} association in C_6H_6 (0.70 to 1.23 wt%, by cryoscopy) indicates an equilibrium between $(Be(CH_3)OC_4H_9\text{-t} \cdot NC_5H_5)_2$, $(Be(CH_3)OC_4H_9\text{-t})_4$, and C_5H_5N C_5H_5N is lost when C_6H_6 solutions are evaporated to dryness at room temperature [7]
4	$Be(CH_3)OCH$-$(C_6H_5)_2 \cdot O(C_2H_5)_2$ II, III [5]	prisms, m.p. 58 to 68° ^1H NMR(C_6H_6): −0.82 (s,CH_3Be) monomeric in C_6H_6 (0.38 to 0.86 wt%, by cryoscopy) removal of ether in a vacuum results in dec. reaction with THF gives No. 5 [5]
5	$Be(CH_3)OCH(C_6H_5)_2 \cdot OC_4H_8$ IV [5]	colorless prisms, m.p. 145° monomeric in C_6H_6 (0.40 to 0.57 wt%, by cryoscopy) [5]
6	$Be(CH_3)OC(C_6H_5)_3 \cdot O(C_2H_5)_2$ II [7]	softens at 82 to 84° (from C_6H_6), the m.p. at 182 to 183° belongs to the ether-free compound ^1H NMR(C_6D_6): −1.53(s,CH_3Be), 1.08(t,CH_3C), 3.30(q,CH_2O), 6.84 to 7.46(m,C_6H_5) IR(Nujol): 1475(m), 1264(s), 1201(w,sh), 1152(w,sh), 1093(s), 1029(s,sh), 1018(s), 869(w,br), 816(m,sh), 800(s), 758(m), 699(m), 666(w), 638(w) monomeric in C_6H_6 (1.15 and 1.50 wt%, by cryoscopy) [7]

Table 26 (continued)

No.	compound method of preparation [Ref.]	properties and remarks [Ref.]
7	$Be(CH_3)OC_6H_5 \cdot O(C_2H_5)_2$ II [7]	needles which soften at 114° (from ether/C_6H_6) dec. at ~160° into a colorless sublimate and a residue which does not melt below 360° ^1H NMR(C_6D_6): −0.48(s, CH_3Be), 0.84(t, CH_3C), 3.52(q, CH_2), 7.24(t, C_6H_5) IR(Nujol): 1597(m), 1263(s), 1193(m), 1153(m), 1094(m), 1075(m), 1038(s), 1023(s), 1002(m), 908(m), 867(s, br), 826(m), 800(s), 765(s), 694(s), 597(w), 509(m, br), 467(w, br) sparingly soluble in ether even at 35°; association in C_6H_6 (0.59 and 1.19 wt%, by cryoscopy) indicates an equilibrium between a monomeric and dimeric adduct and $(Be(CH_3)OC_6H_5)_4$ some ether is lost by evaporation of C_6H_6 to dryness [7]
8	$Be(C_4H_9\text{-}t)OC_4H_9\text{-}t \cdot NC_5H_5$ I [7]	pale yellow crystalline product whose composition is intermediate between those required for No. 8 and for $Be(C_4H_9\text{-}t)_2 \cdot 2NC_5H_5$ ^1H NMR (C_6H_6): 1.28(C_4H_9Be), 1.56(C_4H_9O) in the ratio 4:1 [7]
9	$Be(C_4H_9\text{-}t)OCH\text{-}(C_4H_9\text{-}t)_2 \cdot NC_7H_{13}$ II [3]	colorless needles, m.p. 67 to 69°, 66 to 68° (from C_6H_{14} at −10°) monomeric in C_6H_6 (1.09 and 1.44 wt%, by cryoscopy) hydrolysis gives $(t\text{-}C_4H_9)_2CHOH$ [13]
10	$Be(C_4H_9\text{-}t)OCH\text{-}(C_4H_9\text{-}t)_2 \cdot NC_7H_{10}N$ I [13]	colorless plates, m.p. 110 to 111° (from C_6H_{14}) monomeric in C_6H_6 (1.01 and 1.34 wt%, by cryoscopy) [13]
11	$Be(C{\equiv}CC_6H_5)\text{-}OC_4H_9\text{-}t \cdot OC_4H_8$ II [11]	colorless needles which do not melt when heated to 300° degree of association between 1 and 2 in C_6H_6 (0.60 and 0.91 wt%, by cryoscopy) indicates an equilibrium between the monomeric and dimeric adduct [11]
12	$Be(C{\equiv}CC_6H_5)OC_6H_5 \cdot OC_4H_8$ II [11]	needles, m.p. 203 to 206° with dec. (from C_6H_6/C_6H_{14}) degree of association as for the previous compound (1.04 and 1.54 wt%) [11]
13	$Be(C_6H_5)OCH_3 \cdot O(C_2H_5)_2$ II [6, 7]	m.p. 53 to 55° (from $C_6H_5CH_3$/C_6H_{14} at −78°) [6] identified by elemental analysis monomeric in C_6H_6 (1.27 wt%, by cryoscopy) [7] the product obtained in [6] was assumed to have an indefinite ether content

References on pp. 130/1

1.3.5.3 $Be_3R_2(OC_4H_9\text{-}t)_4$ Type Compounds

Only three compounds of this type have been prepared. They are assumed to have a structure as shown in Formula I:

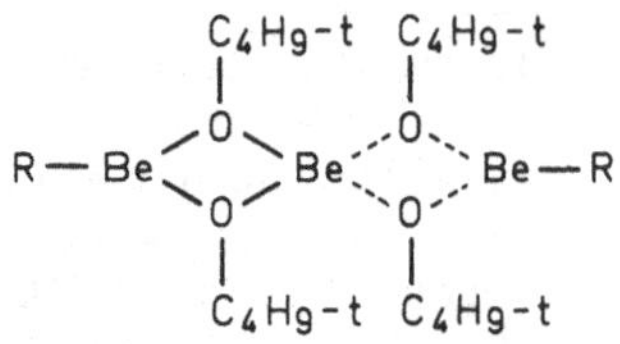

I

$Be_3(CH_3)_2(OC_4H_9\text{-}t)_4$

For the preparation, $t\text{-}C_4H_9OH$ in C_6H_6 is slowly added to a stirred suspension of $Be(CH_3)_2$ in C_6H_6 in the mole ratio 4:3. Methane is evolved. The solvent is evaporated, and the compound crystallizes as colorless plates from C_6H_{14} in nearly quantitative yield, m.p. 188°C.

^1H NMR(C_6D_6): $\delta = -0.41$ (s, CH_3Be) and 1.40 ppm (s, CH_3C). The compound is only sparingly soluble in C_6H_6 and C_6H_{14}. It is monomeric in C_6H_6 (by cryoscopy). Reactions with acids like HN_3, HCN, and HNO_3 give the corresponding compounds $Be_3X_2(OC_4H_9\text{-}t)_4$ with $X = N_3$, CN, or NO_3. No reaction is observed with NH_3, $NH(CH_3)_2$, CH_3SH, $t\text{-}C_4H_9SH$, or $N(CH_3)_2CH_2CH_2N(CH_3)_2$. With CH_3OH, both $Be(OCH_3)_2$ and $Be(OC_4H_9\text{-}t)_2$ are obtained. With $t\text{-}C_4H_9OH$, the products are $Be(OC_4H_9\text{-}t)_2$ and CH_4 [12].

$Be_3(C_4H_9\text{-}t)_2(OC_4H_9\text{-}t)_4$

For the preparation, $LiC_4H_9\text{-}t$ in C_5H_{12} is added to $Be_3Cl_2(OC_4H_9\text{-}t)_4$ in C_6H_6 (mole ratio 2:1). After 1.5 h the precipitated LiCl is separated, and the solvent is removed from the filtrate, leaving a pale yellow oil which solidifies on addition of a little C_6H_{14}. It crystallizes from C_6H_{14} at −78°C, m.p. 98 to 99°C. The yield is not quantitative.

^1H NMR(C_6H_6): $\delta = 1.38$ (s, C_4H_9Be) and 1.48 ppm (s, C_4H_9O). The compound is very soluble in C_6H_6 and C_6H_{14} and dissolves in C_6H_6 as a monomer (by cryoscopy) [12].

$Be_3(C{\equiv}CC_6H_5)_2(OC_4H_9\text{-}t)_4$

The compound is best prepared from $Be(C{\equiv}CC_6H_5)_2 \cdot O(C_2H_5)_2$ (obtained from $C_6H_5C{\equiv}CH$, $LiC_4H_9\text{-}n$, and $BeCl_2$ in the mole ratio 2:2:1 in C_6H_6/ether, and separation of LiCl) and $Be(OC_4H_9\text{-}t)_2$ (mole ratio 1:2) in C_6H_6/ether. The solvent is removed. Another method is the addition of $t\text{-}C_4H_9OH$ in C_6H_6 to $Be(C{\equiv}CC_6H_5)_2 \cdot N(C_2H_5)_3$ in C_6H_6 (mole ratio 4:3). A precipitate is formed.

The product crystallizes as colorless needles from C_6H_{14}, m.p. 210 or 212°C. It dissolves as a monomer in C_6H_6 (by cryoscopy) [11].

References:

[1] Coates, G. E., Glockling, F., Huck, N. D. (J. Chem. Soc **1952** 4512/5).
[2] Coates, G. E., Glockling, F. (J. Chem. Soc. **1954** 22/7).
[3] Funk, H., Masthoff, R. (J. Prakt. Chem. [4] **22** [1963] 250/4).
[4] Masthoff, R. (Z. Anorg. Allgem. Chem. **336** [1956] 252/8).
[5] Bell, N. A., Coates, G. E. (J. Chem. Soc. A **1966** 1069/73).
[6] Coates, G. E., Tranah, M. (J. Chem. Soc. A **1967** 236/9).

[7] Coates, G. E., Fishwick, A. H. (J. Chem. Soc. A **1968** 477/83).
[8] Coates, G. E., Fishwick, A. H. (J. Chem. Soc. A **1968** 640/2).
[9] Coates, G. E., Robert, P. D. (J. Chem. Soc. A **1968** 2651/5).
[10] Mootz, D., Zinnius, A., Böttcher, B. (Angew. Chem. **81** [1969] 398/9; Angew. Chem. Intern. Ed. Engl. **8** [1969] 378/9).

[11] Coates, G. E., Francis, B. R. (J. Chem. Soc. A **1971** 160/4).
[12] Andersen, R. A., Bell, N. A., Coates, G. E. (J. Chem. Soc. Dalton Trans. **1972** 577/82).
[13] Andersen, R. A., Coates, G. E. (J. Chem. Soc. Dalton Trans. **1974** 1171/80).
[14] Andersen, R. A., Coates, G. E. (J. Chem. Soc. Dalton Trans. **1974** 1729/36).
[15] Bell, N. A., Coates, G. E., Fishwick, A. H. (J. Organometal. Chem. **198** [1980] 113/20).

1.3.6 Organoberyllium Oximates

Only one compound of the type Be(R)ON=CR′$_2$ is described in the literature:

$Be(CH_3)ON{=}C(CH_3)_2$

The compound is prepared by addition of an ether solution of $(CH_3)_2C{=}NOH$ to $Be(CH_3)_2$ in ether at ~0°C. To avoid side reactions, a slight excess of acetoxime has to be used. Methane is evolved, and filtration, evaporation to dryness, and crystallization from $C_6H_5CH_3/C_6H_{14}$ (1:3) give the compound as small plates, which decompose from ~180 to 230°C to a brown mass.

The ^1H NMR spectrum in C_6H_6 shows a triplet for CH_3Be at $\delta = 1.92$, 2.12, and 2.24 ppm, and a multiplet for CH_3C at 4.60 to 4.78 ppm. The compound is tetrameric in C_6H_6 (by cryoscopy). In analogy to the corresponding Zn compound, a cage-like structure, as shown in Formula I, is proposed. It is insoluble in c-$C_6D_{11}CD_3$ and decomposes in CCl_4.

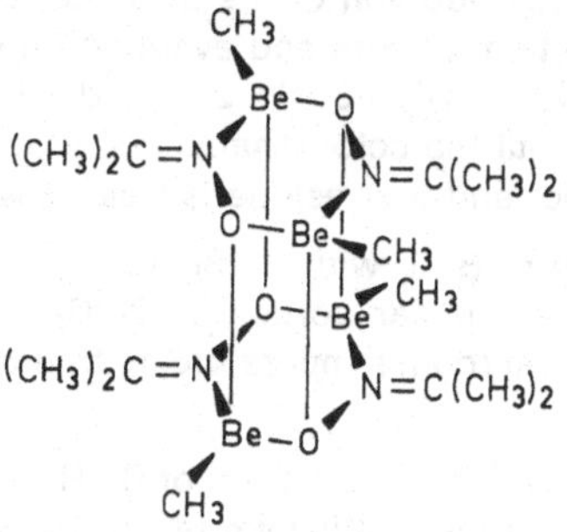

I

Reference:

Coates, G. E., Fishwick, A. H. (J. Chem. Soc. A **1968** 640/2).

1.3.7 Organoberyllium Thiolates, Selenolates, and Their Adducts

General Remarks. The parent compounds of the type Be(R)SR′ are described in Table 27, pp. 133/4, as Nos. 1 to 10. Nos. 11 to 20 in Table 27, pp. 134/5, are adducts of the thiolates with O- and N-donors. Only two selenolates (Nos. 21 and 22, p. 135) have been isolated as adducts.

The compounds that were found to be tetrameric in solutions are assumed to have a cubic structure (Formula I). They generally have a higher tendency to disproportionate than the Be(R)OR′ type compounds, and they vary greatly in their reactions with bases. With an increase in steric congestion (R and/or R′ = i-C_3H_7 or t-C_4H_9), the isolation of adducts is eased. The 1:1 adducts are dimeric in solution, and the structure shown in Formula II is suggested [2].

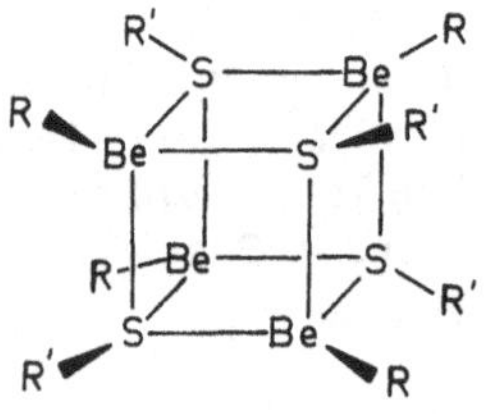

I

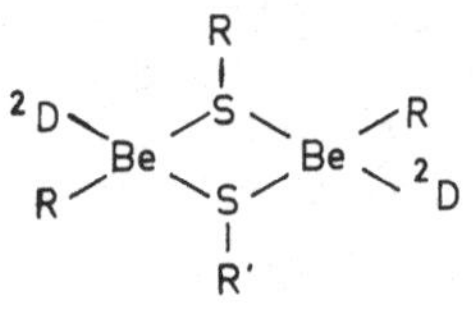

II

The compounds in Table 27 are prepared by the following methods:

Method I: Reaction of BeR_2 with R′SH or R′SeOH in a 1:1 mole ratio.
The appropriate thiol or selenophenol (No. 22) is added to BeR_2, dissolved in ether (Nos. 3 [2], 4 [3], 7, 9, 11, 12, 17, 22 [2]), C_6H_{14} (Nos. 6 to 8 [2]), or C_6H_{14}/ether (No. 10 [4]), at −196°C (Nos. 6, 17), ~−80°C (Nos. 10, 11), ~−50°C (Nos. 3, 4, 9), −40°C (Nos. 8, 22), or ~−20°C (Nos. 7,12). The mixtures are allowed to warm, whereby the product precipitates (No. 4) after boiling for a few minutes (No. 10). Nos. 3, 6 to 9, 11, 12, and 17 remain after evaporation of the solvent. The residues are left in a vacuum for 12 h (No. 3) or are boiled with C_6H_{14} (Nos. 11, 22). No. 20 [5] is prepared from t-C_4H_9SH and Be(C≡CC_6H_5)$_2$·2OC_4H_8 in C_6H_6 by heating for 1 h at 50°C in a closed vessel. The solvent is evaporated.

Method II: Reaction of R′SSR′ or R′SeSeR′ with BeR_2 in a 1:1 mole ratio.
No. 6 is prepared by addition of the disulfide to Be(C_2H_5)$_2$ in ether at −196°C. Warming to room temperature and evaporation of the volatiles gives a residue which is washed with C_5H_{12}. For No. 21, the diselenide and Be(C_2H_5)$_2$ are allowed to react in ether until the color disappears. Excess pyridine is then added. The ether is evaporated, and the residue is treated with boiling C_6H_{14} [2].

Method III: Exchange reaction of BeR_2 with Be(SR′)$_2$.
Slightly less than a 1 molar portion of Be(C_2H_5)$_2$ is added to a suspension of Be(SC_2H_5)$_2$ in ether at room temperature. Filtration and evaporation of the solvent leaves No. 6 [2].

Method IV: The donor ligand is added to Be(R)SR′ or Be(R)SR′·^{2}D. The ligand in excess (Nos. 13, 15, 19), or in a 1:1 mole ratio (No. 14, 16, 18), is added to the organoberyllium thiolate (Be(C_2H_5)SC_4H_9-t·OC_4H_8 for No. 16 and Be(C_3H_7-i)SC_2H_5·O(C_2H_5)$_2$ for No. 18) in C_6H_{14} (Nos. 13, 15, 19) or in ether (Nos. 14, 16, 18). The adduct precipitates (No. 16) after 5 min (Nos. 15, 19), upon concentration (No. 13), or by evaporation of the solvent (Nos. 14, 18) [2].

Table 27

Be(R)ER′ Type Compounds and Their Adducts (E = S, Se).

Further information on numbers preceded by an asterisk is given at the end of the table. Abbreviations: OC_4H_8 = tetrahydrofuran, NC_5H_5 = pyridine, $N_2C_{10}H_8$ = 2,2′-bipyridine. Other explanations, abbreviations, and units on p. X.

No.	compound method of preparation [Ref.]	properties and remarks [Ref.]
*1	$Be(CH_3)SCH_3$ see further information [1]	—
*2	$Be(CH_3)SC_3H_7$-i see further information [2]	—
3	$Be(CH_3)SC_4H_9$-t I [2]	dec. at 142° (from C_6H_6) to give some $Be(CH_3)_2$ ^1H NMR(C_6H_6): −0.23(s, CH_3Be), 1.54(s, CH_3C) tetrameric in C_6H_6 (by cryoscopy) $Be(CH_3)_2$ is slowly deposited from C_6H_6 solutions [2]
*4	$Be(CH_3)SCH_2CH_2N(CH_3)_2$ I [3]	dec. from 194° (washed with ether) ^1H NMR(C_6H_6): 2.11(s, CH_3Be), 4.68 and 4.75 (unequal d, ratio 2:1, CH_3N) trimeric in C_6H_6 (by cryoscopy) readily soluble in C_6H_6 [3]
*5	$Be(C_2H_5)SCH_3$ see further information [2]	—
6	$Be(C_2H_5)SC_2H_5$ I, II, III [2]	white, m.p. 75 to 77° (washed with cold C_5H_{12}) ^1H NMR(C_6H_5): 0.29(q, CH_2Be), 1.04, 1.72(CH_3), 2.74(q, CH_2S) tetrameric in C_6H_6 (by cryoscopy) stable towards disproportionation no adduct could be isolated with ether or pyridine, but complex formation with pyridine in solution is probable [2]
7	$Be(C_2H_5)SC_3H_7$-i I [2]	white, m.p. 58 to 60° (washed with cold C_5H_{12}) ^1H NMR(C_6H_6): 0.44(q, CH_2Be), 1.49(d, CH_3CS), 3.54(quint, CH), CH_3CBe is obscured tetrameric in C_6H_6 (by cryoscopy) stable towards disproportionation no adduct could be isolated with pyridine, but some complex formation in solution is probable exists as a dimeric ether adduct in dilute solution (by ebullioscopy) [2]
8	$Be(C_2H_5)SC_4H_9$-t I [2]	white solid (washed with C_5H_{12}), softens at 65° due to formation of $Be(C_2H_5)_2$ ^1H NMR(C_6H_6): 0.32(q, CH_2), 1.56(s, C_4H_9), CH_3CBe is obscured tetrameric in C_6H_6 (by cryoscopy) solution in C_6H_{14} deposits $Be(SC_4H_9\text{-t})_2$ when warmed to 50° adducts are formed with THF, pyridine, and 2,2′-bipyridine (see Nos. 13 to 16) [2]

References on p. 136

Table 27 (continued)

No.	compound method of preparation [Ref.]	properties and remarks [Ref.]
9	$Be(C_3H_7\text{-}i)SC_3H_7\text{-}i$ I [2]	viscous liquid which could not be crystallized from C_6H_{14} even at −60° gives No. 19 upon addition of pyridine [2]
10	$Be(C_4H_9\text{-}t)SCH_2CH_2N(CH_3)_2$ I [4]	white crystals, m.p. 99 to 101° (from $C_6H_{14}/C_6H_5CH_3$ 3:1) dimeric in C_6H_6 (by cryoscopy) structure as shown in Section 1.3.5.1, Formula III, p. 135, is proposed ($R = t\text{-}C_4H_9$) [4]
adducts of Be(R)ER′ (E = S, Se)		
11	$Be(CH_3)SC_6H_5 \cdot O(C_2H_5)_2$ I [2]	white crystals, dec. at 66 to 96° (from $C_6H_5CH_3/C_6H_{14}$ 1:2) ^1H NMR (C_6H_6): −0.38 (s, CH_3Be) and C_6H_5 as expected dimeric in C_6H_6 (by cryoscopy) no significant disproportionation in C_6H_6 [2]
12	$Be(C_2H_5)SC_3H_7\text{-}i \cdot O(C_2H_5)_2$ I [2]	not isolated identified in dilute ether solutions (0.046 to 0.101 mol/L) by ebullioscopy as a dimer evaporation of the solvent gives the ether-free compound No. 7 [2]
*13	$Be(C_2H_5)SC_4H_9\text{-}t \cdot OC_4H_8$ IV [2]	crystals, m.p. 88° (from C_5H_{12}/THF) ^1H NMR (C_6H_6): 0.32 (q, CH_2Be), 1.64 (s, C_4H_9) gives with 2,2′-bipyridine No. 16 [2]
*14	$Be(C_2H_5)SC_4H_9\text{-}t \cdot NC_5H_5$ IV [2]	white crystals, m.p. 99 to 107° with dec. (from $C_6H_5CH_3/C_6H_{14}$ 1:2 at ~ −30°) [2]
15	$Be(C_2H_5)SC_4H_9\text{-}t \cdot 2NC_5H_5$ IV [2]	lemon-colored needles, m. p. 129 to 130° (from boiling $C_6H_5CH_3/C_6H_{14}$ 2:1) monomeric in C_6H_6 (by cryoscopy) turns white on exposure to air [2]
16	$Be(C_2H_5)SC_4H_9\text{-}t \cdot N_2C_{10}H_8$ IV [2]	orange-red plates, m. p. 133 to 135° (dec.) is completely dec. (turns white) after exposure to air for 3 h [2]
17	$Be(C_3H_7\text{-}i)SC_2H_5 \cdot O(C_2H_5)_2$ I [2]	colorless crystals (from C_6H_{14}), dec. from ca. 160° with loss of ether dimeric in C_6H_6 (by cryoscopy) gives with pyridine No. 18 [2]
18	$Be(C_3H_7\text{-}i)SC_2H_5 \cdot NC_5H_5$ IV [2]	pale yellow plates, m.p. 96 to 98° (recrystallized from ?) dimeric in C_6H_6 [2]
19	$Be(C_3H_7\text{-}i)SC_3H_7\text{-}i \cdot 2\ NC_5H_5$ IV [2]	bright yellow needles, m.p. 75 to 81° with dec. (from C_6H_{14}) monomeric in C_6H_6 (by cryoscopy) [2]

Table 27 (continued)

No.	compound method of preparation [Ref.]	properties and remarks [Ref.]
20	$Be(C{\equiv}CC_6H_5)SC_4H_9$-t$\cdot OC_4H_8$ I [5]	colorless needles, m.p. 150 to 155° with dec. (from C_6H_{14}) ^1H NMR(C_6D_6): 2.09(s, C_4H_9) and the resonances for C_4H_8O and C_6H_5 as expected dimeric in C_6H_6 (by cryoscopy) [5]
21	$Be(C_2H_5)SeC_2H_5\cdot 2NC_5H_5$ II [2]	pale yellow crystals, m.p. 56 to 57° (from $C_6H_5CH_3/C_6H_{14}$ by stirring for 30 min) monomeric in C_6H_6 (by cryoscopy) [2]
22	$Be(C_2H_5)SeC_6H_5\cdot O(C_2H_5)_2$ I [2]	white solid, m.p. 150 to 152° (dec.) dimeric in C_6H_6 (by cryoscopy) exposure to air gives $Se_2(C_6H_5)_2$ heating to 70° in a vacuum gives $Be(SeC_6H_5)_2$ [2]

* Further information:

$Be(CH_3)SCH_3$ (Table **27**, No. **1**) is expected to be formed by addition of CH_3SH to $Be(CH_3)_2$ (1:1 mole ratio) at −183°C. The mixture is allowed to warm to room temperature. At −10°C, CH_4 begins to evolve. A white solid residue is formed which does not sublime at temperatures up to 100°C. Above this temperature, $Be(CH_3)_2$ is evolved. The compound was not characterized [1].

$Be(CH_3)SC_3H_7$-i (Table **27**, No. **2**). Propane-2-thiol in C_6H_6 is added to $Be(CH_3)_2$ in C_6H_6 (mole ratio 1:1) at room temperature. Filtration and evaporation of the solvent leaves a semi-solid mass which is only partially soluble in C_6H_6. The ^1H NMR spectrum of the filtrate had a $CH_3C:CH_3Be$ ratio of ~3:1 that increased to 4.4:1 with deposition of a solid after 16 h. It is assumed that the solutions must have contained $Be(CH_3)SC_3H_7$-i. The precipitate consisted largely of $Be(CH_3)_2$ [2].

$Be(CH_3)SCH_2CH_2N(CH_3)_2$ (Table **27**, No. **4**) is assigned the structure in solution shown in Formula III (R = CH_3) [3].

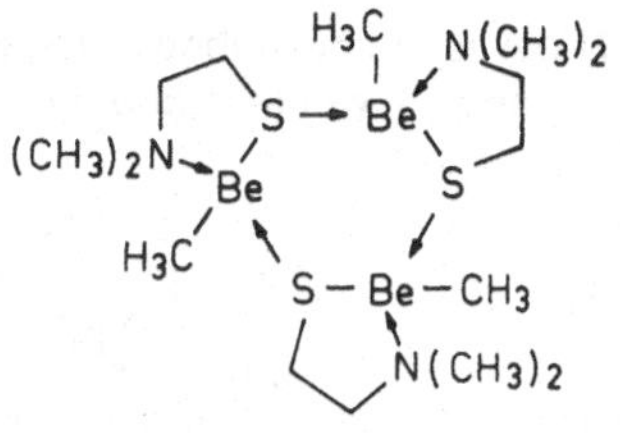

III

$Be(C_2H_5)SCH_3$ (Table **27**, No. **5**) was prepared by addition of CH_3SH to $Be(C_2H_5)_2$ (1:1 mole ratio) in C_6H_{14} at −196°C. A white precipitate was formed upon warming to room temperature. Analysis indicates a deficiency of C_2H_5Be with respect to $Be(C_2H_5)SCH_3$. The product was not further characterized [2].

$Be(C_2H_5)SC_4H_9$-t$\cdot$²D (²D = OC_4H_8, NC_5H_5, Table **27**, Nos. **13**, **14**). For No. 13, the degree of association in C_6H_6 was found to be a little over 1 in C_6H_6 (by cryoscopy), and for No. 14 a little less than 2 was found. The data are consistent with equilibria between the dimeric and monomeric adducts and the tetrameric parent compound: ½ $(Be(C_2H_5)SC_4H_9$-t$\cdot$²D$)_2$

References on p. 136

$\rightleftharpoons Be(C_2H_5)SC_4H_9\text{-t} \cdot {}^2D \rightleftharpoons \frac{1}{4}(Be(C_2H_5)SC_4H_9\text{-t})_4 + {}^2D$. The equilibria are shifted to the right with ${}^2D = OC_4H_8$ and to the left with ${}^2D = NC_5H_5$ [2].

References:

[1] Coates, G. E., Glocking, F., Huck, N. D. (J. Chem. Soc. **1952** 4512/5).
[2] Coates, G. E., Fishwick, A. H. (J. Chem. Soc. A **1968** 635/40).
[3] Coates, G. E., Fishwick, A. H. (J. Chem. Soc. A **1968** 640/2).
[4] Coates, G. E., Roberts, P. D. (J. Chem. Soc. A **1968** 2651/5).
[5] Coates, G. E., Francis, B. R. (J. Chem. Soc. A **1971** 160/4).

1.3.8 Organoberyllium Amides and Their Adducts

General Remarks. In this chapter compounds of the type $Be(R)NR'_2$ and their adducts are described. The degree of association of the parent compounds varies from dimeric and trimeric to polymeric, depending mainly on the steric crowding of the groups R and R′. Polymers (Formula I) are obtained with $NR'_2 = NH_2$ or $NHCH_3$ and $R = CH_3$ or C_2H_5. Small groups R and R′ result in trimers (Formula II), whereas large groups give dimers (Formula III); see the individual entries.

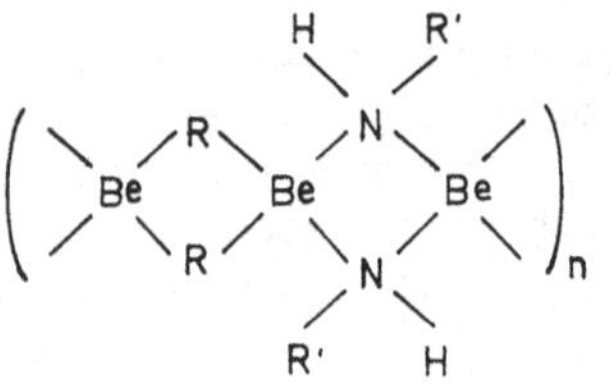

I

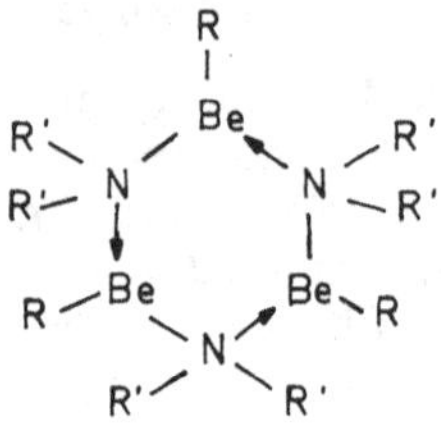

II

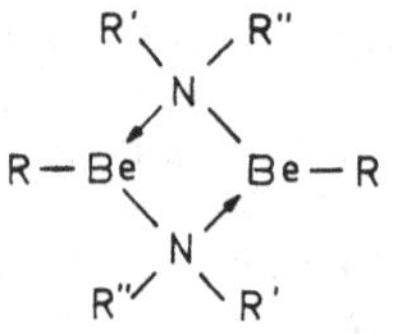

III

Only 1:1 or 1:2 adducts with N-bases are described in the literature. The 1:1 adducts with ${}^2D$ donors are dimeric (Formulas IV and V), whereas adducts with ${}^2D$-${}^2D$ and 1:2 adducts are monomeric.

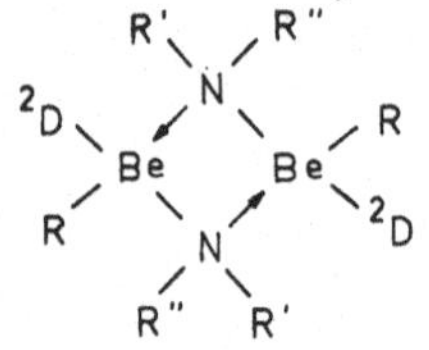

IV

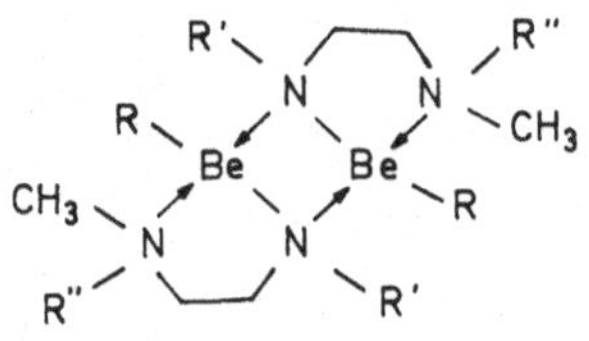

V

At the end of the chapter two amides with a special composition are described, these are a radical cation $[C_6H_5BeNC_5H_4{=}C_5H_4NBeC_6H_5]^{\bullet+}$ and $Be_3(CH_3)_2(N(CH_3)_2)_4$. The two compounds of the composition $Be(R)NC_5H_6 \cdot 2NC_5H_5$ ($R = CH_3$, C_2H_5) are described in Section 4.2, p. 201.

Most of the compounds are prepared by reaction of NHR'_2 with BeR_2. In some cases special methods are applied. The adducts are obtained from the parent compounds by addition of the donor. Details are given with the individual compounds.

$Be(CH_3)NH_2$

The compound is prepared by condensing NH_3 onto a solution of $Be(CH_3)_2$ in ether at −78°C. The mixture is slowly warmed. At −33.5°C, the excess of NH_3 is evaporated, and the temperature is slowly raised to 0 and then to 10°C.

The colorless product is insoluble in ether and has been assigned the polymeric structure shown in Formula I ($R=CH_3$, $R'=H$). Heating to 50°C in a high vacuum leads to liberation of CH_4 and formation of a polymer of the composition $CH_3Be(NHBe)_nNH_2$. The same type of polymer is formed by reaction with an excess of NH_3 [5].

$Be(CH_3)NHCH_3$

The compound is obtained by condensing NH_2CH_3 onto $Be(CH_3)_2$ in a 1:1 mole ratio at −183°C. The mixture is warmed and kept at 40°C until the pressure is constant. One equivalent of CH_4 is pumped off from each mole of reactants. The nonvolatile residue is thus considered to be $(Be(CH_3)NHCH_3)_n$, but it was not further characterized. Further heating induces decomposition which becomes rapid at 110°C and is complete at 200°C; 1.47 mol CH_4 per mole of $Be(CH_3)_2$ is evolved. The residue is formulated as a mixture of the title compound and $Be{=}NCH_3$ [1].

$Be(CH_3)N(CH_3)_2$

$Be(CH_3)_2$ and excess of $NH(CH_3)_2$ form an adduct at −183°C which decomposes to give CH_4 on warming to room temperature. After a few hours at 40 to 55°C, a white crystalline solid is obtained which melts at 55 to 56°C and can be distilled in a high vacuum [1] or sublimed at 50 to 80°C under reduced pressure [6]. The sublimed material melts at 51 to 54°C.

The vapor pressure between 60 and 100°C is given by the equation $\log p = 8.605-2889/T$ (p in Torr, T in K). The latent heat of vaporization is 13.2 kcal/mol. Vapor density measurements at 180 and 190°C gave degrees of association of 3.13 and 3.04, respectively. The molecular mass in C_6H_6, determined by ebullioscopy, also corresponds to a trimer. The cyclic structure shown in Formula II ($R=R'=CH_3$) is suggested [1]. The sublimed material is isotropic when examined with polarized light and an X-ray diffraction gives no lines. It is therefore amorphous, and a polymeric structure may be present in the condensed phase [6].

The ^1H NMR spectrum in c-C_6H_{12} shows resonances at $\delta=-0.87$ (s, $BeCH_3$) and 1.06 ppm (s, CH_3N). The spectrum is unchanged at −90°C in $C_6D_5CD_3$. IR absorptions (in c-C_6H_{12}) are observed at 1479, 1457, 1434 (δCH_3N); 1228, 1169 (ϱCH_3N); 1116, 1033 (νCN); 1206 (δCH_3Be); 920, 894, 841, 816, and 867 cm^{-1} (skeletal vibrations) [6].

The compound fumes in air and is rapidly hydrolyzed by H_2O. Prolonged heating at 180°C yields CH_4 and traces of H_2. With HCl at 140°C, CH_4 is formed. No reaction is observed with $N(CH_3)_3$ between −20 and 50°C [1]. A 1:1 and 1:2 adduct is formed with pyridine (see p. 138), whereas 2,2′-bipyridine gives $Be(CH_3)_2$ [9].

The deuterated analog $Be(CH_3)N(CD_3)_2$ is prepared by condensing $NH(CD_3)_2$ onto $Be(CH_3)_2$ in a 1:1 mole ratio at −196°C. The mixture is warmed to room temperature. When the gas evolution is finished, the volatile matter is removed under reduced pressure, and the product is separated by sublimation. It softens at 50 to 53°C. The IR absorptions in Nujol are at 1205 (δCH_3), 1101, 1086, 1064 (δCD_3); 1064, 966 (ϱCD_3); 1117, 1029 (νCN); 876, 849, 835, 818, 789, 682, and 667 cm^{-1} (skeletal vibrations) [6].

$Be(CD_3)N(CH_3)_2$ is prepared like $Be(CH_3)N(CD_3)_2$ from $Be(CD_3)_2$ and $NH(CH_3)_2$. It softens at 48 to 54°C. The IR absorptions in Nujol are observed at 1475, 1466, 1445 (δCH_3N); 1233,

References on pp. 145/6

1175 (CH_3N); 1117, 1037 (CN), 990 (CD_3); 920, 894, 841, 816, and 687 cm^{-1} (skeletal vibrations) [6].

$Be(CH_3)N(CH_3)_2 \cdot NC_5H_5$ (NC_5H_5 = pyridine)

An equimolar amount of pyridine is added to $Be(CH_3)N(CH_3)_2$ in ether. Colorless crystals are formed after 10 min at room temperature and are recrystallized from ether.

IR absorptions (in Nujol) are observed at 1585 (m), 1550 (w), 1387 (w), 1035 (s), 1006 (m), 936 (s), 909 (m), 858 (m), 833 (m, sh), 784 (s, br), 749 (s), 692 (s), 662 (m), and 646 (w) cm^{-1}. Cryoscopic measurements in C_6H_6 indicate the presence of dimers; see Formula IV, p. 136 (R, R', R'' = CH_3, ^2D = NC_5H_5). The compound decomposes at 225°C with evolution of pyridine [9].

$Be(CH_3)N(CH_3)_2 \cdot 2NC_5H_5$ (NC_5H_5 = pyridine)

An excess of pyridine is added to $Be(CH_3)N(CH_3)_2$ in ether. The solvent is removed under reduced pressure, and the pale cream-colored adduct crystallizes from C_5H_{12} at ~ −50°C.

The IR spectrum (in Nujol) contains absorptions at 1585 (m), 1550 (w), 1387 (w), 1245 (w), 1225 (m), 1193 (w), 1157 (m), 1129 (s), 1101 (s), 1057 (s), 1035 (s), 1006 (m), 936 (s), 909 (m), 858 (m), 833 (m, sh), 784 (s, br), 749 (s), 692 (s), 662 (m), and 646 (w) cm^{-1}. The compound is monomeric in C_6H_6 (by cryoscopy). It decomposes at ~175°C with evolution of pyridine [9].

$Be(CH_3)N(C_2H_5)_2$

For the preparation, $NH(C_2H_5)_2$ is condensed onto a solution of $Be(CH_3)_2$ in a 1:1 mole ratio at −196°C in ether. The mixture is warmed, and CH_4 is rapidly evolved at room temperature. An oil remains after evaporation of the solvent which cannot be distilled without decomposition.

^1H NMR (C_6H_6 vs. C_6H_6): δ = 7.83, 7.77 (d, CH_3Be), 6.10 (q, CH_2), 4.84 to 3.81 ppm (m, CH_3). The unexpected splitting of the signals is not observed in c-C_6H_{12}. Only one line occurs for CH_3Be at 2.44 ppm (relative to c-C_6H_{12}). IR absorptions (contact film) are given at 2916 (s, br), 2809 (s, sh), 1454 (s), 1374 (s), 1337 (w), 1304 (m), 1284 (w), 1252 (m), 1196 (s), 1168 (s), 1142 (s), 1126 (s), 1099 (s), 1077 (s), 1051 (s, sh), 1036 (s, sh), 1014 (s, br), 896 (s, br), 797 (s, br), 676 (s, br), 639 (m, sh), 611 (w, sh), and 525 (w) cm^{-1}. The compound is trimeric in C_6H_6 (by cryoscopy); see Formula II, p. 136 (R = CH_3, R' = C_2H_5) [9].

$Be(CH_3)N(C_3H_7)_2$

The compound is obtained from an equimolar mixture of $NH(C_3H_7)_2$ and $Be(CH_3)_2$ in ether. Gas is evolved during 15 min at room temperature. Removal of solvent leaves a colorless viscous oil. It is dissolved in C_6H_{14}, but no crystals are obtained after filtration and evaporation of the solvent. The oil becomes glassy below ~ −25°C.

The CH_3Be protons give a single resonance at δ = 7.79 ppm in C_6H_6 relative to the solvent, and at 2.47 ppm in c-C_6H_{12} relative to the solvent. The signals due C_3H_7 are complex. The IR spectrum (contact film) contains absorptions at 2958 (s), 2937 (s), 2874 (s), 1468 (m), 1379 (m), 1324 (w), 1298 (w), 1259 (s), 1202 (m), 1166 (w), 1082 (s, br), 1015 (s), 959 (m), 920 (m), 881 (m, sh), 863 (m), 797 (s, br), 749 (m), 726 (w), and 680 (m, br) cm^{-1}. The compound is a dimer in C_6H_6 (by cyroscopy); see Formula III, p. 136 (R = CH_3, R' = R'' = C_2H_5). Reaction with pyridine in C_6H_6 gives $Be(CH_3)_2 \cdot 2NC_5H_5$ [9].

Be(CH$_3$)NC$_5$H$_{10}$ (NC$_5$H$_{10}$ = piperidinyl)

The compound is obtained during a reaction of Be(CH$_3$)$_2$ with excess NHC$_5$H$_{10}$ in ether in a high vacuum system. Below 0°C, the adduct Be(CH$_3$)$_3$·NHC$_5$H$_{10}$ precipitates. Between 0 and 20°C, the title compound is formed, as determined by the amount of CH$_4$ evolved. Cooling the solution below 0°C gives the compound as a white precipitate. Warming to 38°C and subsequent standing for 20 h at room temperature gives Be(NC$_5$H$_{10}$)$_2$ [4].

Be(CH$_3$)NC$_4$H$_8$O (NC$_4$H$_8$O = morpholinyl)

This compound is obtained like the previous one from Be(CH$_3$)$_2$ and morpholine. It precipitates as crystals. Warming to 40°C for several hours gives Be(NC$_4$H$_8$O)$_2$ [4].

Be(CH$_3$)NHCH$_2$CH$_2$N(CH$_3$)$_2$

Excess diamine is condensed onto Be(CH$_3$)$_2$ in ether, cooled in liquid N$_2$. A vigorous exothermic reaction begins just below room temperature, and colorless needles of the compound deposit. It is dimeric in C$_6$H$_6$ (by cryoscopy), Formula V, p. 136, being assigned to it (R = R″ = CH$_3$, R = H). The compound sublimes slowly in a vacuum at 115°C and gradually evolves CH$_4$ at 138°C. At 170°C it melts with vigorous CH$_4$ production to give a polymer [3].

Be(CH$_3$)N(CH$_3$)CH$_2$CH$_2$N(CH$_3$)H

The compound is prepared from the components like the previous one. CH$_4$ evolves as soon as the reactants melt. Removal of the ether and excess diamine gives colorless needles of the compound.

It dissolves in C$_6$H$_6$ as a dimer; see Formula V, p. 136 (R = R′ = CH$_3$, R″ = H). It sublimes at 90°C in a vacuum and suddenly decomposes at 145°C to give more CH$_4$ gas and a polymer (BeN(CH$_3$)CH$_2$CH$_2$NCH$_3$)$_n$. Heating in a tetralin solution also gives CH$_4$ from 140 to 150°C [3].

Be(CH$_3$)N(CH$_3$)CH$_2$CH$_2$N(CH$_3$)$_2$

The diamine (in excess or in a 1:1 mole ratio) is condensed onto Be(CH$_3$)$_2$ in ether and cooled with liquid N$_2$. The mixture is allowed to warm to room temperature. At 18°C colorless needles deposit. In a larger scale preparation, the product is separated by cooling the ether solution, filtration, and pumping off the excess ether and diamine. The compound melts at 116 to 118°C and is dimeric in C$_6$H$_6$; see Formula V, p. 136 (R = R′ = R″ = CH$_3$). Heating a reaction mixture to 60°C for 0.5 h gives no further gas evolution [3].

Be(CH$_3$)N(CH$_3$)CH$_2$CH$_2$C$_5$H$_4$N (C$_5$H$_4$N = pyridin-2-yl)

The amine NH(CH$_3$)CH$_2$CH$_2$C$_5$H$_4$N in ether is added to Be(CH$_3$)$_2$ in ether (1:1 mole ratio) at liquid air temperature. A pale yellow precipitate is formed on warming to room temperature. The precipitate is extracted with ether yielding colorless crystals, which turn yellow at 155°C and melt to a red liquid at 177°C. The compound is assumed to be dimeric (Formula V, p. 136), but the solubility was too low to determine the molecular mass [13].

Be(CH$_3$)N(C$_6$H$_5$)CH$_2$C$_6$H$_5$

The compound is prepared by addition of C$_6$H$_5$CH=N(C$_6$H$_5$) in C$_6$H$_6$ to a solution of (Be(CH$_3$)H·O(C$_2$H$_5$)$_2$)$_2$ in C$_6$H$_6$. Evaporation of the solvent leaves a viscous liquid which could not be distilled. The product can be crystallized, but melts well below room temperature.

References on pp. 145/6

The ^{1}H NMR spectrum in C_6H_6 shows a singlet for CH_3Be at $\delta = -0.36$ ppm. The compound is a dimer in C_6H_6; see Formula III, p. 136 ($R = CH_3$, $R' = C_6H_5$, $R'' = CH_2C_6H_5$). Hydrolysis gives $C_6H_5NHCH_2C_6H_5$. A 1:1 adduct is formed with 2,2'-bipyridine (see below) [7].

$Be(CH_3)N(C_6H_5)CH_2C_6H_5 \cdot N_2C_{10}H_8$ ($N_2C_{10}H_8$ = 2,2'-bipyridine)

2,2'-Bipyridine in C_6H_6 is added to a solution of the previous compound in C_6H_{14}. The orange adduct precipitates. It decomposes from 160°C and is only sparingly soluble in C_6H_6. Attempts to recrystallize by Soxhlet extraction with boiling C_6H_6 results in extensive decomposition [7].

$Be(CH_3)N(C_6H_5)CH(CH_3)C_6H_5$

The compound is prepared from $N(C_6H_5){=}CHC_6H_5$ and $Be(CH_3)_2$ (1:1 mole ratio) in ether at −78°C. The product precipitates, and the suspension is warmed to room temperature and stirred for 2h. Ether is removed in vacuum. Recrystallization from $C_6H_5CH_3$ gives colorless needles which shrink at 140°C and melt at 153 to 154°C. The solubility of the compound in C_6H_6 is too low to determine the degree of association; but in analogy to the following compound, it is assumed to be dimeric; see Formula III, p. 136 ($R = CH_3$, $R' = C_6H_5$, $R'' = CH(CH_3)C_6H_5$). Hydrolysis with dilute H_2SO_4 gives $[NH_2(C_6H_5)CH(C_6H_5)CH_3]_2SO_4$ [12].

$Be(CH_3)N(C_6H_4CH_3\text{-}4)CH(CH_3)C_6H_5$

The compound is prepared from $N(C_6H_4CH_3\text{-}4){=}CHC_6H_5$ and $Be(CH_3)_2$ in ether at −78°C. The mixture is stirred at room temperature for 1 h, and the solvent is evaporated. Recrystallization from C_6H_{14} gives colorless prisms in 28 % yield that melt at 155 to 156°C with decomposition. The compound is dimeric in C_6H_6 (by cryoscopy); see Formula III, p. 136 ($R = CH_3$, $R' = 4\text{-}CH_3C_6H_4$, $R'' = CH(CH_3)C_6H_5$)[12].

$Be(CH_3)N(C_6H_5)_2$

The compound is prepared by addition of $NH(C_6H_5)_2$ in ether to $Be(CH_3)_2$ in ether (mole ratio 1:1) at −78°C. The mixture is warmed to room temperature. The solvent is removed under reduced pressure, and the product crystallizes from C_6H_{14}. It cannot be crystallized from ether, ether/C_6H_{14}, or C_6H_6.

When heated, it turns yellow at 115°C and melts with decomposition at 141°C. The ^{1}H NMR spectrum in C_6H_6 shows a singlet for CH_3Be at $\delta = -0.43$ ppm. The observed IR absorptions in Nujol are 1585(s), 1289(s), 1258(w), 1074(s), 1019(s), 952(s,sh), 922(s,br), 867(s,br), 797(s), 741(s), 686(s), 668(s), 649(m,sh), 599(m), 575(m), 523(m,br), 500(m,br), and 481(m,br) cm^{-1}. The compound is crystalline, as indicated by X-ray powder diffraction. It dissolves in C_6H_6 as a dimer (by cryoscopy) see Formula III, p. 136 ($R = CH_3$, $R' = R'' = C_6H_5$). The compound fumes when exposed to air becoming yellow and then green, and is very sensitive to H_2O. Adducts are formed with pyridine and 2,2'-bipyridine (see below) [9].

$Be(CH_3)N(C_6H_5)_2 \cdot NC_5H_5$ (NC_5H_5 = pyridine)

For its preparation, pyridine is added to a solution of $Be(CH_3)N(C_6H_5)_2$ in C_6H_6 in a 1:1 mole ratio at ~70°C. The pale yellow adduct crystallizes upon cooling and is recrystallized from hot C_6H_6. It has no definite melting point and softens at 158°C, becoming a yellow wax at ~200°C. IR absorptions in Nujol are given at 1589(s), 1577(s), 1477(s), 1329(m), 1302(s), 1282(s), 1216(w), 1207(w), 1176(w), 1165(w), 1148(w), 1079(m), 1065(s), 1049(m), 1027(w), 1016(w), 986(w), 929(s), 886(m), 875(m), 791(m), 774(m), 763(m,sh), 746(s), 699(s), 687(s,sh),

674(m,sh), 651(w,sh), 642(m), 596(m), 572(w), and 526(w) cm^{-1}. It is only sparingly soluble in C_6H_6 such that the molecular mass could not be determined, but it is presumed to be dimeric as shown in Formula IV, p. 136 (R = CH_3, R′=R″ = C_6H_5, ^{2}D = NC_5H_5). It is decomposed by air, but not hydrolyzed by cold H_2O, and reacts only slowly with cold dilute H_2SO_4 [9].

$Be(CH_3)N(C_6H_5)_2 \cdot 2NC_5H_5$ (NC_5H_5 = pyridine)

Pyridine, in excess, is added to $Be(CH_3)N(C_6H_5)_2$ in C_6H_6. A bright yellow product separates when C_6H_{14} is added. The melting point is 83 to 85°C (dec.) after recrystallization from hot C_6H_6/C_6H_{14}. The IR spectrum in Nujol contains absorptions at 1587(s), 1572(s), 1557(s), 1300(m), 1285(s), 1228(w), 1202(s), 1164(m), 1137(m), 1073(w), 1059(m), 1035(m), 1019(m), 1006(m), 978(m), 940(w), 920(s), 877(m), 861(m), 854(m), 784(m), 741(s), 726(w), 694(s), 678(s), 670(s), 662(m), 645(w), 637(m), 628(w), 612(w), 602(w), 590(w), 578(m), 567(m), 554(w), 527(m), and 504(m) cm^{-1}. The compound is monomeric in C_6H_6 [9].

$Be(CH_3)N(C_6H_5)_2 \cdot N_2C_{10}H_8$ ($N_2C_{10}H_8$ = 2,2′-bipyridine)

An excess of 2,2′-bipyridine in C_6H_6 is added to $Be(CH)_3N(C_6H_5)_2$ in C_6H_6 at room temperature. Standing overnight, concentration under reduced pressure, and addition of C_6H_{14} give a brick-red precipitate, which is washed with C_6H_6.

The compound decomposes at ~110°C. The IR absorptions in Nujol are given at 1607(m), 1582(s), 1377(m), 1333(w), 1297(s), 1263(w), 1215(w), 1186(w), 1167(w), 1152(w), 1082(m), 1068(m), 1054(m), 1044(m), 1031(m), 1019(m), 986(m), 945(m), 894(m), 876(m), 862(w), 819(w,sh), 802(m), 770(s), 756(s), 744(s), 709(s), 693(m), 646(w), 540(w), and 523(w) cm^{-1}. The UV spectrum in C_6H_6 has maxima at λ = 3750 Å (ε = 3923 $L \cdot cm^{-1} \cdot mol^{-1}$) and 5520 Å (ε = 4010 $L \cdot cm^{-1} \cdot mol^{-1}$). The compound is not sufficiently soluble in C_6H_6 to determine the degree of association, but it is assumed to be monomeric with $N_2C_{10}H_8$ acting as a ^{2}D-^{2}D ligand in analogy to $Be(C_2H_5)N(C_6H_5)_2 \cdot N_2C_{10}H_8$. It is only slowly hydrolyzed by H_2O and $CH_3OCH_2CH_2OH$ [9].

$Be(C_2H_5)NH_2$

The compound is prepared by condensing NH_3 onto a solution of $Be(C_2H_5)_2$ in ether at −78°C. At −33.5°C, excess NH_3 is evaporated. Warming to 0 to 10°C gives the product as a colorless solid.

The compound is insoluble in ether and has been assigned the polymeric structure shown in Formula I, p. 136 (R = C_2H_5). Thermal decomposition in a high vacuum at 50°C gives a polymer of the type $C_2H_5Be(NHBe)_nNH_2$. The same type of polymer is obtained by reaction with an excess of NH_3 [5].

$Be(C_2H_5)N(CH_3)_2$

The compound is prepared by condensing $NH(CH_3)_2$ onto a frozen (−196°C) solution of $Be(C_2H_5)_2$ in ether (mole ratio 1:1). Warming to room temperature and evaporation of the solvent gives a colorless viscous oil which becomes glassy below −35°C.

^{1}H NMR ($C_6D_{11}CD_3$ vs. C_6H_6): δ = 7.41 (q, CH_2), 6.21 (t, CH_3C), and 4.90 ppm (s, CH_3N). IR (contact film): 2878(s), 2849(s), 2809(s), 2762(m), 2688(w), 1484(s), 1466(s), 1414(m), 1393(m), 1367(w), 1302(w), 1233(m), 1205(s), 1173(m), 1137(m,sh), 1116(m), 1102(m,sh), 1050(s,sh), 1038(s), 897(s,br), 870(s,sh), 804(s,br), and 626(m,br) cm^{-1}. The compound is trimeric in C_6H_6 (by cryoscopy) and is assumed to have the structure shown in Formula II, p. 136 (R = C_2H_5, R′ = R″ = CH_3) [9].

References on pp. 145/6

$Be(C_2H_5)N(C_2H_5)_2$

The compound is prepared by the same procedure as the previous one from $NH(C_2H_5)_2$ and $Be(C_2H_5)_2$.

The nonvolatile oil shows signals in the ^1H NMR spectrum in C_6H_6 at $\delta = -0.10$ (q, CH_2Be) and 2.67 ppm (q, CH_2N). The CH_3 signals are obscured. IR absorptions (contact film) are at 2958(s), 2857(s), 1454(m,sh), 1366 (m,sh), 1337(w), 1310(m), 1282(w), 1262(m), 1225(m), 1199(m), 1185(m), 1150(m), 1135(m), 1105(s), 1091(s,sh), 1066(m), 1013(s), 961(m), 911(s,sh), 882(s), 809(m,sh), 789(s), 732(m), and 630(m) cm^{-1}. The compound is dimeric in C_6H_6 (by cryoscopy) [9].

$Be(C_2H_5)N(CH_3)CH(C_2H_5)C_6H_5$

The compound is obtained by reaction of $N(CH_3)=CHC_6H_5$ with $Be(C_2H_5)_2 \cdot O(C_2H_5)_2$ (1:1 mole ratio) in C_6H_{14} for 2 h at room temperature. Evaporation of the solvent and pumping overnight at 10^{-2} Torr gives a solid which crystallizes as colorless needles from C_6H_{14} at 5°C; yield 73 %; m.p. 82 to 84°C. The compound is dimeric in C_6H_6; see Formula III, p. 136 ($R = C_2H_5$, $R' = CH_3$, $R'' = CH(C_2H_5)C_6H_5$) [12].

$Be(C_2H_5)N(C_6H_5)CH(C_2H_5)C_6H_5$

An equivalent of $N(C_6H_5)=CHC_6H_5$ in ether is added to $Be(C_2H_5)_2$ in ether at −78°C. The mixture is warmed to room temperature, stirred for 30 min, and the ether is evaporated. Crystallization from C_6H_{14}/C_6H_6 (6:1) gives colorless prisms in 28 % yield which soften at ~90°C and melt at 134 to 136°C. The compound is dimeric in C_6H_6 (by cryoscopy); see Formula III, p. 136 ($R = C_2H_5$, $R' = C_6H_5$, $R'' = CH(C_2H_5)C_6H_5$). Hydrolysis with dilute HNO_3 gives $[H_2N(C_6H_5)CH(C_2H_5)C_6H_5]NO_3$ [12].

$Be(C_2H_5)N(C_6H_5)_2$

The compound is prepared from $NH(C_6H_5)_2$ and $Be(C_2H_5)_2$ by the same procedure as described for $Be(CH_3)N(C_6H_5)_2$. The crystals obtained from C_6H_{14} melt at 154°C.

IR (Nujol): 1572(m), 1364(w), 1295(w), 1251(w), 1209(m), 1164(m), 1151(m), 1067(s), 1009(s), 917(m), 902(m), 859(s), 793(m), 754(m,sh), 742(s), 717(m), 687(s), 659(w), 621(w,br), 598(w), 545(w,br), and 526(w,sh) cm^{-1}. The compound is dimeric in C_6H_6 (by cryoscopy); see Formula III, p. 136 ($R = C_2H_5$, $R' = R'' = C_6H_5$). Adducts are formed with pyridine and 2,2'-bipyridine (see below). A 1:1 adduct with pyridine could not be isolated [9].

$Be(C_2H_5)N(C_6H_5)_2 \cdot 2NC_5H_5$ (NC_5H_5 = pyridine)

Addition of pyridine to $Be(C_2H_5)N(C_6H_5)_2$ in hot C_6H_6 and evaporation of ~⅔ of the solvent gives crystals that decompose at 85°C and become a yellow wax at ~190°C.

IR absorptions in Nujol are observed at 1587(s), 1572(s), 1557(s), 1300(m), 1285(s), 1228(w), 1202(s), 1164(m), 1137(m), 1073(w), 1059(m), 1035(m), 1019(m), 1006(m), 978(m), 940(w), 920(s), 877(m), 861(m), 854(m), 784(m), 741(s), 726(w), 694(s), 678(s), 670(s), 662(m), 645(w), 637(m), 628(w), 612(w), 602(w), 590(w), 578(m), 567(m), 554(w), 527(m), and 504(m) cm^{-1}. Solutions in C_6H_6 contain monomers (by cryoscopy) [9].

$Be(C_2H_5)N(C_6H_5)_2 \cdot N_2C_{10}H_8$ ($N_2C_{10}H_8$ = 2,2'-bipyridine)

The adduct is prepared from an excess of 2,2'-bipyridine and $Be(C_2H_5)N(C_6H_5)_2$ in C_6H_6. After concentration, C_6H_{14} is added. Recrystallization from hot C_6H_6/C_6H_{14} gives a brick-red solid which decomposes at ~135°C and turns dark purple.

The IR spectrum in Nujol contains absorptions at 1607(s), 1575(s), 1333(w), 1307(s), 1218(m), 1185(w), 1167(m), 1154(m), 1082(w,sh), 1073(w), 1052(m), 1043(w), 1031(m), 1022(m,sh), 988(m), 946(m), 934(m), 888(m), 877(w), 858(w), 790(w), 770(s), 752(s), 743(s), 719(m), 706(m), 691(m), 676(m), 652(w,sh), 616(w), 605(m), and 572(w) cm^{-1}. The UV spectrum in C_6H_6 has absorptions at $\lambda = 3780$ Å ($\varepsilon = 2109\ L \cdot mol^{-1} \cdot cm^{-1}$) and 5540 Å ($\varepsilon = 4100\ L \cdot mol^{-1} \cdot cm^{-1}$). It dissolves readily in C_6H_6 at 50°C and is monomeric in this solvent (by cryoscopy) [9].

$Be(C_3H_7)N(CH_3)CH_2CH_2N(CH_3)_2$

The diamine $NH(CH_3)CH_2CH_2N(CH_3)_2$ in C_6H_{14} is added to $Be(C_3H_7)_2 \cdot 2O(C_2H_5)_2$ in C_6H_{14} at −78°C in a 1:1 mole ratio. Warming to room temperature and removal of the solvent gives the adduct, which is recrystallized from c-$C_6H_{11}CH_3$ and decomposes at 60°C. Solutions in C_6H_6 contain dimeric molecules (by cryoscopy); see Formula V, p. 136 ($R = C_3H_7$, $R' = R'' = CH_3$) [10].

$Be(C_3H_7\text{-}i)N(CH_3)_2$

For the preparation, $NH(CH_3)_2$ is condensed onto a frozen (−196°C) solution of $Be(C_3H_7\text{-}i)_2$ in ether in a 1:1 mole ratio. A colorless liquid remains after the C_3H_8 evolution has ceased at room temperature and the solvent has been evaporated [2, 9]. It is mentioned without details that $BeHC_3H_7$-i and $NH(CH_3)_2$ give H_2 and $Be(C_3H_7\text{-}i)N(CH_3)_2$ [2].

^1H NMR (C_6H_6 vs. C_6H_6): $\delta = 5.09$ (s, CH_3N), 5.99 (d, CH_3C) and 6.75 ppm (sept, CH). IR (contact film): 2907(s), 2833(s), 1484(m,sh), 1462(s), 1439(m,sh), 1377(w), 1264(w), 1222(m), 1172(w), 1153(w), 1136(w), 1116(m), 1053(w,sh), 1035(s), 980(m), 919(s,sh), 893(s), 827(s), 769(s,br), 583(m,br), and 553(m,br) cm^{-1}. The compound is trimeric in C_6H_6 (by cryoscopy) [9].

The compound decomposes above 100°C to give C_3H_6 and $BeHN(CH_3)_2$. The decomposition is complete within 15 min at 220°C. Reaction with an excess of $NH(CH_3)_2$ gives $Be(N(CH_3)_2)_2$ [2].

$Be(C_3H_7\text{-}i)N(CH_3)CH_2CH_2N(CH_3)_2$

Equimolar amounts of $Be(C_3H_7\text{-}i)_2$ and $NH(CH_3)CH_2CH_2N(CH_3)_2$ in C_6H_{14} and ether are mixed at −78°C and then allowed to warm. The precipitate is recrystallized from $C_6H_{14}/C_6H_5CH_3$ (3:1). On heating, it shrinks at 120 to 124°C, becomes pale yellow at ~150°C, and melts with decomposition at 165 to 167°C. The ^1HNMR spectrum in $C_6D_5CD_3$ shows two doublets at $\delta = 1.32$ and 1.41 ppm for CH_3C at 20°C (J(H,H) = 7 Hz). At 80°C, the two doublets collapse to a single doublet at $\delta = 1.26$ ppm (J(H,H) = 7 Hz). The behavior is ascribed to a restricted Be–C bond rotation. The compound is dimeric in C_6H_6 (by cryoscopy) [10].

$Be(C_4H_9\text{-}t)N(CH_3)_2$

An equimolar mixture of $Be(C_4H_9\text{-}t)_2 \cdot O(C_2H_5)_2$, $NH(CH_3)_2$, and $N(CH_3)_3$ in C_5H_{12} is stirred overnight, after which all volatile material is removed under reduced pressure leaving a colorless oil. The t-C_4H_9:Be ratio in the product was 0.97:1, but the IR and ^1HNMR spectra are consistent with a mixture of the title compound and $Be(C_4H_9\text{-}t)_2NH(CH_3)_2$. ^1HNMR(C_6D_6): $\delta = 1.06$, 1.14 (d, C_4H_9), 1.75 (br, CH_3 of $NH(CH_3)_2$), 2.22 ppm (s, CH_3 of $N(CH_3)_2$). The IR spectrum contained weak absorptions due to NH. Distillation at 60°C/~0.1 Torr results in disproportionation [9].

References on pp. 145/6

$Be(C_4H_9\text{-}t)N(CH_3)CH_2CH_2N(CH_3)_2$

An equimolar amount of $NH(CH_3)CH_2CH_2N(CH_3)_2$ is added to $Be(C_4H_9\text{-}t)_2 \cdot O(C_2H_5)_2$ in C_5H_{12} at ~ −100°C. Warming to room temperature and evaporation of the solvent after the gas evolution has ceased gives a liquid which is distilled at 45°C/1 Torr and collected at −40°C.

IR absorptions (liquid film) are given as 2994(m,sh), 2907(vs), 2857(vs), 2809(vs), 2778(vs,sh), 2770(vs,sh), 2717(m,sh), 2681(m), 1468(s), 1449(m,sh), 1408(w), 1379(m), 1370(m), 1348(s), 1290(s), 1252(vs), 1222(m), 1188(vs), 1174(s,sh), 1161(m,sh), 1120(m), 1103(m), 1065(s), 1059(vs), 1022(s), 1012(m,sh), 961(vs), 946(vs), 899(w), 878(w), 837(w), 799(m), 787(s), 756(w,br), 707(s,br), 629(w,br), 595(w), 524(s), and 506 (w,br,sh) cm^{-1}. The compound is monomeric in C_6H_6 (by cryoscopy), and Formula VI is suggested [9].

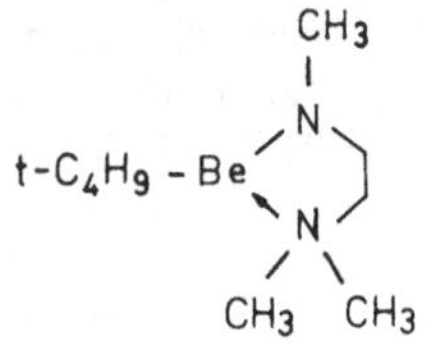

VI

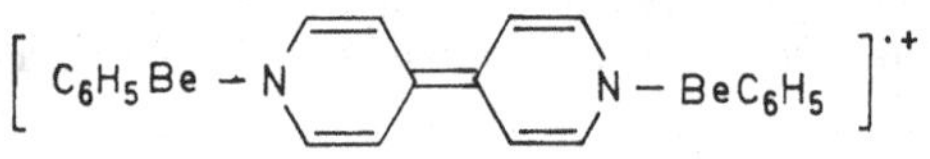

VII

$Be(C_4H_9\text{-}t)N(CH_3)CH_2C_6H_5$

The compound is prepared by addition of $N(CH_3){=}CHC_6H_5$ to $Be(C_4H_9\text{-}t)_2 \cdot O(C_2H_5)_2$ in C_6H_{14} at −78°C (mole ratio 1:1). Stirring for 2h at room temperature, evaporation of the solvent, and pumping for 3h at 10^{-2} Torr gives a solid that crystallizes as colorless prisms from C_6H_{14}.

It shrinks at ~170°C and melts at 325 to 327°C with decomposition. Sublimation occurs at 120 to 130°C/10^{-2} Torr. The compound is dimeric in C_6H_6 (by cryoscopy); see Formula III, p. 136 ($R = t\text{-}C_4H_9$, $R' = CH_3$, $R'' = CH_2C_6H_5$). Hydrolysis gives $NH(CH_3)CH_2C_6H_5$. No reaction occurs with $N(CH_3){=}CHC_6H_5$ [12].

$Be(C_4H_9\text{-}t)N(C_6H_5)CH_2CH{=}CHC_6H_5$

For the preparation, $N(C_6H_5){=}CHCH{=}CHC_6H_5$ in ether is added to $Be(C_4H_9\text{-}t)_2 \cdot O(C_2H_5)_2$ in ether at −78°C (mole ratio 1:1). Stirring for 1 h at room temperature, evaporation of ether, and recrystallization from C_6H_{14} gives colorless prisms that shrink at ~100°C and melt at 130 to 133°C with decomposition. The compound is dimeric in C_6H_6 (by cryoscopy), see Formula III, p. 136 ($R = t\text{-}C_4H_9$, $R' = C_6H_5$, $R'' = CH_2CH{=}CHC_6H_5$). Hydrolysis gives $NH(C_6H_5)CH_2CH{=}CHC_6H_5$ [12].

$Be(C_6H_5)N(CH_3)_2$

The amine $NH(CH_3)_2$ is condensed onto a solution of $Be(C_6H_5)_2$ in ether/$C_6H_5CH_3$ (1:1) at −78°C in an equimolar amount. Warming to 40°C, concentration, and addition of C_6H_{14} give long needles at −78°C, m. p. 154°C. The compound is trimeric in C_6H_6 (by cryoscopy); see Formula II, p. 136 ($R = C_6H_5$, $R' = CH_3$) [8].

$Be(C_6H_5)N(C_6H_5)_2$

The reaction between $NH(C_6H_5)_2$ and $Be(C_6H_5)_2$ in $C_6H_5CH_3$/ether is so exothermic at room temperature that the mixture has to be cooled. The solvent is evaporated, and crystals are obtained from C_6H_{14} at −78°C. The compound melts at 260 to 262°C with decomposition and is dimeric in C_6H_6 (by cryoscopy); see Formula III, p. 136 ($R = R' = R'' = C_6H_5$).

When exposed to air, the color changes to green and then dark blue. Hydrolysis is vigorous. Addition of 2,2′-bipyridine gives cream-colored crystals of $Be(C_6H_5)_2 \cdot N_2C_{10}H_8$ and red-colored crystals presumably of $\mathbf{Be(C_6H_5)N(C_6H_5)_2 \cdot N_2C_{10}H_8}$ [8].

$Be(C_6H_5)N(CH_3)CH_2CH_2N(CH_3)_2$

Equimolar amounts of $NH(CH_3)CH_2CH_2N(CH_3)_2$ react with $Be(C_6H_5)_2$ in ether at room temperature. The compound precipitates. Recrystallization from C_6H_6/C_6H_{14} (1:1) gives small needles, m.p. 255 to 256°C (dec.). As determined by cryoscopy, the compound is dimeric in C_6H_6; see Formula V, p. 136 ($R = C_6H_5$, $R' = R'' = CH_3$). Hydrolysis with cold water is very slow, and warm dilute H_2SO_4 is necessary for complete decomposition [8].

$[Be(C_6H_5)NC_5H_4{=}C_5H_4NBeC_6H_5]^{\bullet +}[Be(C_6H_5)_3]^-$

A solution of 4,4′-bipyridine and a ~5-fold excess of $Be(C_6H_5)_2$ in THF reacts with a K mirror to yield the very persistent and intensely blue-colored title compound (Formula VII). The radical cation is characterized by its ESR spectrum. The coupling constants in mT are a(H) = 0.086 and 0.0205, $a(^{14}N) = 0.350$, and $a(^9Be) = 0.029$. The g-factor is 2.0031 [14].

$Be_3(CH_3)_2(N(CH_3)_2)_4$

The compound is prepared by reaction of $Be(N(CH_3)_2)_2$ and $Be(CH_3)_2$ in ether. The mixture is stirred for 15 min at room temperature, and the ether is evaporated. Recrystallization from C_6H_{14} at 5°C gives colorless prisms that melt at 38 to 39°C.

The compound is monomeric in C_6H_6 (by cryoscopy). It is believed to have the structure shown in Formula VIII, but the ^1H NMR spectrum is anomalous. Though the area ratio of the CH_3N resonances to that of the CH_3Be resonance at $\delta = -0.62$ ppm is 4:1, the CH_3N signal is split into two peaks at 2.10 and 2.19 ppm in the area ratio 5:1 [11].

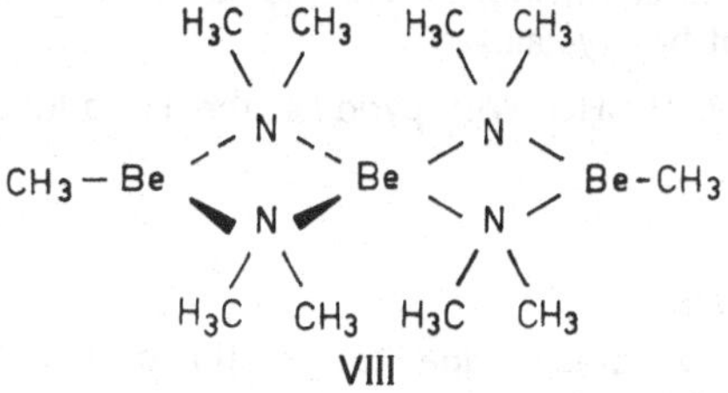

VIII

References:

[1] Coates, G. E., Glockling, F., Huck, N. D. (J. Chem. Soc. **1952** 4512/5).
[2] Coates, G. E., Glockling, F. (J. Chem. Soc. **1954** 22/7).
[3] Coates, G. E., Green, S. I. E. (J. Chem. Soc. **1962** 3340/8).
[4] Funk, H., Masthoff, R. (J. Prakt. Chem. [4] **22** [1963] 255/8).
[5] Masthoff, R., Vieroth, C. (Z. Chem. [Leipzig] **5** [1965] 142).
[6] Bell, N. A., Coates, G. E., Emsley, J. W. (J. Chem. Soc. A **1966** 49/52).
[7] Bell, N. A., Coates, G. E. (J. Chem. Soc. A **1966** 1069/73).
[8] Coates, G. E., Tranah, M. (J. Chem. Soc. A **1967** 236/9).
[9] Coates, G. E., Fishwick, A. H. (J. Chem. Soc. A **1967** 1199/204).
[10] Coates, G. E., Roberts, P. D. (J. Chem. Soc. A **1968** 2651/5).

[11] Andersen, R. A., Bell, N. A., Coates, G. E. (J. Chem. Soc. Dalton Trans. **1972** 577/82).
[12] Andersen, R. A., Coates, G. E. (J. Chem. Soc. Dalton Trans. **1974** 1171/80).
[13] Bell, N. A., Coates, G. E., Fishwick, A. H. (J. Organometal. Chem. **198** [1980] 113/20).
[14] Kaim, W. (J. Organometal. Chem. **241** [1983] 157/69).

1.3.9 Organoberyllium Imides and Their Adducts

General Remarks. The imides of the general formula Be(R)N=CHR′ are obtained by addition of Be(R)H to nitriles R′CN, except for $Be(C_4H_9\text{-}i)N{=}CHC_6H_5$ which is formed by reaction of $Be(C_4H_9\text{-}i)_2$ with C_6H_5CN. When Be(R)H is used in the form of its ether complex, the products could not be purified, and the compounds are further characterized as adducts with $N(CH_3)_3$ or pyridine (NC_5H_5). These adducts are dimeric with bridging N=CHR′ groups, as shown in Formula I [2].

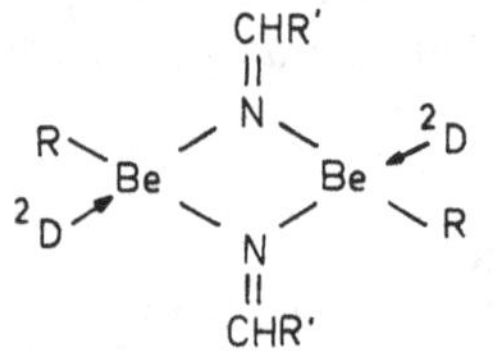

I

$Be(CH_3)N{=}CHC_4H_9\text{-}t$

An equimolar amount of $t\text{-}C_4H_9CN$ in C_6H_6 is added to $Be(CH_3)H$ etherate in C_6H_6 at −78°C. Warming to room temperature, stirring for 2 h, and removal of all volatile material leaves a colorless oil that could not be crystallized.

Acid hydrolysis gives $t\text{-}C_4H_9CHO$. With pyridine, the 1:1 adduct can be crystallized (see below) [2].

$Be(CH_3)N{=}CHC_4H_9\text{-}t \cdot NC_5H_5$

An equimolar amount of pyridine is added to a solution of $Be(CH_3)N{=}CHC_4H_9\text{-}t$, prepared as described above, at −78°C. Stirring for 2 h at room temperature and evaporation of the volatile material gives a residue that crystallizes from C_6H_6 as yellow prisms, m.p. 127 to 129°C.

IR absorptions (Nujol) for bridging ν(C=N) are observed at 1682(s), 1650(sh), and 1603(s) cm^{-1}. The adduct is dimeric in C_6H_6 (by cryoscopy); see Formula I ($R = CH_3$, $R' = t\text{-}C_4H_9$, $^2D = NC_5H_5$). Hydrolysis with dilute acid gives $t\text{-}C_4H_9CHO$. No reaction is observed with another molar equivalent of $Be(CH_3)H$ [2].

$Be(CH_3)N{=}CHC_6H_5$

The compound can be prepared by slow addition of C_6H_5CN in ether to ethereal $Be(CH_3)H$ at −96°C (printing error, −196°C?) (mole ratio 1:1). Warming to room temperature, stirring for 1 h, and evaporation of the solvent gives a product that crystallizes from C_6H_6/C_6H_{14} at −30°C as a glassy solid that is dried in a vacuum. The glassy solid is X-ray amorphous. It shrinks at 80°C before melting in the range 90 to 130°C and is tetrameric in C_6H_6 (by cryoscopy). No reaction is observed with an excess of $Be(CH_3)H$ [3].

A preparation from the components in $C_6H_5CH_3$, as described for $Be(CH_3)N{=}CHC_4H_9$-t, gave only an oil that could not be crystallized. It may be the etherate. The molecular mass of the residue obtained after heating to 100°C in a vacuum for 1 h was >2000 (cryoscopically in C_6H_6). It analyzed correctly for Be, but the CH_3 content was too low. Hydrolysis gave C_6H_5CHO [2, 3].

The adducts with $N(CH_3)_3$ and pyridine are described below [2].

$Be(CH_3)N{=}CHC_6H_5 \cdot N(CH_3)_3$

The adduct is prepared by addition of C_6H_5CN to $Be(CH_3)H \cdot N(CH_3)_3$ (mole ratio 1:1) in $C_6H_5CH_3$ at −196°C. Stirring for 2 h at room temperature, evaporation of the solvent, and crystallization from C_6H_6 gives yellow crystals that melt at 127 to 131°C with decomposition.

IR absorptions in Nujol are found at 2728(m), 1966(w), 1890(w), 1822(w), 1690(m), 1653(s), 1648(s), 1618(m), 1593(m), 1577(s), 1488(m), 1474(s), 1308(m), 1284(m), 1244(s), 1228(m), 1180(s), 1105(m), 1100(m), 1070(m), 1021(w), 1005(s), 1000(s), 960(m), 910(w), 856(s), 827(m), 772(s), 720(s), and 692(s) cm^{-1}. The compound is dimeric in C_6H_6 (Formula I, R = CH_3, R′ = C_6H_5, $^2D = N(CH_3)_3$), and acid hydrolysis gives C_6H_5CHO. The $N(CH_3)_3$ is partly removed by heating with formation of oligomers [2].

$Be(CH_3)N{=}CHC_6H_5 \cdot NC_5H_5$

Pyridine is added to $Be(CH_3)N{=}CHC_6H_5$ in toluene (1:1 mole ratio) at −78°C. Reaction and workup as described for the preceding compound gives cream-colored crystals (from C_6H_6), m.p. 162 to 163°C.

The IR spectrum in Nujol contains absorptions for bridging ν(C=N) at 1688(m), 1653(s), and 1603(s) cm^{-1}. Formula I (R = CH_3, R′ = C_6H_5, $^2D = NC_5H_5$) is proposed. Acid hydrolysis gives C_6H_5CHO [2].

$Be(CH_3)N{=}CHC_6H_4CH_3\text{-}2 \cdot N(CH_3)_3$

This adduct is prepared from $Be(CH_3)H \cdot N(CH_3)_3$ and 2-$CH_3C_6H_4CN$ as described above for $Be(CH_3)N{=}CHC_6H_5$.

Yellow prisms are formed (from C_6H_6); m.p. 137 to 139°C with decompositon; IR(Nujol): 2798(m), 2770(w), 1970(w), 1926(w), 1952(w), 1837(w), 1693(w), 1655(sh), 1649(s), 1640(s), 1478(s), 1406(m), 1375(s), 1282(m), 1244(m), 1230(w), 1208(m), 1174(s), 1151(m), 1108(sh), 1101(m), 1048(w), 1031(w), 1004(s), 967(m), 871(m), 855(s), 830(m), 818(s), 783(s), 758(s), 728(s), and 710(m) cm^{-1}. The compound is dimeric in C_6H_6 (by cryoscopy); see Formula I (R = CH_3, R′ = 2-$CH_3C_6H_4$, $^2D = N(CH_3)_3$). Hydrolysis gives 2-$CH_3C_6H_4CHO$. Thermolysis for 30 min at 150°C in a vacuum gives 1.75 mol $N(CH_3)_3$ per mol of dimer [2].

$Be(CH_3)N{=}CHC_6H_4CH_3\text{-}3 \cdot N(CH_3)_3$

The adduct is prepared from $Be(CH_3)H \cdot N(CH_3)_3$ and 3-$CH_3C_6H_4CN$ as described for $Be(CH_3)N{=}CHC_6H_5$ above.

Yellow prisms are formed (from C_6H_6); m.p. 133 to 136°C with decomposition; IR(Nujol): 2710(m), 1900(w), 1825(w), 1750(w), 1647(s), 1600(w), 1580(m), 1278(s), 1238(s), 1225(sh), 1171(s), 1148(s), 1092(s), 1080(sh), 990(s), 956(m), 900(m), 870(m), 830(m), 760(m), 700(m), and 680(w) cm^{-1}. The adduct is dimeric in C_6H_6 (by cryoscopy) see Formula I (R = CH_3, R′ = 3-$CH_3C_6H_4$, $^2D = N(CH_3)_3$). Hydrolysis gives 3-$CH_3C_6H_4CHO$. Reaction with 2,2′-bipyridine yields $Be(CH_3)_2 \cdot N_2C_{10}H_8$ [2].

References on p. 148

$Be(C_4H_9\text{-}i)N{=}CHC_6H_5$

$Be(C_4H_9\text{-}i)_2$ is added at −30°C to C_6H_5CN in a 1:1 or 2:1 mole ratio. Heating for 4 h at 67°C results in the evolution of i-C_4H_8. The product is assumed to be the title compound, since acid hydrolysis gives C_6H_5CHO. Another product formed is i-$C_4H_9(C_6H_5)CO$, which dominates at reaction temperatures of ≦47°C. The title compound was not further characterized [1].

$Be(C_4H_9\text{-}t)N{=}CHC_4H_9\text{-}t$

An equimolar amount or an excess of t-C_4H_9CN in C_6H_{14} is added to $Be(C_4H_9\text{-}t)H$ etherate in C_6H_{14} at −196 or −78°C. Warming to room temperature and evaporation of the solvent give a colorless oil that is miscible with C_6H_6 but could not be crystallized. No crystalline adduct is obtained by addition of $N(CH_3)_3$, but the 1:1 adduct with pyridine can be isolated [2].

$Be(C_4H_9\text{-}t)N{=}CHC_4H_9\text{-}t \cdot NC_5H_5$

For the preparation, an equimolar amount of pyridine is added to $Be(C_4H_9\text{-}t)N{=}CHC_4H_9\text{-}t$ in $C_6H_5CH_3$ at −78°C. Stirring for 2 h at room temperature and evaporation of the solvent give the adduct that crystallizes from C_6H_{14} as yellow prisms, m.p. 124 to 126°C.

The IR spectrum in Nujol shows bands at 1670(s), 1655(s), and 1610(s) cm^{-1} for the bridging ν(C=N) groups. The compound is dimeric in C_6H_6 (by cryoscopy); see Formula I, p. 146 (R = R′ = t-C_4H_9, $^2D = NC_5H_5$). Acid hydrolysis gives t-C_4H_9CHO [2].

$Be(C_4H_9\text{-}t)N{=}CHC_6H_5$

A 10% excess of C_6H_5CN is added to $Be(C_4H_9\text{-}t)H$ etherate in C_6H_{14} at −196°C. Stirring for 2 h at room temperature gives a yellow solid. Attempts to crystallize the compound from C_6H_6 gave only an oil. A crystalline adduct is formed with pyridine (see below) [2].

$Be(C_4H_9\text{-}t)N{=}CHC_6H_5 \cdot NC_5H_5$

Pyridine is added to $Be(C_4H_9\text{-}t)N{=}CHC_6H_5$ in $C_6H_5CH_3$ at −78°C (mole ratio 1:1). The yellow prisms (from C_6H_6) melt at 177 to 179°C with decomposition. The IR spectrum in Nujol shows absorptions at 2738(m), 1640(s), 1601(m), 1590(w), 1575(m), 1482(m), 1441(s), 1302(w), 1208(m), 1190(m), 1150(w), 1063(m), 1046(m), 1024(w), 1011(w), 1001(w), 992(w), 981(w), 955(w), 948(w), 838(m), 809(m), 764(w), 741(s), 718(s), 700(s), 690(s), 677(m), 644(m), 630(w), 610(w), and 595(m) cm^{-1}. The adduct is dimeric in C_6H_6 (by cryoscopy); see Formula I (R = t-C_4H_9, R′ = C_6H_5, $^2D = NC_5H_5$). Acid hydrolysis gives C_6H_5CHO [2].

References:

[1] Giacomelli, G. P., Lardicci, L. (Chem. Ind. [London] **1972** 689/90).
[2] Coates, G. E., Smith, D. L. (J. Chem. Soc. Dalton Trans. **1974** 1737/40).
[3] Bell, N. A., Coates, G. E., Fishwick, A. H. (J. Organometal. Chem. **198** [1980] 113/20).

1.3.10 Organoberyllium Hydrazides

General Remarks. Only three hydrazides with the general formula $Be(C_2H_5)N(CH_3)NRR'$ (R = R′ = H or CH_3, R = H and R′ = CH_3) are described in the literature. They are prepared by transferring $NH(CH_3)N(CH_3)_2$ under vacuum to a solution of $Be(C_2H_5)_2$ or $Be(C_2H_5)_2 \cdot N(CH_3)_3$ in C_8H_{18} at −196°C and subsequent warming. The compounds appear to be highly associated or polymeric as indicated by molecular weight data, low volatility, and very high melting temperatures (usually decomposition occurs before melting). A structure as shown in Formula I is proposed.

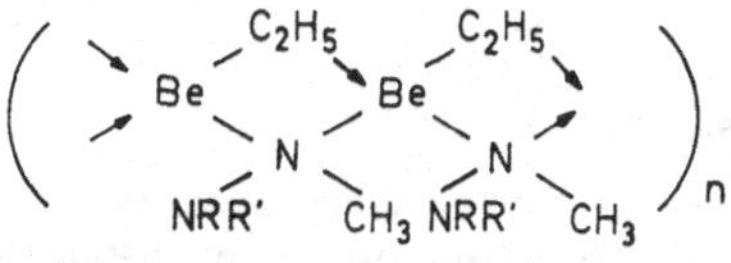

I

$Be(C_2H_5)N(CH_3)NH_2$

Reaction of $Be(C_2H_5)_2 \cdot N(CH_3)_3$ with $NH(CH_3)NH_2$ in C_8H_{18} in the mole ratio 1:1.56 for 2 h at room temperature gives 1.61 equivalents of C_2H_6. Removal of the solvent leaves a powdery white solid which does not sublime or dissolve in hydrocarbons. The results of the elemental analysis agree with the formula of the title compound. Methanolysis gives the expected amount of C_2H_6. An analogous reaction of $Be(C_2H_5)_2$ with $NH(CH_3)NH_2$ in the mole ratio 1:2.35 for 2 days at room temperature gives 1.5 equivalents of C_2H_6. The powdery white solid obtained analyzes correctly for Be and N, but a sample taken for the C and H analysis exploded before a result could be obtained. Methanolysis gives only 55.2 % of the expected C_2H_6. If the reaction of $Be(C_2H_5)_2$ with $NH(CH_3)NH_2$ is performed at 65°C in C_9H_{20} for 14 h, obviously increased formation of $Be(N(CH_3)NH_2)_2$ occurs.

The crude product exhibits a strong IR absorption at 3100 cm^{-1} for ν(NH) and a weak absorption at 1700 cm^{-1} for $\delta(NH_2)$, which supports the given formula instead of $Be(C_2H_5)$-$NHNHCH_3$.

$Be(C_2H_5)N(CH_3)NHCH_3$

Reaction of $NH(CH_3)NH(CH_3)$ with $Be(C_2H_5)_2$ (mole ratio 1:1) by heating at 75°C for 3 h gives a little more than 1 equivalent of C_2H_6. A nonvolatile gummy yellow product is formed, the elemental analysis of which agrees with the given formula. Methanolysis of a sample gives 91 % of the expected volume of C_2H_6. The IR spectrum contains a strong absorption at 3120 cm^{-1} for ν(NH).

The compound is poorly soluble in C_6H_6. The measured degrees of association in C_6H_6 (by cryoscopy) are 15.1 (0.142 M solution) and 7.95 (0.216 M solution).

$Be(C_2H_5)N(CH_3)N(CH_3)_2$

The compound is prepared from $NH(CH_3)N(CH_3)_2$ and $Be(C_2H_5)_2$ (mole ratio 1:1). During 2 h at room temperature, nearly 1 equivalent of C_2H_6 is evolved. The solvent is removed in a vacuum and ~15 % of the crude product consists of $Be(C_2H_5)_2 \cdot NH(CH_3)N(CH_3)_2$, which is removed at 25°C/0.05 Torr. The remaining nonvolatile viscous liquid gives the expected elemental analysis for the title compound, and methanolysis yields the correct amount of C_2H_6. Reaction of $Be(C_2H_5)_2$ and $NH(CH_3)N(CH_3)_2$ in the mole ratio 1:2 in C_7H_{16} at 55°C for 4 h still mainly gives the title compound, and only partial formation of $Be(N(CH_3)N(CH_3)_2)_2$ is observed. Reaction of $Be(C_2H_5)_2 \cdot N(CH_3)_3$ with $NH(CH_3)N(CH_3)_2$ (mole ratio 1:1) at 25°C for 1 h gives a crude product that contains also ~50% of $Be(C_2H_5)_2 \cdot NH(CH_3)N(CH_3)_2$. The latter is removed in a vacuum as described before.

According to cryoscopic measurements in C_6H_6, the degree of association in this solvent is 4.18 for 0.107 M and 0.143 M solutions.

Reference on p. 150

Reference:

Fetter, N. R. (Can. J. Chem. **42** [1964] 861/6).

1.3.11 Organoberyllium Phosphides

Organoberyllium phosphides are not explicitly described in the literature; but from the stoichiometry of the following reaction, the formation of **$Be(CH_3)P(CH_3)_2$** could be implied.

Heating equimolar amounts of $Be(CH_3)_2$ and $PH(CH_3)_2$ to 190°C for 5 h results in the formation of ~0.5 mol CH_4, and a nonvolatile, evidently polymeric material that is insoluble in C_6H_6 is obtained [1]. A reaction of the components in ether (mole ratio 1:1.13) in a sealed tube at 65°C for 12 h results in 0.82 equivalents of CH_4. The residue, obtained after evaporation of all volatile material, could not be characterized. Sublimation at 65°C in a vacuum gave $Be(CH_3)_2$ as a sublimate [2].

References:

[1] Coates, G. E., Glockling, F., Huck, N. D. (J. Chem. Soc. **1952** 4512/5).
[2] Bell, N. A., Coates, G. E., Fishwick, A. H. (J. Organometal. Chem. **198** [1980] 113/20).

1.3.12 Monoorganylberyllates and Their Adducts

$Li[BeC_2H_5(OC_4H_9\text{-}t)_2]$

The compound is prepared by addition of LiC_2H_5 in C_6H_6 to $Be(OC_4H_9\text{-}t)_2$ in C_6H_{14} (mole ratio 1:1), stirring for 1 h, and removal of the solvent in a vacuum. The residue crystallizes from C_6H_{14}/C_6H_6 (4:1) as white needles in quantitative yield, m.p. 97 to 98°C.

The compound is easily soluble in C_6H_6 and dimeric in this solvent (by cryoscopy). The structure shown in Formula I ($R=C_2H_5$) is proposed.

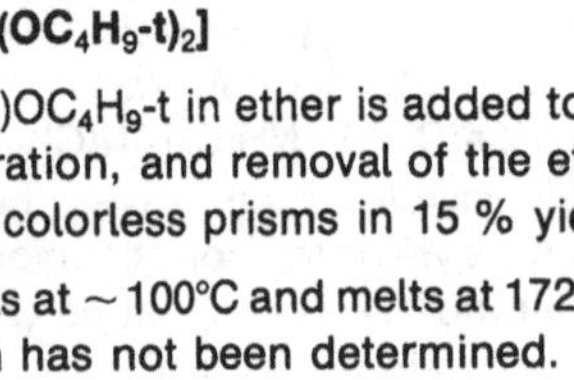

I

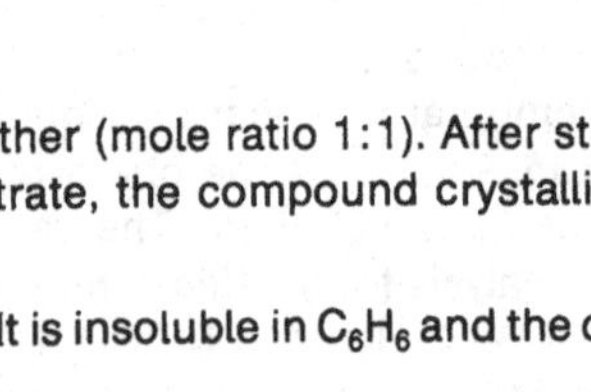

II

$Na[BeC_2H_5(OC_4H_9\text{-}t)_2]$

$Be(C_2H_5)OC_4H_9$-t in ether is added to $NaOC_4H_9$-t in ether (mole ratio 1:1). After stirring for 30 min, filtration, and removal of the ether from the filtrate, the compound crystallizes from toluene as colorless prisms in 15 % yield.

It shrinks at ~100°C and melts at 172 to 174°C (dec.). It is insoluble in C_6H_6 and the degree of association has not been determined.

$Li[BeC_4H_9\text{-}t(OC_4H_9\text{-}t)_2]$

LiC_4H_9-t in C_5H_{12} is added to $Be(OC_4H_9\text{-}t)_2$ in C_6H_{14}. The white precipitate is stirred for 1 h and the solvent is removed in a vacuum. The residue crystallizes from C_6H_6 as large white plates that melt at 174 to 175°C to a turbid liquid.

^1H NMR (C_6H_6): $\delta = 1.30$ (s, C_4H_9Be) and 1.34 ppm (s, C_4H_9O). Dimers are present in C_6H_6 (by cryoscopy). This is consistent with the structure shown in Formula I ($R = t\text{-}C_4H_9$).

An addition compund is formed with $(CH_3OCH_2CH_2)_2O$, which is described next.

$Li[BeC_4H_9\text{-}t(OC_4H_9\text{-}t)_2] \cdot O(CH_2CH_2OCH_3)_2$

The preceding compound is dissolved in an excess of $(CH_3OCH_2CH_2)_2O$. After stirring for 15 min the volatile matter is removed in a vacuum, and the residue crystallizes as waxy plates from C_6H_{14}.

The complex melts at 60°C. The ^1H NMR spectrum in C_6H_6 shows resonances at $\delta = 1.51$ (s, C_4H_9O), 1.64 (s, C_4H_9Be), 3.07 (s, CH_2), and 3.23 ppm (s, CH_3O). Monomers are present in C_6H_6 (by cryoscopy). The structure in Formula II is tentatively suggested.

$Li[BeC_4H_9\text{-}t(OC(C_2H_5)_3)_2]$

Equimolar amounts of LiC_4H_9-t and $Be(OC(C_2H_5)_3)_2$ in C_5H_{12}/C_6H_{14} are stirred for 4 h. Then the volatile material is removed in a vacuum, and the residue crystallizes from C_6H_6 as colorless needles that shrink at ~90°C and melt at 99°C.

The degree of association in C_6H_6 is between 1 and 2 (by cryoscopy). From the values obtained for 1.04, 1.41, and 1.87 wt% solutions, the equilibrium constant for the reaction dimer $\rightleftharpoons$2 monomers is calculated as 0.0030, 0.0028, and 0.0035 mol/L.

$Li_2[BeC_4H_9\text{-}t(OC_4H_9\text{-}t)_3]$

An equimolar amount of LiC_4H_9-t in C_5H_{12} is added to $Li[Be(OC_4H_9\text{-}t)_3]$ in C_6H_{14}. The mixture is stirred for 8 h, and all volatile material is removed in a vaccum. Crystallization of the residue from C_6H_6 gives colorless prisms, m.p. 170 to 172°C.

The ^1H NMR spectrum in C_6H_6 shows three singlets of the relative intensity 1:3:3 at $\delta = 1.40$, 1.42, and 1.43 ppm. Since the molecular weight (in C_6H_6 by cryoscopy) is found to be concentration-dependent and smaller than calculated for the dimer, it is suggested that an equilibrium between monomers and dimers exists.

$Li[BeC_4H_9\text{-}t(N(CH_3)_2)_2]$

LiC_4H_9-t in C_5H_{12} is added to $Be(N(CH_3)_2)_2$ in C_6H_{14} (mole ratio 1:1). After stirring for 2 h the solvent is evaporated, and the compound crystallizes from C_6H_6 as colorless prisms which decompose at ~275°C.

Absorptions in the IR spectrum (in Nujol) are observed at 2760(s), 1248(s), 1240(m,sh), 1167(m), 1140(m), 1120(s), 1093(w), 1045(s), 995(m), 980(w,sh), 950(s), 825(s), 810(s), 753(m), 715(w), and 550(s) cm^{-1}.

The compound is only sparingly soluble in C_6H_6 which suggests an oligomeric or salt-like constitution.

Reference:

Andersen, R. A., Coates, G. E. (J. Chem. Soc. Dalton Trans. **1974** 1729/36).

2 Compounds with Ligands Bonded by Two or Three Carbon Atoms (Beryllocarbaboranes)

3,1,2-$BeC_2B_9H_{11}$

To 1,2-$C_2B_9H_{13}$ in C_6H_6 is added an equimolar amount of $Be(C_2H_5)_2 \cdot 3.3 O(C_2H_5)_2$. C_2H_6 is evolved, and a white precipitate is formed in a small yield. The precipitate is separated by filtration, washed with C_6H_6, and dried at 50°C.

The solid does not melt below 350°C. The IR spectrum in Nujol exhibits five distinct absorptions in the BH region at 2850, 2640, 2610, 2590, and 2550 cm^{-1}. A bridge-type hydrogen absorbs at 2150 cm^{-1}. Absorptions between 1200 and 700 cm^{-1} (not given) can be ascribed to cage absorptions. The same pattern as for the following two compounds is observed, shifted only by 15 cm^{-1} to lower frequency. The insolubility (see below) suggests a high molecular weight. It is therefore suggested that the compound is composed of repeating units of icosahedral geometry having BeHB bridge linkages.

The compound is insoluble in common organic solvents like C_6H_6, C_5H_{12}, CH_2Cl_2, $CH_3C(O)CH_3$, THF, and $CHCl_3$. It is extremely sensitive to air and hygroscopic. Reaction with $N(CH_3)_3$ gives the adduct described on p. 153 [2].

3,1,2-$BeC_2B_9H_{11} \cdot O(C_2H_5)_2$

To a solution of 1,2-$C_2B_9H_{13}$ in C_6H_6/ether is added an approximately equimolar amount of $Be(C_2H_5)_2 \cdot 2 O(C_2H_5)_2$ at 35°C. The solvent is removed at 50°C on a vacuum line. After ~12 h evolution of C_2H_6 ceases [2]; see also [1]. The remaining solid is recrystallized from CH_2Cl_2/C_6H_{14} [2] or CH_2Cl_2/C_5H_{12} [1] by slow evaporation of the solvent mixture. The procedure is repeated three times. The yield is 55% [2]. The reaction can also be performed with $Be(CH_3)_2$ in ether/C_6H_6 [1,2]. The product is obtained as a yellow oil, which gives by treatment with C_6H_6 and subsequent evaporation of C_6H_6 a semisolid that was reacted further with $N(CH_3)_3$ (see below) [2]. If the reaction is performed with $Be(C_2H_5)_2 \cdot 0.33 O(C_2H_5)_2$ and 1,2-$C_2B_9H_{13}$ in C_6H_6, some insoluble (3,1,2-$BeC_2B_9H_{11})_n$ precipitates (see above), and the title compound can be isolated from the filtrate [2].

The white [1,2], nonvolatile [1] crystals melt at 120 to 121°C. The ^1H NMR spectrum in CH_2Cl_2 contains a triplet and a quartet characteristic of two equivalent C_2H_5 groups at 1.80 and 4.62 ppm, and a broad singlet at $\delta = 2.75$ ppm for the hydrogens on the carborane carbon atoms [1]; see also [2]. The ^{11}B NMR spectrum in CH_2Cl_2 contains five unresolved resonances [1] which are not reported due to their uninformative nature [2]. IR absorptions between 1200 and 700 cm^{-1} in Nujol (not given) can be ascribed to cage absorptions [2]. The structure shown in Formula I ($^2D = O(C_2H_5)_2$) is proposed [1,2].

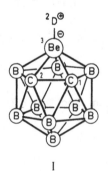

I

The compound is very sensitive to air and moisture [1,2]. It liquifies if exposed to the atmosphere and is decomposed by H_2O [2]. Degradation with ethanolic KOH gives $[1,2\text{-}C_2B_9H_{12}]^-$, which is isolated as the $[N(CH_3)_4]^+$ salt [2]. Reaction with $N(CH_3)_3$ gives the following compound [1,2]. The ether is only incompletely displaced by $S(C_2H_5)_2$. The product obtained with NH_3 is unstable. No products were characterized [2].

$3,1,2\text{-}BeC_2B_9H_{11} \cdot N(CH_3)_3$

The crude product, obtained by reaction of $1,2\text{-}C_2B_9H_{13}$ with $Be(CH_3)_2$ etherate (see above), is dissolved in CH_2Cl_2, and $N(CH_3)_3$ is passed through the solution [2]; see also [1]. The precipitate is separated and washed with C_6H_6. Recrystallization is performed from CH_2Cl_2/C_6H_{14} [2] or CH_2Cl_2/C_5H_{12} [1] by passing N_2 over the surface of the solution. The yield is 51% [2]. The compound is also obtained by treatment of $(3,1,2\text{-}BeC_2B_9H_{11})_n$ (see above) with liquid $N(CH_3)_3$ for 5 h; yield 67% [2].

The white solid melts with decomposition at 221 to 223°C. ^1H NMR(CH_2Cl_2): $\delta = 2.55$ (br s, CH) and 2.90 ppm (s, CH_3). The ^{11}B NMR spectrum in CH_2Cl_2 shows five unresolved resonances [1,2] at $\delta = 6.0$, 12.0, 15.9, 19.5, and 25.0 ppm [1]. IR absorptions between 1200 and 700 cm^{-1} in Nujol (not given) can be assigned to cage absorptions. The structure shown in Formula I ($^2D = N(CH_3)_3$) is proposed [2].

The mass sprectrum shows the parent ion. The compound is less sensitive to air than the etherate [1, 2], but is decomposed by H_2O. Degradation by ethanolic KOH gives $[1,2\text{-}C_2B_9H_{12}]^-$, which was isolated as the $[N(CH_3)_4]^+$ salt [2].

$Be(C_8H_{15}BN)_2$ ($C_8H_{15}BN$ = 1-t-butyldihydro-2-methyl-1H-azaborolyl)

The compound is prepared by reaction of $LiC_8H_{15}BN$ ($C_8H_{15}BN$ = 1-t-butyldihydro-2-methyl-1H-azaborolyl) with $BeCl_2$ in the mole ratio 2:1 in ether at −100°C. After being warmed to room temperature the mixture is heated at reflux for 24 h, and the LiCl is filtered. The filtrate is evaporated, and the residue is extracted with petroleum ether. The solution is evaporated to half the volume and again filtered. Evaporation to dryness yields the compound in 99% yield. Due to the prochiral character of the $C_8H_{15}BN$ ligand, two diastereoisomers are formed in a 1:1 ratio. If the two rings in one molecule use the same site for the coordination, the atomic sequence is BN/NB; different sites result in the BN/BN isomer. The BN/NB isomer crystallizes at −30°C from ether. The crystals become soft on warming above −20°C. The BN/BN isomer remains in solution and is oily when the solvent is evaporated.

NMR spectra of the BN/BN isomer: ^1H NMR($CDCl_3$): $\delta = 0.60$ (s, CH_3B), 1.46 (s, C_4H_9), 3.20 (dd, H-3; J(H-3,5) = 4.2 Hz, J(H-3,4) = 1.3 Hz), 5.78 (dd, H-4; J(H-5,4) = J(H-5,3) = 4.2 Hz), 6.47 ppm (dd, H-5; J(H-4,5) = 3.3 Hz, J(H-4,3) = 1.6 Hz). ^{11}B NMR($CDCl_3$): $\delta = 34.2$ ppm. ^{13}C NMR(C_6D_6): $\delta = 31.5$ (CH_3 in C_4H_9), 55.1 (C in C_4H_9), 61.9 (C-3), 112.2 (C-4), 116.7 (C-5) ppm.

NMR spectra of the BN/NB isomer: ^1H NMR($CDCl_3$): $\delta = 0.68$ (s, CH_3B), 1.47 (s, C_4H_9), 3.16 (dd, H-3; J(H-3,5) = 4.1 Hz, J(H-3,4) = 1.4 Hz), 5.65 (dd, H-5; J(H-5,4) = J(H-5,3) = 3.8 Hz), 6.40 ppm (dd, H-4; J(H-4,5) = 3.3 Hz, J(H-4,3) = 1.6 Hz). ^{11}B NMR($CDCl_3$): $\delta = 34.2$ ppm. ^{13}C NMR(C_6D_6): $\delta = 31.5$ (CH_3 in C_4H_9), 55.2 (C in C_4H_9), 61.9 (C-3), 111.5 (C-4), 116.6 (C-5) ppm.

The ^1H NMR spectrum of a mixture of BN/BN and BN/NB isomer in $CDCl_3$ or $C_6D_5CD_3$ at room temperature shows the signals of both compounds simultaneously. The range of the chemical shifts for H-3, -4, and -5 is typical for a sandwich complex. The room temperature ^1H NMR spectrum gives no indication of a σ-bonded C-3 atom. ^1H NMR experiments between 20 and −55°C in $CDCl_3$, and at −70°C in $C_6D_5CD_3$, show no change in the spectra.

References on p. 154

An X-ray structure determination of the BN/NB isomer was carried out at −150°C. No phase change could be detected between −150 and −20°C. The crystals are monoclinic, space group $P2_1/c-C^5_{2h}$ (No. 14) with the cell dimensions a = 8.606(8), b = 10.121(9), c = 20.582(20) Å, β = 92.78°; Z = 4, d_c = 1.61 g/cm³. The structure was refined to R = 0.077 and R_w = 0.078. It is shown in **Fig. 13** together with selected bond lengths and bond angles. The Be atom is η^5-coordinated by one $C_8H_{15}BN$ ring and η^1-bonded to C(3) of the second ligand. A comparison of the bond distances in both rings shows that the η^1-coordinated ring behaves like a diene, as the C(3)–B and C(3)–C(4) distances are longer than in the η^5 ring, but C(4)–C(5) and BN are significantly shorter. The rings are not coplanar. The interplanar angle is 14.5°. The distance of Be to the η^5 ring is 1.47 Å and somewhat shorter than in $Be(C_5H_5)_2$ (1.505 Å). Also the BeC(3) distance is with 1.755 Å significantly shortened compared with $Be(C_5H_5)_2$ (1.826 Å). The bond angles at C(3) characterize this atom as not fully sp^3 hybridized. The investigated BN/NB diastereoisomer should exist in enantiomeric forms (R,R and S,S). The crystal structure indeed shows the two forms alternating in the unit cell.

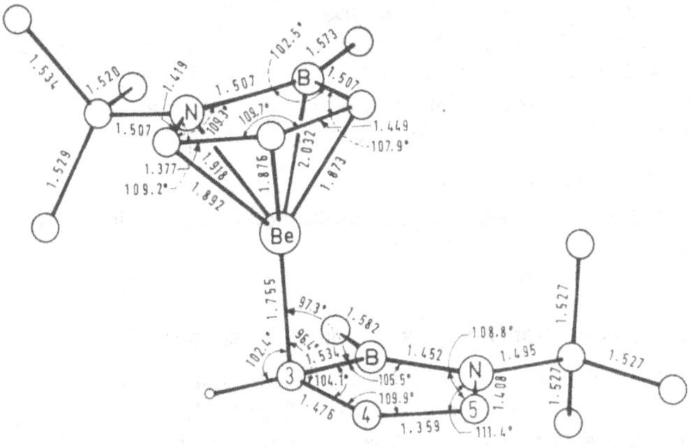

Fig. 13. Molecular structure of $Be(C_8H_{15}BN)_2$ [3].

On dissolving the BN/NB isomer in C_6D_6 or $C_6D_5CD_3$, the original 1:1 mixture of the BN/NB and BN/BN isomers is formed again within 5 days. This suggests a dissociation of one ring into $[BeC_8H_{15}BN]^+$ and $[C_8H_{15}BN]^-$, enabling the flip. The dissociated fraction must be very small, since no remarkable conductivity is found in C_6H_6, $C_6H_5CH_3$, $CHCl_3$, petroleum ether, and ether. More basic solvents like THF decompose the complex.

The compound reacts with $Be(C_5H_5)_2$ in ether at room temperature to give $Be(C_5H_5)C_8H_{15}BN$ in quantitative yield [3].

References:

[1] Popp, G., Hawthorne, M. F. (J. Am. Chem. Soc. **90** [1968] 6553/4).

[2] Popp, G., Hawthorne, M F. (Inorg. Chem. **10** [1971] 391/3).

[3] Schmid, G., Boltsch, O., Boese, R. (Organometallics **6** [1987] 435/9).

3 Compounds with Ligands Bonded by Five Carbon Atoms

3.1 Cyclopentadienylberyllium Hydride and Borohydrides

$Be(C_5H_5)H$

$Be(C_5H_5)H$ is prepared by the reaction of $Be(C_5H_5)BH_4$ with $P(C_6H_5)_3$. A 2-fold excess of $P(C_6H_5)_3$ is condensed onto $Be(C_5H_5)BH_4$, and the flask is evacuated at −196°C. The mixture is heated for 1 h at 50°C, and the product is vapor-transferred as a mixture of a white solid and colorless crystals. $Be(C_5H_5)D$ is prepared analogously from $Be(C_5H_5)BD_4$. The reaction of $Be(C_5H_5)BH_4$ with a large excess of $N(CH_3)_3$ or $P(CH_3)_3$ at room temperature gave $Be(C_5H_5)_2$, BeH_2, and $BH_3 \cdot N(CH_3)_3$ or $BH_3 \cdot P(CH_3)_2$, respectively. The mass spectrum of $Be(C_5H_5)_2$ indicated the presence of some $Be(C_5H_5)H$. $Be(C_5H_5)H$ could not be obtained by reactions of LiH or $LiAlH_4$ with $Be(C_5H_5)X$ (X = Cl, Br); BeH_2 with $Be(C_5H_5)_2$; $Na[BH(C_2H_5)_3]$ with $Be(C_5H_5)X$ (X = Cl, Br); $Sn(CH_3)_3H$ with $Be(C_5H_5)R$ (R = alkyl); or pyrolysis of $Be(C_5H_5)R$ (R = alkyl or BH_4) [3].

$Be(C_5H_5)H$ and $Be(C_5H_5)D$ do not melt. They appear wet over a range of temperatures and then appear dry as if they are going through phases. At about 270°C, $Be(C_5H_5)H$ starts to darken; and at about 325°C, it is completely decomposed. It also appears to shrink above about 100°C. The compounds are easily sublimed at room temperature/10^{-3} Torr over a period of several days to give small, colorless crystals. The vapor pressure at 23°C is 0.24 Torr [3].

$Be(C_5H_5)H$ and $Be(C_5H_5)D$ are monomeric in the vapor state and in solution, but are associated to an unknown extent in the solid state [3]. A dipole moment of 2.08 D for $Be(C_5H_5)H$ is derived from the microwave spectrum [3,4].

The ^1H NMR spectrum shows a single resonance for the C_5H_5 protons at $\delta = 5.72$ ppm in C_6H_6 and at $\delta = 5.75$ ppm in c-C_6H_{12}. The C_5H_5 peak is not concentration-dependent in C_6H_6. The hydride proton was not detected [3].

IR spectra of $Be(C_5H_5)H$ and $Be(C_5H_5)D$ were measured for the vapor phase, in Nujol, and at −196°C. The Raman spectrum of $Be(C_5H_5)H$ was measured in C_6H_6 solution and in the solid state. The vapor phase IR and solution Raman spectra are assigned and compared with the computer-calculated spectra for the monomeric species with C_{5v} symmetry. The vapor phase IR absorptions for $Be(C_5H_5)H$ are observed at 3130(w), 3095(vw), 2082(vw), 2048(sh), 2032(s, sh), 2023(sh), 2018(sh), 1945(vvw), 1813(vw), 1734(w), 1683(vw), 1436(w), 1390 to 1340(w, br), 1083(w, sh), 1017(m, sh), 875(s, sh, PQR), 730(vw), 547(s), 435(m), and 310(vw) cm^{-1}. The solution Raman spectrum has bands at 3123(vs, sh), 3100(vs), 3070(s), 1412(m), 1356(m), 1247(w), 1085(vs, sh), 1068(w), 1028(m), 870(m), 845(w), 836(m), 622(w), 550(m), 435(m), and 310(s, sh) cm^{-1}. The BeH stretching frequency in the vapor is found at ~2030 cm^{-1} and shifts to ~1535 cm^{-1} for BeD. The solid state IR and Raman spectra are more complex and indicate that the C_{5v} local symmetry about the C_5H_5 ring is reduced and that a mixture of monomeric and associated species is present. Some possible structures for the solid state are discussed. The BeH stretching mode is found at ~1720 cm^{-1}, and ν(BeD) at ~1275 cm^{-1}. Calculated force constants (dyn/cm) for the BeH bond are 2.20×10^5 (2.28×10^5 for BeD) in the vapor and 1.58×10^5 (1.58×10^5 for BeD) in the solid state [3].

$Be(C_5H_5)H$ has been studied by gas phase electron diffraction analysis and is shown to have a BeH bond length of 1.306(6) Å, BeC = 1.929(7) Å, BeC_5(center) = 1.502(7) Å, and CC = 1.422(2) Å [3], but a quantitative structure determination by this method is not possible due to the instability of the molecule [4].

It was not possible to record a He(I) photoelectron spectrum because of disproportionation to $Be(C_5H_5)_2$ [18].

References on pp. 164/5

The microwave spectra of $Be(C_5H_5)H$, $Be(C_5H_5)D$, $Be(^{13}CC_4H_5)H$, and $Be(^{13}CC_4H_5)D$ are reported. They are typical of symmetric tops, and the C_{5v} symmetry (Formula I) is confirmed. The H···ring distance in 2.806±0.003 Å. The other structural parameters for $Be(C_5H_5)H$ and $Be(C_5H_5)D$ are included in Table 28.

Four different vibrationally excited normal modes are identified and their wave numbers were determined assuming A(E) symmetry: $\tilde{\nu}_1$=325(452), $\tilde{\nu}_2$=516(643), $\tilde{\nu}_3$=487(614), and $\tilde{\nu}_4$=717(843) cm^{-1}; $\tilde{\nu}_4$ is tentatively assigned to δ (ring BeH). The other three modes presumably involve ν(Be ring) and $\delta(C_5H_5)$. The excited state lines are weak in comparison with the ground state lines, indicating a rigid molecule [3,4].

Several structures are probed by MO calculations with the MNDO [7,17], PRDDO [13], and ab initio [8] methods. The models I to IV are considered in the ab initio calculations [8].

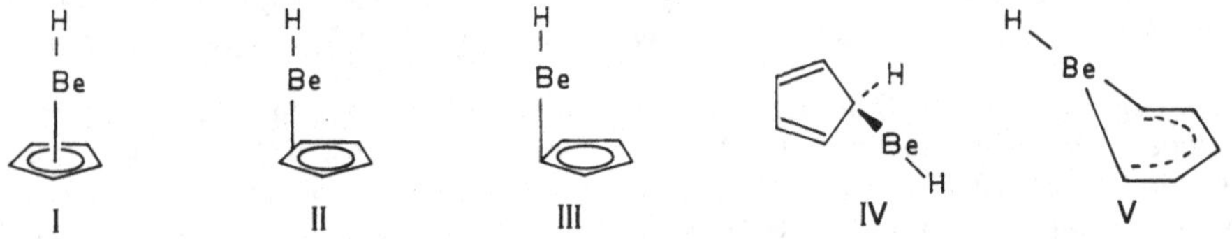

In agreement with the experimental results, the C_{5v} geometry (model I) was found to have the lowest energy [7,8,13,18] (total energy = −204.970 a.u. [8]). The calculated and experimental parameters for the C_{5v} geometry are summarized in Table 28. The structural data obtained with the STO-3G basis set are in remarkable agreement with the experimental data. It follows from the ab initio calculations that the CH bonds in C_5H_5 are bent towards Be [8]. This contrasts with the PRDDO [13] and MNDO results [7]. The bending is explained with a rehybridization of the π orbitals in C_5H_5 to achieve maximum overlap with the two Be 2p orbitals [8]. The Be is sp hybridized, with two empty 2p orbitals acting as acceptors for the C_5H_5 e_1 orbitals. The HOMO is a degenerate e_1 type [7,8]. A schematic interaction diagram is shown in [8].

SCF-X_α-scattered wave calculations give a sequence of valence orbitals that is in agreement with the qualitative interaction diagram. The HOMO ($3e_1$) is mostly a ring orbital with a small amount of Be(2p) character. The next MO ($4a_1$) is the BeH σ bond. The charge densities demonstrate only a very small contribution from the a_1 ring π orbital. The $3a_1$ MO derives principally from the interaction of the Be(2s) AO with the totally symmetric ring MO. The computed ionization energies are (in eV): 8.70 ($3e_1$), 12.02 ($4a_1$), 12.91 ($3a_1$), 15.26 ($2e_2$), 15.75 ($2e_1$), 19.17 ($2a_1$), 19.52 ($1e_2$), 23.74 ($1e_1$), and 28.94 ($1a_1$) [19]. Other calculated ionization potentials are (in eV): 9.88 [7], 9.64 [14] for model I and 8.82 for model IV [7].

It follows from the ab initio (STO-3G) calculations that model II is 43.5 kcal/mol less stable than model I. Model III is 58.5 kcal/mol less stable, and model IV 51 kcal/mol [8]. But in the MNDO calculation model IV is a stable form and only 9 kcal/mol less stable than model I [7]. A beryllaheterocycle with C_s symmetry (Formula V) was found to be ~60 kJ/mol (~15 kcal/mol) more stable than the C_{5h} structure (Formula I) by MNDO calculations [17].

The calculated dipole moments are in good agreement with the experimental value. The negative end of the dipole moment is directed towards the BeH group. This argues against ionic bonding and implies considerable electron donation from the ring into the Be p orbitals. The Mulliken charge on Be is 0.222 and −0.045 on the BeH hydrogen [8].

The isodesmic reaction $C_5H_6 + Be(CH_3)H \rightarrow CH_4 + Be(C_5H_5)H$ ($\Delta H = -54.8$ kcal/mol) indicates the large stabilization of $Be(C_5H_5)H$ relative to $Be(CH_3)H$. The equation $LiC_5H_5 + BeH^+ \rightarrow Li^+ + Be(C_5H_5)H$ ($\Delta H = -92.6$ kcal/mol) compares the relative stability of LiC_5H_5 and $Be(C_5H_5)H$.

The difference between Li and BeH bonding to C_5H_5 is expressed by the isodesmic reaction $LiC_5H_5 + BeH \rightarrow Li + Be(C_5H_5)H$ ($\Delta H = -35.8$ kcal/mol) [8].

Table 28
Structural Parameters of $Be(C_5H_5)H$ (C_{5v}).
Distances in Å, μ in D.

parameter	experiment [3,4]	STO-3G [8]	4-31G [8]	PRDDO [13]	MNDO [7]
CC	1.423±0.001	1.419	1.421	1.40	1.458
CH	1.09[a)]	1.079	1.079	1.06	1.084
BeC	1.920	1.901	1.976	1.83	1.991
Be ring	1.49[a)]	1.461	1.563	1.39	1.557
BeH	1.32±0.01	1.280	1.280	1.26	1.285
∢ring CH	180°	179.1°[b)]	179.2°[b)]	−177.6°[b)]	−172.0°[b)]
dipole moment μ	2.08	2.80	1.792		

[a)] Assumed values, [b)] positive values indicate CH bending towards Be, negative values away from Be.

It appears from PRDDO calculations that $C_5H_5BeH_2BeC_5H_5$ dimers with either σ- or π-bonded C_5H_5 groups are not energetically favorable [18].

$Be(C_5H_5)H$ and $Be(C_5H_5)D$ are moderately soluble in C_6H_6 and c-C_6H_{12}, in which they are monomeric (by cryoscopy). The hydride is not decomposed in refluxing C_6H_6 in the presence of a Pt sheet within 24 h. Especially, no C_5H_5Be–BeC_5H_5 is formed. The hydride and deuteride slowly decompose over a large temperature range up to 300°C [3]. One of the decomposition products is C_5H_6, which was identified by its microwave spectrum [3, 4]. The compounds are sensitive towards air [3].

The mass spectra of $Be(C_5H_5)H$ and $Be(C_5H_5)D$ were obtained at ionizing voltages of 70, 30, 20, and 10 eV. The major species in the 70, 30, and 20 eV spectra is $[BeC_5H_5]^+$, while the major species in the 10 eV spectrum is $[Be(C_5H_5)H]^+$. Other fragments observed in the 70 eV spectrum are $[BeC_5H_n]^+$ (n = 4 to 1), $[BeC_4H_n]^+$ (n = 5 to 3), $[BeC_3H_n]^+$ (n = 4 to 0), $[Be(C_5H_5)H]^{2+}$, $[BeC_2H_3]^+$, $[BeCH_n]^+$ (n = 2, 1), and fragments of the type $[C_nH_m]^+$. No associated species were observed. The ionization potential for the hydride is 10.33 eV and for the deuteride 10.87 eV. The appearance potentials for $[BeC_5H_5]^+$ are 11.76 and 11.72 eV, respectively. The BeH bond dissociation energy is 0.93 eV (= 21.45 kcal/mol), and the BeD bond dissociation energy is 0.94 eV (= 21.68 kcal/mol) [3].

No reactions were observed with $Be(C_5H_5)H$ and CH_3I, C_5H_5I, 2-pentene, 1-Br-2-butene, 2-CH_3-2-butene, 3-Br-propene, and 3-I-propene in C_6H_6 at room temperature. Acetone, $(C_6H_5)_2CO$, and pyridine are reduced by the hydride, but only $(C_6H_5)_2CH_2$ is identified in the IR spectrum for the second case. The reaction of $NH(CH_3)_2$ with $Be(C_5H_5)H$ in the mole ratio 3.42:1.42 gives $Be(N(CH_3)_2)_2$. With the same reactants in a ~1:1 mole ratio, $Be(C_5H_5)N(CH_3)_2$ is obtained. With $NH(C_2H_5)_2$ in a ~2:1 mole ratio, a mixture of $Be(C_5H_5)N(C_2H_5)_2$ and $Be(N(C_2H_5)_2)_2$ is obtained. The mole ratio 1.14:1.73 gives $Be(C_5H_5)N(C_2H_5)_2$ and some unreacted $Be(C_5H_5)H$ [3].

$Be(C_5H_5)BH_4$

The compound and its deuterated analog $Be(C_5H_5)BD_4$ are prepared by the reaction of $Be(C_5H_5)X$ (X = Cl [3, 5], Br [3]) with $LiBH_4$ [3, 5] in a 1:1 mole ratio or with an excess of $LiBD_4$

References on pp. 164/5

[3, 5] without a solvent at room temperature. The mixture is subsequently heated to 60°C for 30 min [5] to 1 h [3] to ensure completion of the reaction. The flask is evaporated, first at −196°C, then at −80°C, and the product is vapor-transferred into a storage vessel [3, 5], also [1, 6]. If the reaction is performed with $NaBH_4$ instead of $LiBH_4$, heating to higher temperatures is required, which leads to more decomposition [3].

The borohydride and borodeuteride are colorless liquids [1, 3, 5], m.p. −22°C [3]. The vapor pressure is 2.7 mm at 23°C [3] (~3 mm [5]). The compound is monomeric in solution and in the gas phase [2, 3].

The ^1H NMR spectrum consists of a sharp singlet for the C_5H_5 protons at $\delta = 5.78$ ppm in C_6H_6 and $\delta = 6.13$ ppm in c-C_6H_{12} [3]. The four protons of the BH_4 group are equivalent at room temperature [2, 3, 6, 12] in C_6H_6, c-C_6H_{12} [3], and $C_6H_5CH_3$ [2]. The signal is split by ^{11}B into a quartet of four equally intense peaks. Each peak of the quartet is again split by ^9Be into a quartet with four equally high peaks, J(H, Be) = 10.2 Hz [3, 6, 12]. The ^9Be decoupled ^1H NMR spectrum shows the $^1H,^{11}B$ quartet (J = 84 Hz) and the $^1H,^{10}B$ septet (J = 28 Hz) [3, 12, 15]. In C_6H_6, the quartet of quartets centers at $\delta = -0.24$ ppm and in c-C_6H_{12} at $\delta = -0.66$ ppm [3]. Cooling to −80°C does not give separate signals for terminal and bridging H's [2].

The chemical shift of $\delta = -53.6$ ppm (solvent?) for the boron atom in the ^{11}B NMR spectrum is unusually high for a tetrahydroborate species [15]. The observed quintet of quartets ($J(^{11}B,H) = 84$ Hz, $J(^{11}B,^9Be) = 3.7$ Hz) [12, 15] indicates that rapid positional exchange of the hydrogens takes place within the BH_4 group, but intermolecular exchange of BH_4 groups is not observed. The exceedingly narrow lines in the ^1H and ^{11}B NMR spectra (half-height widths = 2 and 1.5 Hz, respectively) suggest that the electric field gradient at the boron nucleus is effectively zero [15].

The ^9Be NMR spectrum in C_6F_6 exhibits a multiplet at $\delta = -22.1$ ppm, resulting from the ^9Be,H coupling (J = 10.2 Hz) and the ^9Be, ^{11}B coupling (1:1:1:1 quartet, J = 3.6 Hz). The resolution of this pattern is complicated due to overlap and a 20% contribution of an unresolved ^9Be, ^{10}B septet coupling of 1.2 Hz (with equally intense peaks) [12]. Figures of all NMR spectra are given in the original literature [3, 12, 15].

IR spectra are reported for solid $Be(C_5H_5)BH_4$ and $Be(C_5H_5)BD_4$, and Raman spectra are reported for solid and liquid $Be(C_5H_5)BH_4$ und $Be(C_5H_5)BD_4$. Vibrational assignments for most of the C_5H_5, $BeBH_4$, $BeBD_4$, and skeletal modes are made. The observed vibrations of $Be(C_5H_5)BH_4$ and their assignment are listed in Table 29. The C_5H_5 vibrations are in close accord with those of $Fe(C_5H_5)_2$ while the vibrations of the $BeBH_4$ group are analogous to those reported for $[BeBH_4]^+$ in solid $Be(BH_4)_2$. The observed $BeBH_4$ vibrations give conclusive evidence for a double hydrogen bridge configuration (Formula VI, p. 161) [5]. For $Be(C_5H_5)BH_4$ and $Be(C_5H_5)BD_4$, gas phase IR spectra are also given which are very similar to the solid phase spectra [3]. Comparison of the liquid and solid Raman spectra of $Be(C_5H_5)BH_4$ and $Be(C_5H_5)BD_4$ also shows only small frequency shifts between these phases, and thus it may be concluded that the covalently bonded structure VI is retained in all three phases. Observation of several transitions for the skeletal torsional motion for $Be(C_5H_5)BH_4$ implies that free internal rotation about the symmetry axis does not occur, and calculations yield a value of 500 cm^{-1} (1.4 kcal/mol) for the tenfold torsional barrier. Raman polarization data indicate that the $BeH_{2b}B$ bridging plane and the C_s molecular symmetry plane are coincident for the most probable orientation [5].

Table 29
IR and Raman Vibrations (in cm^{-1}) of $Be(C_5H_5)BH_4$ [5].

IR solid 14 K	Raman solid 15 K	Raman liquid 50°C	ϱ[a)]	C_{5v}[b)]	C_s[b)]	C_{2v}[b)]	assignment[c)]
3125(w)	3123(vvs)	3123(vvs)	0.12	A_1			ν_1, ν(CH)
3118(m) 3110(m)				E_1			ν_5, ν(CH)
3102(w)	3103(vs)	3104(vs)	0.8	E_2			ν_9, ν(CH)
2938(w)	2930(vvw)	2938(vvw)	0.00			A_1	$2\nu'_{11}$
	2855(vw)	2858(vw)	0.16	A_1			$2\nu_6$
	2777(vw)	2778(vw)	0.80	E_2			$\nu_6+\nu_{10}$
	2696(vvw)	2694(vvw)	0.10	A_1			$2\nu_{10}$
2477(vs)	2473(s)	2479(s)	0.6			B_1	ν'_7, $\nu(BH_t)$
2426(vs)	2422(vs)	2428(vs)	0.06			A_1	ν'_1, $\nu(BH_t)$
2251(w)	2252(vw)			A_1			$2\nu_2$
2170(vs)						B_2	ν'_{10}, $\nu_{as}(BH_b)$
	2166(vs)	2162(vs)	0.05			A_1	ν'_2, $\nu_s(BH_b)$
	1964(vw)	1969(vw)	0.12	A_1			$\nu_3+\nu_{11}$
1750(w, br)					A″		$\nu'_5+\nu_8$
		1685(vvw)	0.19	A_1			$2\nu_{13}$
1477(vs)	1478(m)					A_1	ν'_3, $\nu_s(BeH_b)$
1470(vs)	1472(m)	1476(w)	0.76			B_2	ν'_{11}, $\nu_{as}(BeH_b)$
1434(m)	1435(w)	1433(vw)	0.8	E_1			ν_6, ν(CC)
	1353(s)	1355(s)	0.83	E_2			ν_{10}, ν(CC)
1262(vw) 1235(vw)	1239(w)	1238(w)	0.06	E_2			ν_{11},$2\nu_{14}$
	1173(m)	1176(vvw, sh)	dp,?			A_2	ν'_6, $\delta(BH_{2t})$ torsional
1131(vs)	1131(vvw)					A_1	ν'_4, $\delta(BH_{2t})$
1124(s, sh)	1123(vvs)	1124(vvs)	0.05	A_1			ν_2, ν(CC)
	1070(s)	1067(s)	0.76	E_2			ν_{12}, δ(CH) out-of-plane
1065(vw)					A″		$\nu''_4+\nu_{13}$
1016(vs)	1019(m)	1016(vw)	0.72	E_1			ν_7, δ(CH) in-plane
	996(vw, sh)					A″	$\nu'_9+\nu'_{12}$
990(m)	990(m)	990(vvw)	0.6			B_1	ν'_8, $\varrho(BH_{2t})$
973(m, sh)					A′		$2\nu''_2$
935(m, sh)	936(w)				A′		$2\nu''_3$
921(m, sh)	921(vw, sh)	920(vw, sh)	dp,?		A″		$\nu_3+\nu''_4$
904(vs)	913(vw)	906(vw)	dp,?		A″		$\nu_3+\nu'_9$
889(vs)	888(m)	896(m)	0.15			A_1	ν'_5, ν(BeB)
863(vw, sh)	863(m)	864(w)	0.72	E_1			ν_8, δ(CH) out-of-plane
	843(vw)	834(w)	0.65	E_2			ν_{13}, δ(ring) in-plane

References on pp. 164/5

Table 29 (continued)

IR solid 14 K	Raman solid 15 K	Raman liquid 50°C	$\varrho^{a)}$	$C_{5v}{}^{b)}$	$C_s{}^{b)}$	$C_{2v}{}^{b)}$	assignment[c)]
823(vs)	824(vw)					E_2	ν'_{12}, $\delta(BH_{2t})$ wagging
735(s)	739(m)	735(w)	0.34	A_1			ν_3, δ(CH) out-of-plane
	622(w)	619(vw)	0.54	E_2			ν_{14}, δ(ring) out-of-plane
	497(m)				A′		ν''_2
486(vw)	488(m)	471(m, sh)	dp,?		A″		ν''_3
416(vw)	418(s)	418(m, sh)	dp,?				ν''_6 (A→A)
	401(m)						ν''_6 (A→E_1)
391(vw)	392(s)	389(vs)	0.14		A′		ν''_1
	332(m)	314(vw, sh)	dp,?				ν''_6 (A→E_2)
	324(m)						ν''_6 (E_2→E_2)
	220(s)	213(s)	0.79		A, A″		ν''_4, ν''_5
	182(m)	171(m, sh)	0.8			B_1	ν'_9, $\delta(BeH_2B)$

a) Polarization coefficient, b) C_s = total molecular symmetry, C_{5v} = local symmetry of C_5H_5, C_{2v} = local symmetry of $BeBH_4$, c) ν_1 to ν_{14} ring vibrations, ν'_1 to ν'_{12} = $BBeH_4$ vibrations, ν''_1 to ν''_6 = skeletal vibrations.

The gas phase molecular structure was determined by electron diffraction at 60°C. The radial distribution curves confirm the monomeric nature of the compound in the gas phase and the C_{5v} symmetry of the $Be(C_5H_5)B$ trunk. But the refinements could not distinguish between a double-bridged (Formula VI, C_{2v} symmetry) or a triple-bridged (Formula VII, C_{3v} symmetry) $BeBH_4$ group. Both models can be brought into satisfactory agreement with the electron diffraction data. For both models, the same structural parameters for the BeC_5H_5 fragment are obtained (distances in Å): CH = 1.116(8), CC = 1.422(1), BeC = 1.915(5), Be ring center = 1.484(7). Interatomic distances (in Å) and angles for the $BeBH_4$ part in Formulas VI and VII are as follows [2]:

	Formula VI	Formula VII
BeB	1.88(1)	1.89(1)
BH_t	1.17(3)	1.16(2)
BH_b	1.29(5)	1.28(3)
BeH_b	1.78(9)	1.70(5)
∢$BeBH_b$	65(4)°	62(2)°
∢H_tBH_b	123(9)°	—

An MNDO calculation shows that the double-bridged form VI is lower in energy than the triple-bridged form VII by 0.5 kcal/mol (ΔH_f(VI) = −6.4 kcal/mol). The calculated molecular geometries (in Å) for structures VI and VII (in parentheses) are: CC = 1.456(1.455), CH = 1.084(1.084), BeC = 2.019(2.021), BeH_b = 1.611(1.818), BH_b = 1.271(1.230), and BH_t = 1.166(1.159). The ionization potentials obtained with Koopmans' theorem are 9.67 eV (Formula VI) and 9.70 eV (Formula VII), and the calculated dipole moments are 2.58 D and 3.46 D, respectively [7].

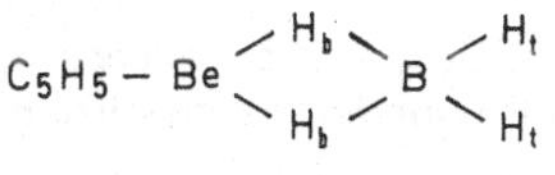

VI

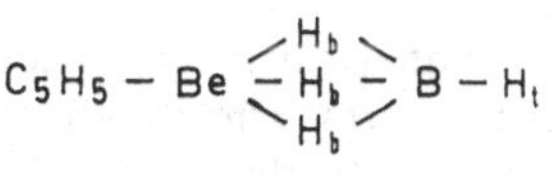

VII

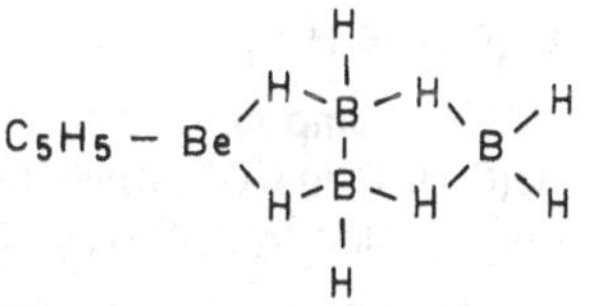

VIII

The borohydride and borodeuteride dissolve as monomers in C_6H_6 [2,3] and c-C_6H_{12} (by cryoscopy) [3]. Mass spectra of $Be(C_5H_5)BH_4$ and $Be(C_5H_5)BD_4$ are obtained at ionizing voltages of 70, 30, 20, and 10 eV. All fragments containing boron are observed as two species, one mass unit apart due to the two isotopes ^{11}B and ^{10}B. For the borohydride, the major species at 10 eV is the parent cation $[Be(C_5H_5)BH_4]^+$. The major species at 70, 30, and 20 eV is $[Be(C_5H_6)B]^+$. The other fragments observed at 70 eV are $[BeC_5H_nB]^+$ (n = 8 to 5), $[BeC_5H_n]^+$ (n = 6 to 1), $[BeC_3H_nB]^+$ (n = 4 to 1), $[BeC_3H_nB]^+$ (n = 3 to 1), $[BeCH_2B]^+$, $[BeBH_n]$ (n = 4 to 0), $[BeH_n]^+$ (n = 2 to 0), as well as $[C_nH_m]^+$ type and $[BH_n]^+$ type fragments. The major species for the borodeuteride are at 70 and 30 eV $[BeC_5H_5]^+$, at 20 eV $[BeC_5H_5D]^+$, and at 10 eV the parent cation $[BeC_5H_5BD_4]^+$. The other fragments observed are analogous to the borohydride. The ionization potential for $Be(C_5H_5)BH_4$ is 10.03 eV; the appearance potential of $[BeC_5H_5]^+$ is 14.03 eV; and the BeH bond dissociation energy is 4.00 eV (= 92.24 kcal/mol). The corresponding values for $Be(C_5H_5)BD_4$ are 10.01, 14.05, and 4.04 eV (= 93.17 kcal/mol) [3].

The compound is pyrophoric [1,5]. No decomposition is observed by pyrolysis at ~100°C in an atmosphere of N_2. The borohydride reacts with excess of $N(CH_3)_3$ and $P(CH_3)_3$. The products observed are $Be(C_5H_5)_2$, BeH_2, and $BH_3 \cdot N(CH_3)_3$ or $BH_3 \cdot P(CH_3)_3$, respectively. $Be(C_5H_5)H$ is an intermediate in these reactions and disproportionates to $Be(C_5H_5)_2$ and BeH_2. $Be(C_5H_5)BH_4$ and $Be(C_5H_5)BD_4$ give, with a 2-fold excess of $P(C_6H_5)_3$, the hydride $Be(C_5H_5)H$ or the deuteride $Be(C_5H_5)D$, respectively [3].

$Be(C_5H_5)B_3H_8$

Reaction of $Be(B_3H_8)_2$ with an excess of NaC_5H_5 for several hours at room temperature produces a moderate yield of a slightly volatile liquid which is isolated by vacuum condensation into a trap cooled to 0°C. The mass spectrum showed no parent peak, but the $[M-2H]^+$ peak confirmed the proposed formula.

The ^1H NMR spectrum in $C_6D_5CD_3$ shows at 23°C a broad resonance at $\delta = 0.46$ ppm with some fine structure due to B_3H_8, which shifts to 1.16 ppm at −75°C. A sharp singlet for the C_5H_5 protons is observed at 5.87 or 6.50 ppm at 23°C or at 5.08 or 5.59 ppm at −75°C. $Be(C_5H_5)_2$ contamination prevented precise assignment. The line width of the BH resonance decreases from 164 Hz at room temperature to 85 Hz at −75°C. The CH resonance remains a sharp singlet at −75°C.

The ^{11}B NMR spectrum in CD_2Cl_2 shows a nonet resonance at $\delta = -25.5$ ppm with $J(^{11}B,H) = 34$ Hz at 23°C. At −87°C the coupling in the ^{11}B spectrum is no longer visible, but the same overall width as at room temperature is observed.

The NMR results indicate that the B_3H_8 moiety undergoes hydrogen positional exchange which is rapid on the NMR time scale even at −87°C. For static conditions, the B_3H_8 ligand is proposed to be a bidentate species (Formula VIII) as observed for most B_3H_8 complexes.

The IR spectrum of a solid film at −196°C shows the following bands (in cm^{-1}): 3100(w), 2545(m), 2470(m), 2190(w), 1640(s,br), 1440(m), 1170(m), 1130(m), 1090(w), 1020(m), 985(s), 900(s), 835(m), 800(m,sh), 790(m), 755(m), and 680(w) [15].

References on pp. 164/5

$Be(C_5H_5)B_5H_8$

The compound is formed in a high yield in the reaction of KB_5H_8 and an excess of $Be(C_5H_5)Cl$ in C_5H_{12} while the mixture is warmed from −40°C to room temperature with stirring. It is purified by high vacuum trap-to-trap fractionation [10, 11].

The product is a colorless solid of low volatility (vapor pressure ≪1 Torr at room temperature) which melts at ~38°C. The mass spectrum shows the parent ion [10, 11].

The ^1H NMR spectrum (solvent?) contains a sharp singlet at $\delta = 5.4$ ppm for the C_5H_5 protons, quartets at $\delta = 1.1$, 1.8, and 2.7 ppm for the terminal BH protons, and two broad singlets of the intensity 1:2 at $\delta = -2.5$ and −3.7 ppm for the bridging BH hydrogens [10, 11].

The ^9Be NMR spectrum (C_5H_{12}/CF_3Br) shows a signal at $\delta = -21.0$ ppm (half-height width 28 Hz) [12].

The ^{11}B NMR spectrum is appropriate for a μ-substituted B_5H_9 derivative. It consists of three doublets of the intensity 2:2:1 at $\delta = -13.4$ (J(B,H) = 161 Hz), −21.8 (J(B,H) = 141 Hz), and −54.5 ppm (J(B,H) = 170 Hz). With ^1H decoupled, the ^{11}B(apex)/^{11}B (base) coupling is observed [10, 11]. The large upfield shift of the resonances associated with two of the basal borons suggests a higher electron density at these centers [11].

The crystals formed from the melt are monoclinic, space group $P2_1/c\text{-}C^5_{2h}$ (No. 14), with a = 10.266(8), b = 5.616(4), c = 16.187(7)Å, β = 98.50(5)°, V = 923.0(5) Å^3; Z = 4, and $d_c = 0.978$ g/cm^3. The crystal structure, determined at −100°C, was refined to an R value of 0.0551 ($R_w = 0.0831$). It is shown in **Fig.14**. As indicated by the NMR spectra, a bridging position on $[B_5H_8]^-$ is occupied by $[BeC_5H_5]^+$. The B(2)–B(3) bond is shorter than the rest of the basal B–B distances. H(1) is tilted by 5° from the fourfold axis of the pyramid toward Be. The BeC_5H_5 moiety is tilted up toward H(1). H(4-5) is tilted further under the base of the B_5 framework compared to H(3-4) or H(5-2). The C_5H_5 ring is nearly planar with local C_{5v} symmetry. The CH hydrogens are tilted slightly inward towards the Be. The distortions cannot be explained by crystal packing phenomena as the intermolecular distances are much too large [10, 11].

The thermal stability of the compound is surprisingly high. It is stable at room temperature in the solid state and in nonbasic solvents such as C_6H_6 or CH_2Cl_2 [10]. In C_6H_6, thermal decomposition occurs slowly at 80°C and rapidly at 140°C [10, 11] to uncharacterized insoluble materials [11]. No decomposition is observed in $O(C_4H_9)_2$ even after several hours at 140°C. It decomposes readily in the presence of O_2 or H_2O [10]. The compound does not react or rearrange with $O(CH_3)_2$ at ambient temperature or at 100°C, or with $O(C_2H_5)_2$ at ambient temperature in C_6H_6 in a sealed NMR tube. On the other hand, $N(CH_3)_3$ reacts rapidly, producing $BH_3 \cdot N(CH_3)_3$. In contrast, 1,3-bis(dimethylamino)naphthalene does not react upon heating to 100°C in C_6D_6 solution or in ether at room temperature [11].

Reaction with $Zn(CH_3)_2$ in ether appears to result in deprotonation to form CH_4 and what is tentatively characterized as $Be(C_5H_5)B_5H_7ZnCH_3$ (Formula IX). The reaction of $Zn(CH_3)_2$ with the title compound in C_6D_6 solution is very different and produces $Be(CH_3)C_5H_5$ and presumably $^3\mu\text{-}Zn(CH_3)B_5H_8$. The reaction requires more than 1 month to reach completion in the dark. It is much faster in the presence of light, but then also significant quantities of metal, presumably Zn, are formed [11].

Brønsted acids HX (X = Cl, Br, OH) react rapidly and cleanly with the compound in nonbasic solvents to produce B_5H_9 and $Be(C_5H_5)X$. In aromatic solvents, such as C_6D_6, insoluble orange-brown solids form, suggested to be $[Be(C_5H_5)C_6D_6]^+X^-$. Upon standing for 1 week, B_5H_9 had undergone exchange with C_6D_6 to form $1\text{-}DB_5H_8$, presumably catalyzed by a Be complex [11].

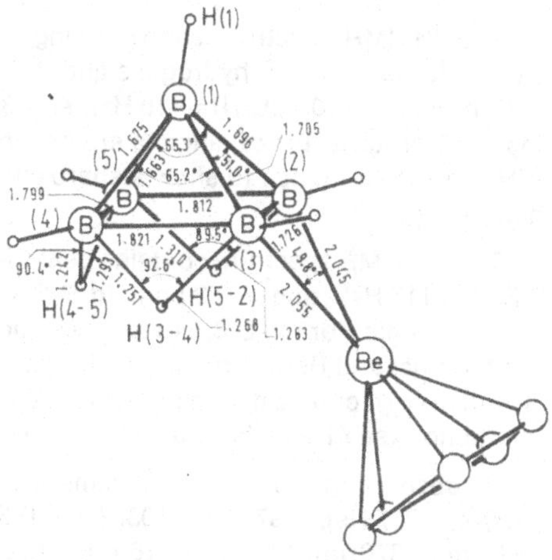

Fig. 14. Molecular structure of $Be(C_5H_5)B_5H_8$ [10, 11].

Other selected bond distances (in Å) and bond angles (in °):

distances		angles	dihedral[a)]	from basal plane
CC	1.391(3) to 1.401(3)	Be	33.82	56.18
CH	0.90(3) to 1.05(3)	H(3-4)	25.92	64.08
BeC	1.877(3) to 1.894(3)	H(5-2)	27.27	62.73
$BH_{terminal}$	0.98(3) to 1.10(2)	H(4-5)	21.35	68.65
intermolecular	3.670(4) (minimum)			

a) Angle between the normal to the B_4 plane and the vector from the center of each basal edge to its bridging group.

$Be(C_5H_5)B_5H_7ZnCH_3$

Equimolar amounts of $Be(C_5H_5)B_5H_8$ and $Zn(CH_3)_2$ in ether react immediately upon warming to room temperature with evolution of CH_4. The product is tentatively assigned the stoichiometry of the title compound (Formula IX) on the basis of the ^{11}B NMR spectrum which contains five equally intense doublets at $\delta = -47.4, -17.8, -14.0, -12.4,$ and -10.0 ppm [11].

$Be(C_5H_5)B_5H_{10}$

Reaction of $Be(B_5H_{10})Br$ with a slight excess of NaC_5H_5 at room temperature for 8 to 10 h produces a slightly volatile liquid in 81 % yield. The mass spectrum shows the parent ion.

As demonstrated for $Be(B_5H_{10})BH_4$ and $Be(B_5H_{10})_2$ by structural characterization, Be in this case is expected also to occupy a cage position, as shown in Formula X.

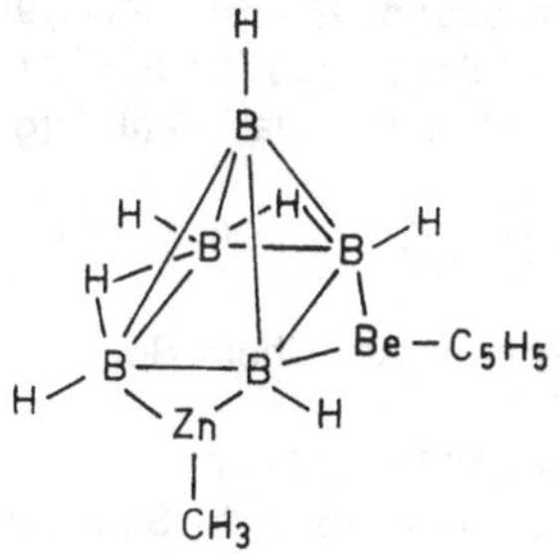

IX

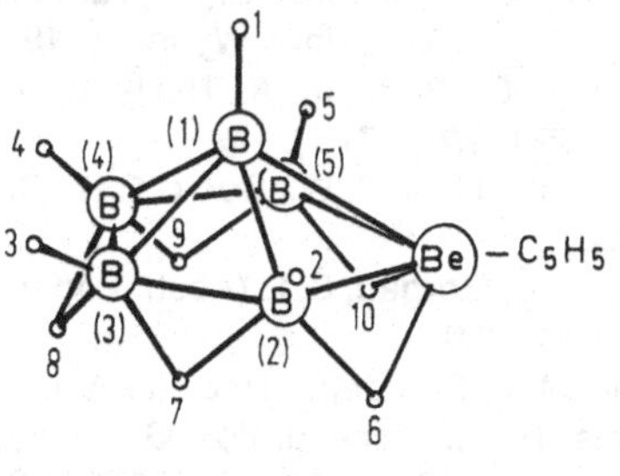

X

References on pp. 164/5

The ^1H NMR spectrum shows a singlet at $\delta = 5.77$ ppm for the C_5H_5 protons, and three signals for terminal BH hydrogens and BeHB hydrogens at $\delta = -1.45$ (q, H-1; J(B, H) = 115 Hz), 2.32 (m, H-2,5,6,10; J(B, H) = 115 Hz), and 3.23 ppm (q, H-3,4; J(B, H) = 156 Hz). For the bridging BHB protons, broad singlets are observed at $\delta = -1.09$ (H-7,9) and -3.20 (H-8) ppm. The C_5H_5 protons continue to show a sharp singlet down to $-75°C$ (see Formula X, p. 163, for the numbering).

The ^{11}B NMR spectrum contains peaks at $\delta = -64.0$ (d, B-1; J(B, H) = 117 Hz), -3.6 (t, B-2,5; J(B, H) = 117 Hz), and -0.3 ppm (d, B-3,4; J(B, H) = 156 Hz) in C_6D_6 or $C_6D_5CD_3$. The well-resolved triplet for the B-3,5 resonance suggests nearly equivalent coupling of B-2 and B-5 to the terminal and BeHB bridging hydrogens. The high chemical shift of B-1 and the lower B-1, H-1 coupling constant compared to other BeB_5H_{10} compounds suggest a strong Be–B-1 interaction (see Formula X, p. 163, for the numbering).

IR absorptions for the solid films are observed at 3140 (w), 2610 (s), 2600 (s), 2520 (s), 2490 (s), 2470 (s), 1970 (w), 1930 (w), 1850 (w, br), 1750 (w), 1490 (m), 1420 (m), 1410 (m), 1390 (m), 1370 (m), 1340 (m), 1160 (m), 1120 (w), 1090 (m), 1080 (m), 1050 (s), 1020 (s), 970 (s), 940 (m, sh), 890 (m), 850 (w), 840 (w), and 790 (w) cm^{-1} [16].

Molecular orbital studies of the title compound and of other $Be(B_5H_{10})X$ ($X = BH_4$, B_5H_{10}, CH_3) compounds with the PRDDO program are presented. The bonding in these molecules is analyzed in terms of charge stability, static reactivity indices, degrees of bonding, overlap populations, and fractional bond orders from localized molecular orbitals. A C_s symmetry and a staggered configuration of C_5H_5 relative to the BeH_bB hydrogen bridges with CC = 1.415 and CH = 1.09 Å is assumed. It follows from the calculations that the CC bonds in the C_5H_5 unit have an average degree of bonding (1.257) characteristic of substantial aromatic character. The Be is bonded to B_5H_{10} only through fractional bonds. The charge on Be is slightly negative (−0.008) and that of C_5H_5 positive (0.17), whereas it is the reverse for the other $Be(B_5H_{10})X$ compounds, and Be is more covalently bonded in the title compound than in the three other compounds studied. The relatively large gap of 0.459 a.u. between the HOMO and LUMO is indicative of some resistance to oxidation or reduction. Within the BeB_5H_{10} unit, the bond strengths are in the order B(1)–B(2) > B(1)–(B(3) > B(3)–B(4) > B(2)–B(3) > Be–B(1) > Be-B(2). Thus, the Be–B(1) bonding is stronger than the Be–B(2) and the Be–B(3) bonding, as it is indicated by the ^{11}B NMR spectrum. The $B–H_t$ bonds are stronger than the $B–H_b$ fragments of the $B–H_b–Be$ bonds, and this fragment is stronger than the $B–H_b$ fragment of the $B–H_b–B$ bonds. The extreme relative weakness of the $Be–H_b$ fragments is due to the electronegativity difference between B and Be (see Formula IX, p. 163, for the atom numbering, H_t = terminal hydrogens, H_b = bridging hydrogens) [9].

References:

[1] Drew, D. A., Morgan, G. L. (unpublished results from [2]).

[2] Drew, D. A., Gundersen, G., Haaland, A. (Acta Chem. Scand. **26** [1972] 2147/9).

[3] Bartke, T. C. (Diss. Univ. Wyoming 1975; Diss. Abstr. Intern. B **36** [1976] 6141).

[4] Bartke, T. C., Bjørseth, A., Haaland, A., Marstokk, K. M., Møllendal, H. (J. Organometal. Chem. **85** [1975] 271/7).

[5] Coe, D. A., Nibler, J. W., Cook, T. H., Drew, D. A., Morgan, G. L. (J. Chem. Phys. **63** [1975] 4842/53).

[6] Drew, D. A., Morgan, G. L. (Abstr. Papers 173rd Natl. Meeting Am. Chem. Soc., New Orleans 1977, INOR 21).

[7] Dewar, M. J. S., Rzepa, H. S. (J. Am. Chem. Soc. **100** [1978] 777/84).

[8] Jemmis, E. D., Alexandratos, S., Schleyer, P. von R., Streitwieser, A., Schaefer III, H. F. (J. Am. Chem. Soc. **100** [1978] 5695/700).

[9] Bicerano, J., Lipscomb, W. N. (Inorg. Chem. **18** [1979] 1565/71).
[10] Gaines, D. F., Coleson, K. M., Calabrese, J. C. (J. Am. Chem. Soc. **101** [1979] 3979/80).

[11] Gaines, D. F., Coleson, K. M., Calabrese, J. C. (Inorg. Chem. **20** [1981] 2185/8).
[12] Gaines, D. F., Coleson, K. M., Hillenbrand, D. F. (J. Magn. Resonance **44** [1981] 84/8).
[13] Marynick, D. S. (private communication from [8]).
[14] Goodman, W. (private communication from [7]).
[15] Gaines, D. F., Walsh, J. L., Morris, J. H., Hillenbrand, D. F. (Inorg. Chem. **17** [1978] 1516/22).
[16] Gaines, D. F., Walsh, J. L. (Inorg. Chem. **17** [1978] 1238/41).
[17] Bews, J. R., Glidewell, C. (J. Organometal. Chem. **219** [1981] 279/93).
[18] Böhm, M. C., Gleiter, R., Morgan, G. L., Lusztyk, J., Starowieyski, K. B. (J. Organometal. Chem. **194** [1980] 257/63).
[19] Lattman, M., Cowley, A. H. (Inorg. Chem. **23** [1984] 241/7).

3.2 Cyclopentadienylberyllium Halides

$Be(C_5H_5)F$

The compound could not be prepared by the reactions of BeF_2 with NaC_5H_5 in ether nor from $Be(C_5H_5)Cl$ with NaF in c-C_6H_{12} [19]. Also, the reaction of $Be(C_5H_5)_2$ with BeF_2 in the solid failed to give $Be(C_5H_5)F$ [8]. However, the compound is presumably formed, if the latter reaction is performed with equimolar amounts of reactants in boiling C_6H_6 for several hours. A ^{1}H NMR spectrum of the colorless solid product in C_6H_6 shows a peak at $\delta = 5.75$ ppm which is assigned to $Be(C_5H_5)_2$ and another peak at 5.78 ppm which may belong to the title compound. In c-C_6H_{12}, the peaks are observed at $\delta = 5.63$ and 6.11 ppm, respectively. Mass spectra were obtained at 70, 30, 20, and 10 eV and ambient source temperature. They showed that the sample is highly contaminated with $Be(C_5H_5)_2$. The fragments which are characteristic for the title compound are $[BeC_5H_5F]^+$, $[BeC_3H_3F]^+$, and $[BeF]^+$. The parent ion is observed at all the applied ionizing voltages and is the major species at 30, 20, and 10 eV [4].

$Be(C_5H_5)Cl$

The compound is prepared from an excess of anhydrous $BeCl_2$ and $Be(C_5H_5)_2$ at room temperature [8], warmed a few minutes with hot tap water [4], or at 60°C for 1 to 2 h [6, 14]. The product can be vapor-transferred into a storage vessel [4, 8, 14]. It is also prepared from $BeCl_2$ and an equimolar amount of NaC_5H_5 in ether [7, 15]. After 3 h stirring, the ether is removed and the residue is cooled in an ice bath [15]. The compound is purified by crystallization from C_6H_{14} [4] and vacuum sublimation [14, 15] at room temperature/10^{-2} Torr [4]. It is also formed in the reaction of $Be(C_5H_5)B_5H_8$ with HCl [12]. Reaction of $BeCl_2$ with NaC_5D_5 in ether gives $Be(C_5D_5)Cl$ [7].

It forms volatile [6] colorless needles, melting with decomposition [8] at 43 to 44°C [8, 14] or 42.0 to 42.5°C [4]. The vapor pressure at 27°C is 1.0 Torr [8] or 0.9 Torr at 23°C [4]. Monomers are present in the solid, liquid, and gas phase (see below).

The ^{1}H NMR spectrum shows singlets for the C_5H_5 protons at $\delta = 5.66$ in C_6D_6 (5.38 [4] or 5.88 [8] in C_6H_6), 6.18 [4] or 6.27 [16] in c-C_6H_{12}, and 6.37 ppm in C_6F_6. The high benzene-induced upfield shift is explained with a stereospecific solvent-solute orientation [16]. The ^{9}Be NMR spectrum in C_6H_6 shows a sharp singlet at $\delta = -18.8$ ppm with a half-height width of 3 Hz (5 Hz [4]) [4, 8]. In C_6F_6 solution the signal appears at -19.1 ppm [13].

References on pp. 171/2

The microwave spectrum reveals that $Be(C_5H_5)Cl$ is a symmetrical top with C_{5v} symmetry. The spectrum consists of groups of strong lines separated by constant intervals. Transitions in the range $J=2\rightarrow3$ to $J=9\rightarrow10$ are measured for the species $Be(C_5H_5)^{35}Cl$, $Be(C_5H_5)^{37}Cl$, $Be(C_4{}^{13}CH_5)^{35}Cl$, $Be(C_4{}^{13}CH_5)^{37}Cl$, and $Be(C_5H_4D)^{35}Cl$ in natural abundance. A value of 250 ± 75 cm^{-1} is calculated for the ring–Be–Cl bending vibration. A dipole moment of $\mu=4.26\pm0.16$ D is determined for the $J=4\rightarrow5$ transition, and a ^{35}Cl quadrupole constant of $eqQ=22\pm2$ MHz for the $J=2\rightarrow3$ transition. The following structural parameters are obtained (in Å):

	r_s	r_o
CH (coplanar)	1.09 (3)	1.090
CC	1.424 (3)	1.424
BeCl	1.81 (3)	1.839
Be–ring	1.52 (3)	1.485
Cl···C	3.546 (5)	3.538

They are in good agreement with the data obtained from electron diffraction (see below). The comparatively long BeCl distance (1.75 Å in $BeCl_2$), the high dipole moment, and the small quadrupole coupling constant suggest that the BeCl bond is a polar single bond with little $p\pi\rightarrow d\pi$ back bonding [3].

IR spectra of $Be(C_5H_5)Cl$ as a solid film at 14 K [6] and 77 K [7], in Nujol [4], in CS_2 solution [14], and the vapor phase [4], and the Raman spectra for the solid at 15 K [6], 25°C [7] and for the liquid at 50°C [6, 7] are reported. Figures of the spectra are shown in [6] and [7]. The solid state IR spectrum at 77 K and the Raman spectrum of $Be(C_5D_5)Cl$ at 25°C are given as figures and used to assign the IR and Raman spectra of $Be(C_5H_5)Cl$ [7]. The observed IR vibrations for the solid and the Raman vibrations for the solid and liquid are listed in Table 30. Assignments are made for a C_{5v} symmetry [6]. The IR and Raman spectra taken in [7] are in very good agreement with those reported in Table 30. The assignment is confirmed by the relative positions of the bands in the spectra of $Be(C_5H_5)Cl$ and $Be(C_5D_5)Cl$ [7].

The vapor phase IR spectrum is of only moderate quality due to the low volatility of the compound, but the C_{5v} symmetry of the molecule is confirmed. The bands at 1720 (w) and 1000 (vs, br) cm^{-1} are assigned to a composite of ν(BeCl) and δ(CH). The ν(BeCl) vibrations are observed in Nujol at 1755 (w, m), 910 (s), and 897 cm^{-1} [4]. The IR spectrum in CS_2 resembles the solid state spectra. The characteristic ν(BeCl) vibration is observed at 994 (vs) cm^{-1}, and ν(BeC) is assigned to the very strong band at 898 cm^{-1} [14].

The He(I) photoelectron spectrum has been recorded. To interpret the spectrum the orbital energies (ε_J) were calculated with a modified CNDO procedure. The comparison suggests that the first two peaks in the PE spectrum are to be assigned to four ionization events from two e orbitals with a Jahn-Teller splitting of 0.1 to 0.2 eV [11]:

band	I.P. (in eV)	assignment	$-\varepsilon_J$ (in eV)
1	9.60	$e_n(Cl)$	9.88
2	~9.9		
3	11.15	$e_\pi C_5H_5)$	11.57
4	11.5		
5	12.45	(BeCl)	11.83

Table 30
IR and Raman Vibrations (in cm^{-1}) of $Be(C_5H_5)Cl$ [6].

IR solid 14 K	Raman solid 15 K	Raman liquid 50°C	p[a)]	C_{5v}	assignment[b)]
3120(s)	3120(vs)	3124(vs)	0.13	A_1	ν(CH)
3112(vs)	3112(s)			E_1	ν(CH)
	3103(vs)	3104(s, sh)	0.6	E_2	ν(CH)
2852(vw)	2850(vw)	2853(w)	0.22	A_1	2ν(CC)
2771(vw)	2770(vw)	2774(w)	0.82	E_1	ν(CC) + ν(CC)
2694(w)		2692(vw)	0.17	A_1	2ν(CC)
2409(w)				E_1	ν(CC) + δ(CH) out-of-plane
2275(w)				A_1	ν(CC) + δ(CH) out-of-plane
2081(w)				E_1	δ(ring) in-plane + out-of-plane
1851(m)				E_1	δ(CH) in-plane + δ(ring) in-plane
1764(s)				E_1	ν(BeCl) + δ(CH) out-of-plane
1628(s)				E_1	δ(CH) in-plane + δ(ring) out-of-plane
1489(w)				E_1	δ(CH) out-of-plane + δ(ring) out-of-plane
1459(vw)				E_1	δ(ring) in-plane + δ(ring) out-of-plane
1427(vs)	1427(m)	1430(w)	0.7	E_1	ν(CC)
	1352(s)	1352(s)	0.82	E_2	ν(CC)
1327(w)				E_1	ring tilt + δ(ring) in-plane
1256(m)	1259(m)	1244(m)	0.08	E_2, A_1	δ(CH) in-plane, 2δ(ring) out-of-plane
1249(vw)	1248(m)				
1233(w)	1232(vw)			A_1	ν(BeC) + ν(BeCl)
1126(vs)	1125(vs)	1126(vs)	0.04	A_1	ν(CC)
1118(w, sh)	1119(s)	1120(w, sh)	0.1	A_1	ν(^{13}CC)
1100(s)	1102(m)			E_1	δ(ring BeCl) + ν(BeCl)
1074(s)	1073(s)				
	1066(s)	1066(s)	0.81	E_2	δ(CH) out-of-plane
1064(s)	1063(m)				
1019(vs)	1018(m)	1017(w)	0.79	E_1	δ(CH) in-plane
1000(vs)	1002(m)	1000(w)	0.75	E_1	δ(ring BeCl) + δ(ring) in-plane
974(m)				E_1	2 ring tilts
956(m)				E_1	
939(vs)	940(w)	912(m)	0.23	A_1	ν(BeCl)
919(s)	920(s)				
865(s)	866(m)	865(w)	0.69	E_1	δ(CH) out-of-plane
855(s)					
846(vw)	848(m)	837(w)	0.72	E_1	δ(ring) in-plane
838(w)	842(w, sh)				

References on pp. 171/2

Table 30 (continued)

IR solid 14 K	Raman solid 15 K	Raman liquid 50°C	p[a)]	C_{5v}	assignment[b)]
786(s)	794(vw)	785(vw)	0.63	A_1	δ(CH) out-of-plane
		646(vw)	0.3	A_1	ring tilt + δ(ring BeCl)
623(w)	625(m)	624(vw)	0.54	E_2, A_1	δ(ring) out-of-plane, 2ν(Be ring)
496(vw)	493(m, sh)			E_1	ν(BeC) + δ(ring BeCl)
	485(s)	486(m, sh)	0.7	E_1	
482(m)	479(s)	474(m)	0.80	E_1	ring tilt
	357(vw, br)	347(w, sh)	0.1	A_1	2δ(ring BeC)
314(s)	314(s)	312(vs)	0.12	A_1	ν(BeCl)
309(w, sh)	310(m, sh)				
	186(m, sh) 175(s)[b)]	177(vs)	0.78	E_1	δ(ring BeCl)

[a)] Polarization coefficient, [b)] Raman active lattice modes were also observed at 101 (m), 93 (m), 81 (m), and 41 (m) cm^{-1}.

The calculated net charges are −0.082 at C_5H_5, 0.114 at Be, and −0.276 at Cl [11]. The orbital sequence and computed ionization energies obtained by SCF-X_α-scattered wave calculations are not in agreement with the previous study. The principal difference in the results is that the ring π($4e_1$) and Cl lone-pair ($3e_1$) MO's are reversed. The HOMO is doubly degenerate ($4e_1$) and represents principally a C_5H_5 ring π-orbital with minor contribution from BeCl. Interestingly, the contribution of the Cl(3p) AO's to this MO exceeds that of the Be(2p) AO's. The second MO corresponds to the Cl lone pairs. There is only a minor contribution from the Be(2p) AO and the e_1 ring MO. The $5a_1$ MO corresponds to the BeCl σ bond; the Cl(3p) AO contribution exceeds that of the Be(2p) AO's. The computed ionization energies for the first six MO's are (in eV): 8.25($4e_1$), 10.97($3e_1$), 11.93($5a_1$), 13.09($4a_1$), 14.04($2e_2$), and 14.63($2e_1$). MNDO calculations (not given) produce the same orbital sequence as the X_α-scattered wave calculations [18].

The crystal structure has been determined by X-ray diffraction. The orthorhombic crystals belong to the space group $P2_12_12_1$-D_2^4 (No. 19) with a = 9.539(2), b = 9.549(2), c = 6.513(1) Å; Z = 4, d_c = 1.23 g/cm³. Refinement of the structure converged at R = 0.032 and R_w = 0.029. The molecular structure with important bond distances and bond angles is shown in **Fig. 15**a. Fig. 15b shows the arrangement in the unit cell. Be(C_5H_5)Cl consists in the crystalline state of discrete monomeric molecules, separated by distances >3.0 Å. The shortest intermolecular distance is C···H = 3.05(4) Å. The shortest Be···Cl distance is 4.052(3) Å, and the corresponding Cl···Cl contact is 4.156(1) Å. The molecule possesses a C_{5v} axis of symmetry which passes through the center of the C_5H_5 ring, the Be and Cl atoms. The distance Be–C_5H_5(center) is 1.451 Å, and the angle Cl–Be–C_5H_5(center) is 178.4°. Although not precisely located, the H atoms in the C_5H_5 ring are slightly bent towards the Be atom, as was observed for Be(C_5H_5)H [15].

The molecular structure in the gas phase has been determined by electron diffraction. The C_{5v} symmetry of the solid state is also valid in the gas phase. The bond distances obtained for the gas phase are included in Fig. 15a in parentheses. The distance Be–C_5H_5(center) is 1.484 Å. Since it was not possible to locate the hydrogen atoms precisely, they are assumed to lie in the plane of the C_5 ring [1, 2].

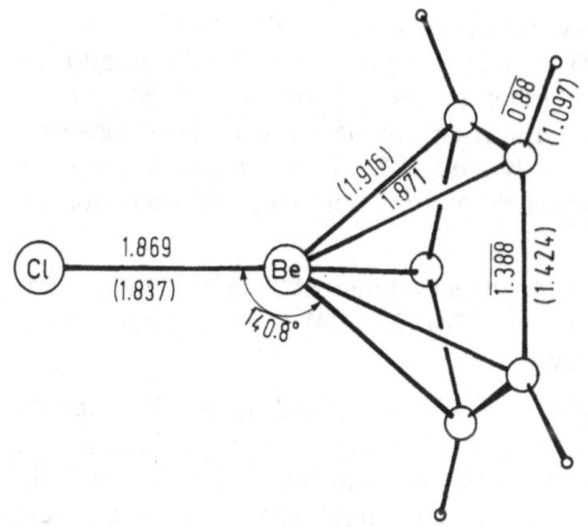

Fig. 15a. Crystal and gas phase structure of $Be(C_5H_5)Cl$. The values in parentheses are for the gas phase [1, 2, 15].

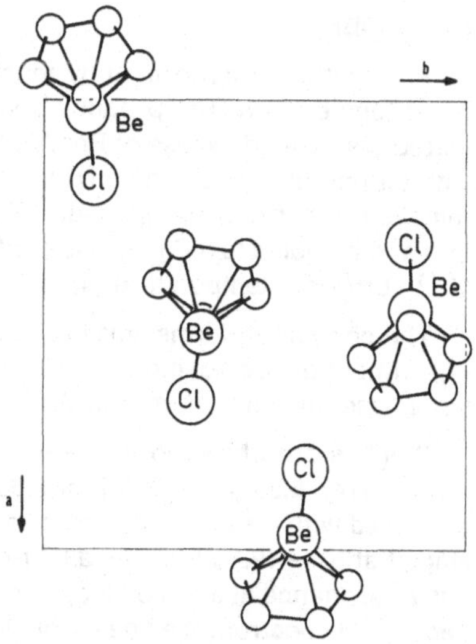

Fig. 15b. Unit cell of $Be(C_5H_5)Cl$ looking down c [15].

The relatively long BeCl bond distance (1.75 Å in $BeCl_2$) [1 to 3, 15], the high dipole moment [3], and the small ^{35}Cl quadrupole coupling constant [3] suggest that the BeCl bond is a polar single bond [3] with little $p\pi \rightarrow d\pi$ back bonding [1 to 3, 15].

The compound dissolves as a monomer in C_6H_6 [4, 8] and in c-C_6H_{12} [4] (by cryoscopy). Mass spectra of $Be(C_5H_5)Cl$ were recorded at 12, 20, and 70 eV [8] and at 10, 20, 30, and 70 eV ionizing voltages [4]. The spectra reflect a large amount of C_5H_6 contamination at a source temperature of 160°C. At 12 eV, no fragmentation of the parent ion is observed. The 20 eV spectrum shows the $[BeC_5H_5]^+$ ion, but no peak for $[BeCl]^+$. The latter is observed in the spectrum recorded at 70 eV [8]. A mass spectrum measured at 35°C/75 eV (with a direct inlet system) contained also peaks indicative of Be_3 and Be_2 species; however, this may be an artifact [7]. The parent ion is observed at 10 to 70 eV and ambient source temperature, and it is the major peak from 10 to 30 eV. The other fragments observed at 70 eV are $[BeC_5H_3Cl]^+$, $[BeC_3H_3Cl]^+$, $[BeC_5H_5]^+$, $[C_5H_5]^+$, $[BeC_3H_3]^+$, $[BeCl]^+$, $[C_3H_3]^+$, and $[Be]^+$. The ionization potential is 11.28 eV; the BeCl bond dissociation energy is 2.75 eV (=63.41 kcal/mol) [4].

The compound is extremely sensitive towards air and H_2O [15]. No reaction is observed with LiH in refluxing C_6H_6 or in ether. Only the formation of $Be(C_5H_5)_2$ and $BeCl_2$ is found [4]. Reactions with $LiBH_4$ [4, 6, 10] or $LiBD_4$ [4, 6] give $Be(C_5H_5)BH_4$ and $Be(C_5H_5)BD_4$, respectively [4, 6, 10]. The reaction can also be performed with $NaBH_4$ or $NaBD_4$, but higher temperatures are needed, which leads to more side products. No reaction is observed with $LiAlH_4$ in ether or in refluxing C_6H_6 or with $NaBH_3CN$, and the only products observed with $NaBH(C_2H_5)_3$ are $Be(C_5H_5)_2$ and $B(C_2H_5)_3$ [4]. With KB_5H_8, $Be(C_5H_5)B_5H_8$ is formed [12, 17]. Reactions with $LiNR_2$ (R = H, CH_3, C_2H_5, C_6H_5) in C_6H_6 give the corresponding $Be(C_5H_5)NR_2$; and with $NaN(Si(CH_3)_3)_2$, $Be(C_5H_5)N(Si(CH_3)_3)_2$ is formed, whereas no reactions occur with $LiPR_2$ (R = CH_3, C_6H_5), $LiAs(CH_3)_2$, and MN_3 (M = Na, K). $Be(C_5H_5)Cl$ reacts with LiR (R = CH_3, C_2H_5, C_4H_9, and C_6H_5), or with $Mg(C_2H_5)Br$, to yield the corresponding $Be(C_5H_5)R$ compounds [4].

References on pp. 171/2

$Be(C_5H_5)Br$

Preparation is performed by reaction of an excess of anhydrous $BeBr_2$ and $Be(C_5H_5)_2$ at room temperature. The product is vapor-transferred into a storage vessel [4, 8]. In another method, a twofold excess of $Be(C_5H_5)_2$ dissolved in ether is added to anhydrous $FeBr_2$ in ether with stirring at −78°C. The mixture is stirred for 2 h at room temperature, and the solvent is removed under reduced pressure. The product is sublimed at 0.1 Torr at room temperature into a trap cooled to 0°C [14]. It is purified by recrystallization from hot C_6H_{14} and sublimation at 10^{-3} Torr/room temperature [4].

The compound forms small [8] colorless [8, 14] needles that melt at 33 to 35°C (35 to 36°C [4]) without decomposition [8]. The vapor pressure is 0.3 Torr/23°C [4]. It is monomeric in the gas phase, in solution, and in the solid (see below).

Singlets are observed in the ^1H NMR spectra at $\delta = 6.20$ ppm in c-C_6H_{12} [4], 6.29 ppm in c-C_6D_{12} [14], 5.86 [8], 5.73 [4], or 5.66 ppm [14, 16] in C_6H_6. The C_6H_6 induced upfield shift is explained with a stereospecific solvation [16]. The ^9Be NMR spectrum in C_6H_6 exhibits a sharp singlet at $\delta = -19.5$ ppm with a half-height width of 4 Hz (7 Hz [4]) [8]. The ^{13}C NMR spectrum has a resonance at $\delta = 105.52$ ppm with J(C,H) = 179.1 Hz in C_6D_6 at 30°C. In the proton noise-decoupled spectrum the line is split into a quartet. In the high resolution spectrum a coupling of J(Be,C) = 1.1 Hz can be observed [9].

The IR spectrum was recorded in the vapor phase [4, 8], in Nujol [4], and in CS_2 and C_6D_{12} [14]. The observed vibrations in solution (in cm^{-1}) are 3980 (vw), 3120 (m), 2700 (vw), 2420 (vw), 2275 (vw), 2065 (vw), 1835 (w), 1742 (m), 1590 (mw), 1430 (m), 1125 (s), 1060 (vw), 1017 (s), 973 (vvs), 899 (vvs), 880 (sh), 860 (sh), 775 (mw), 730 (vw), 610 (vw), 500 (vw), 480 (vw), and 410 (vw). Characteristic are the very strong absorptions at 973 and 899 cm^{-1}, which are assigned to ν(BeBr) and ν(BeC), respectively. The spectrum is largely solvent-independent [14]. The vapor phase IR spectrum at 70°C contains only three fundamentals assignable to the ring: ν(CH) at 3097 (w) cm^{-1}, δ(CH in-plane) at 996 (s, PQR) cm^{-1}, and δ(CH out-of-plane) at 901 (s) cm^{-1}. The band at 970 (m) cm^{-1} (899 (s, br) cm^{-1} in Nujol [4]) is assigned to ν(BeBr). The assignments are based on C_{5v} symmetry. The absence of an absorption in the range 1126 to 1097 cm^{-1} may reflect a higher degree of ionic interaction between Be and the ring [8].

The He(I) photoelectron spectrum shows two peaks (corresponding to 4 bands) which are well separated from the others. The spectrum is interpreted with the aid of calculated molecular orbital energies, $-\varepsilon_J$, using a modified CNDO procedure. The comparison suggests that the first four bands are due to two $^2E_1$ states with a Jahn-Teller splitting of 0.1 to 0.2 eV.

band	I.P. (in eV)	assignment	$-\varepsilon_J$ (in eV)
1	9.52	$e_n(Br)$	9.23
2	9.78		
3	10.79	$e_\pi(C_5H_5)$	11.30
4	~11.0		
5	12.00	σ(BeBr)	11.77

The calculated net charges are −0.086 for C_5H_5, 0.077 for Be, and −0.194 for Br [11].

The gas phase molecular structure was determined by electron diffraction. The structure is analogous to that of $Be(C_5H_5)Cl$. The bond distances (in Å) are CC = 1.424(2), CH = 1.087(7), BeC = 1.950(12), and BeBr = 1.943(15). The Be–ring distance is 1.528(16) Å [5].

The mass spectrum, recorded at 65°C and 70 eV, shows the parent peak [14]. At 160°C and 20 eV, $[BeBr]^+$ and $[BeC_5H_5]^+$ (base peak) are also observed; and at 70 eV, the small fragments like $[BeC_5H_5]^+$, $[Br]^+$, $[C_3H_3]^+$ (base peak), and $[C_3H_2]^+$ are additionally seen [8]. See [4] for the mass spectra at 10, 20, 30, and 70 eV and ambient source temperature. An ionization potential of 10.55 eV is obtained, and a BeBr dissociation energy of 2.23 eV (=51.42 kcal/mol) [4].

The compound dissolves as a monomer in C_6H_6 [4, 8] and c-C_6H_{12} [4] (by cryoscopy). The reactions described in [4] for $Be(C_5H_5)Cl$ can also be performed with $Be(C_5H_5)Br$; see p. 169 [4].

$Be(C_5H_5)I$

The iodide is prepared from $Be(C_5H_5)_2$ and an excess of anhydrous BeI_2 [8] by heating for 1 h at 60°C. The vessel is cooled with liquid N_2 and evacuated [4]. The product is vapor-transferred [4, 8] with little heating [4] into a storage vessel [4, 8]. It is purified by recrystallization from C_6H_6/C_6H_{14} and by sublimation at 10^{-3} Torr/50°C [4].

The colorless needles [8] melt at 32 to 33°C [4, 8] without decomposition [8]. The vapor pressure is 0.007 Torr at 23°C [4].

The ^1H NMR spectrum shows a singlet at $\delta = 5.66$ [4] or 5.85 ppm [8] in C_6H_6 [4, 8], and at $\delta = 6.21$ ppm in c-C_6H_{12} [4]. The IR spectrum was measured in the gas phase [4, 8] and in Nujol [4]. With KBr optics, no spectrum is obtained due to rapid reaction with the cell plates [8]. The gas phase IR spectrum with BaF_2 optics contains two bands assignable to the C_5H_5 ring at 3090(w) cm^{-1} for ν(CH) and at 992(m) cm^{-1} for δ(CH). A strong absorption at 890 cm^{-1} is assigned to C_5H_6 decomposition [8], whereas the strong band at 967 cm^{-1} in the gas phase and at 890 cm^{-1} in Nujol are assigned to ν(BeI) in [4].

The compound is monomeric in C_6H_6 and c-C_6H_{12} (by cryoscopy) [4]. The mass spectra were obtained at 10, 20, 30, and 70 eV ionizing radiation and ambient source temperature. The parent ion is observed in all spectra. It is the major species at 10 eV ionizing radiation. At 20, 30, and 70 eV, the major species is $[BeC_5H_5]^+$. The other peaks observed at 70 eV belong to $[BeC_5H_4I]^+$, $[BeI]^+$, $[I]^+$, $[BeC_5H_5I]^{2+}$, $[BeC_5H_n]$ (n=5,4), $[C_5H_n]$ (n=5,4), $[C_4H_3]^+$, $[BeC_3H_3]^+$, $[C_3H_n]^+$ (n=3,2), $[BeH]^+$, and $[Be]^+$. The ionization potential for the parent species is 9.33 eV, and the BeI bond dissociation energy is 1.70 eV (=39.20 kcal/mol) [4].

References:

[1] Drew, D. A., Haaland, A. (J. Chem. Soc. Chem. Commun. **1971** 1551/2).
[2] Drew, D. A., Haaland, A. (Acta Chem. Scand. **26** [1972] 3351/6).
[3] Bjørseth, A., Drew, D. A., Marstokk, K. M., Møllendal, H. (J. Mol. Struct. **13** [1972] 233/9).
[4] Bartke, T. C. (Diss. Univ. Wyoming 1975; Diss. Abstr. Intern. B **36** [1976] 6141).
[5] Haaland, A., Novak, D. P. (Acta Chem. Scand. A **28** [1974] 153/6).
[6] Coe, D. A., Nibler, J. W., Cook, T. H., Drew, D. A., Morgan, G. L. (J. Chem. Phys. **63** [1975] 4842/53).
[7] Starowieyski, K. B., Lusztyk, J. (J. Organometal. Chem. **133** [1977] 281/4).
[8] Drew, D. A., Morgan, G. L. (Inorg. Chem. **16** [1977] 1704/8).
[9] Fischer, P., Stadelhofer, J., Weidlein, J. (J. Organometal. Chem. **116** [1976] 65/73).
[10] Drew, D. A., Morgan, G. L. (Abstr. Papers 173rd Natl. Meeting Am. Chem. Soc., New Orleans 1977, INOR 21).

[11] Böhm, M. C., Gleiter, R., Morgan, G. L., Lusztyk, J., Starowieyski, K. B. (J. Organometal. Chem. **194** [1980] 257/63).
[12] Gaines, D. F., Coleson, K. M., Calabrese, J. C. (Inorg. Chem. **20** [1981] 2185/8).
[13] Gaines, D. F., Coleson, K. M., Hillenbrand, D. F. (J. Magn. Resonance **44** [1981] 84/8).
[14] Pratten, S. J., Cooper, M. K., Aroney, M. J. (Polyhedron **3** [1984] 1347/50).
[15] Goddard, R., Akhtar, J., Starowieyski, K. B. (J. Organometal. Chem. **282** [1985] 149/54).

[16] Pratten, S. J., Cooper, M. K., Aroney, M. J., Filipczuk, S. W. (J. Chem. Soc. Dalton Trans. **1985** 1761/5).
[17] Gaines, D. F., Coleson, K. M., Calabrese, J. C. (J. Am. Chem. Soc. **101** [1979] 3979/80).
[18] Lattman, M., Cowley, A. H. (Inorg. Chem. **23** [1984] 241/7).
[19] Starowieyski, K. B. (private communication from [4]).

3.3 $Be(C_5H_5)OH$ and Cyclopentadienylberyllium Amides

$Be(C_5H_5)OH$

It is mentioned that the compound is obtained by the reaction of $Be(C_5H_5)B_5H_8$ with H_2O, but no further details are given [2].

$Be(C_5H_5)NH_2$

The compound is probably obtained in the reaction of $LiNH_2$ with $Be(C_5H_5)Cl$. A yellow paste is obtained, which is only slightly soluble in hydrocarbons and nonvolatile at temperatures up to 165°C and pressures down to 10^{-3} Torr. A mass spectrum of the sample, contaminated with $Be(C_5H_5)Cl$, at 70 eV and >165°C source temperature shows the parent ion. Another characteristic fragment is $[Be(C_5H_5)N]^+$. No associated species are observed, but the parent compound is assumed to be associated in the solid state [1].

$Be(C_5H_5)N(CH_3)_2$

The compound is prepared by mixing equimolar amounts of $LiN(CH_3)_2$ and $Be(C_5H_5)X$ (X = Cl, Br) in C_6H_6. The solution is filtered, and the solvent is removed under vacuum. The compound is also obtained by stirring an equimolar mixture of $Be(C_5H_5)_2$ and $Be(N(CH_3)_2)_2$ in C_6H_6 for 1 h. C_6H_6 is removed under vacuum. Both methods give a red paste that is recrystallized from C_6H_6/C_6H_{14}.

The deep red crystals melt at 104 to 105°C or at 105 to 108°C with decomposition. Attempted sublimation results in decomposition above 100°C. The vapor pressure is too low to measure at 23°C/10^{-3} Torr.

The ^1H NMR spectrum consists of two sharp singlets. The C_5H_5 protons are observed at $\delta = 6.11$ ppm in C_6H_6 and at 6.08 ppm in c-C_6H_{12}. The CH_3 protons are found at $\delta = 2.04$ ppm in C_6H_6 and lie under the solvent lock signal in c-C_6H_{12}.

The vapor phase IR spectrum could not be measured because of the low volatility of the compound. The observed IR vibrations in Nujol are (in cm^{-1}) 3085(m), 2795(m), 1772(w), 1750(w-m), 1610(w), 1402(w-m), 1362(m), 1345(m), 1227(s), 1175(s), 1123(s), 1060(w), 1035(m-s), 1020(s), 980(s), 923(s), 795(s), 745(s,sh), 670(m-w), 602(s), and 465(w).

The compound is dimeric in C_6H_6 and c-C_6H_{12} (by cryoscopy), and the structure shown in Formula I ($R = CH_3$) is proposed. Mass spectra were obtained at 10, 20, 30, and 70 eV ionizing voltages and a source temperature of 150°C. The parent ion is observed in all spectra and is the only peak at 10 eV. The other fragments observed at 70 eV are $[BeC_7H_{11}N]^+$, $[BeC_6H_nN]^+$ (n = 8,7), $[BeC_5H_5N]^+$ (base peak), $[BeC_5H_5]^+$, $[BeN(CH_3)_2]^+$, $[BeC_3H_3]^+$, $[N(CH_3)_2]^+$, $[C_3H_3]^+$, $[NCH_3]^+$, $[N]^+$, and $[Be]^+$. No associated species are observed. The ionization potential is 9.45 eV, and the BeN bond dissociation energy is 4.15 eV (= 95.70 kcal/mol) [1].

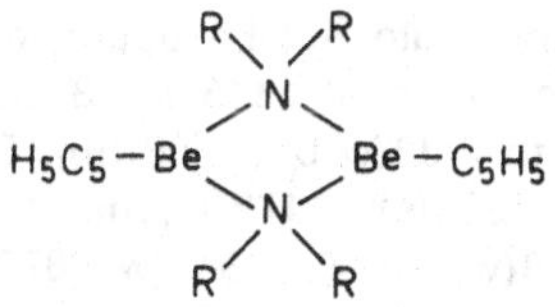

I

$Be(C_5H_5)N(C_2H_5)_2$

The compound is prepared from $LiN(C_2H_5)_2$ and $Be(C_5H_5)X$ (X = Cl, Br) as described for the previous compound. It is also prepared from $Be(C_5H_5)_2$ and $Be(N(C_2H_5)_2)_2$, as described for the previous compound, but with several hours of stirring. Recrystallization from C_6H_6/C_6H_{14} gives yellow-orange crystals that melt at 115°C or 112 to 116°C with decomposition. Attempted sublimation results in decomposition above 110°C. The vapor pressure is too low to be measured at 22°C/10^{-3} Torr.

The ^1H NMR spectrum shows the following signals in C_6H_6: $\delta = 0.96$ to 1.10 (t, CH_3; J = 7 Hz), 2.21 to 2.42 (q, CH_2; J = 7 Hz), and 6.14 ppm (s, C_5H_5). In c-C_6H_{12} as a solvent, the signal for the protons in C_5H_5 is observed at 6.06 ppm, and for CH_2 at 2.30 to 2.51 ppm. The signal for CH_3 is hidden under the solvent peak.

The following vibrations are observed in the IR spectrum of the neat solid (in cm^{-1}): 3080 (m), 2962 (s), 2925 (s), 2860 (s), 2830 (m), 2810 (m), 2720 to 2660 (complex, w), 2604 (w), 1730 (w, br), 1590 (w, br), 1472 (s), 1463 (s), 1448 (s), 1385 (m), 1370 (s), 1362 (s, sh), 1332 (s), 1304 (m), 1275 (m), 1212 (s, sh), 1178 (s), 1165 (s), 1145 (m), 1125 (m), 1095 (s), 1078 to 1055 (s, br), 1022 (m), 972 (m), 902 (s), 880 (s), 858 (w), 845 (vw), 820 (s), 799 (s, br), 745 (m), 730 (m), 702 (m), 658 (w), 648 (w), 620 (w), 605 (w), 593 (w), 578 (w), 550 (w), and 480 (w, br). The vapor phase IR spectrum could not be measured because of the low volatility of the compound.

The compound is dimeric in C_6H_6 and c-C_6H_{12}. The structure shown in Formula I ($R = C_2H_5$) is proposed. Mass spectra were obtained at ionizing voltages of 10, 20, 30, and 70 eV and 150°C source temperature. The parent ion is observed at all voltages, and it is the only peak at 10 eV. The other fragments observed at 70 eV are $[BeC_8H_nN]^+$ (n = 12, 11), $[BeC_7H_{10}N]^+$, $[BeC_6H_nN]^+$ (n = 8, 7), $[BeC_5H_5N]^+$, $[BeN(C_2H_5)_2]^+$, $[BeC_5H_5]^+$, $[N(C_2H_5)_2]^+$, $[C_5H_5]^+$, $[NC_3H_7]^+$, $[BeC_3H_3]^+$, $[NC_2H_5]^+$, $[C_3H_3]^+$, $[N]^+$, and $[Be]^+$. No associated species are observed. The ionization potential is 9.53 eV; the BeN bond dissociation energy is 4.37 eV (= 100.77 kcal/mol) [1].

$Be(C_5H_5)N(C_6H_5)_2$

An equimolar solution of $LiN(C_6H_5)_2$ and $Be(C_5H_5)X$ (X = Cl, Br) in C_6H_6 is stirred overnight. The mixture is filtered, and dry C_6H_{14} is added dropwise until the compound starts to precipitate. After standing overnight, the crystals are filtered and washed with C_6H_{14}. Another method of preparation is the reaction of $Be(C_5H_5)_2$ and $Be(N(C_6H_5)_2)_2$ in equimolar amounts in C_6H_6. The mixture is stirred overnight, the solution is filtered, and about half of the C_6H_6 is evaporated. The compound is crystallized from C_6C_6/C_6H_{14}.

The light yellow crystals melt at 99 to 101°C or 101 to 102°C with decomposition to a red oil. The compound sublimes to give fine yellow crystals at ~70°C/10^{-3} Torr. The vapor pressure is too low to be measured at 22°C/10^{-3} Torr.

The ^1H NMR spectrum consists of a singlet at $\delta = 5.78$ ppm in C_6H_6 and 6.16 ppm in c-C_6H_{12}. In C_6H_6 the phenyl ring proton signals are lost under the solvent resonance, while they are seen as a complex multiplet in c-C_6H_{12} at $\delta = 6.81$ to 6.97 (meta position) and 6.44 to 6.60 ppm (ortho and para positions). The ^9Be NMR spectrum consists of a single broad peak at $\delta = -18.09$ ppm with a half-height width of 59 Hz.

References on p. 175

The vapor phase IR spectrum could not be obtained due to the low volatility of the compound. In Nujol, bands are found at 3118(w), 3102(w), 3080(w), 3060(w), 3030(w), 3010(w), 1970(w), 1840(w), 1755 to 1730(w, br), 1600(m), 1582(s, sh), 1567(m, sh), 1485(s, sh), 1330(m), 1308(w), 1285(s, sh), 1225(s), 1172(m, sh), 1152(vw), 1130(s, sh), 1095(s, sh), 1062(s, sh), 1018(m), 998(w), 968(w), 895(w), 885(w), 870(m), 852(s, sh), 828(vw), 768(w), 768(m), 762(m), 753(s, sh), 742(w), 728(w), 704(m, sh), 699(s, sh), 632(m), 612(w), 585(w), 530(m, sh), 494(w), and 457(w-m) cm^{-1}.

The compound is monomeric in C_6H_6 and c-C_6H_{12} (by cryoscopy). The mass spectra, obtained at 10, 20, 30, and 70 eV ionizing voltages and 100°C source temperature, all contain the parent ion as a base peak. The other peaks observed in the 70 eV spectrum belong to $[BeC_{17}H_{14}N]^+$, $[BeC_{15}H_nN]^+$ (n = 13, 12), $[BeC_{13}H_nN]^+$ (n = 11, 10), $[BeC_{12}H_{10}N]^+$, $[BeC_{11}H_nN]^+$ (n = 10 to 8), $[BeC_{10}H_nN]^+$ (n = 9, 7), $[BeC_9H_nN]^+$ (n = 7, 6), $[BeC_8H_nN]^+$ (n = 6, 5), $[BeC_7H_nN]^+$ (n = 6, 5), $[BeC_6H_nN]^+$ (n = 6,5), $[BeC_5H_5N]^+$, $[BeC_4N_nN]^+$ (n = 3 to 1), $[BeC_5H_n]^+$ (n = 4, 3), $[BeC_4H_n]^+$ (n = 4 to 1), $[BeC_2HN]^+$, $[BeC_3H_3]^+$. No associated species are observed. The ionization potential is 9.78 eV; the BeN bond dissociation energy is 5.33 eV (= 122.91 kcal/mol) [1].

$Be(C_5H_5)N(Si(CH_3)_3)_2$

For the preparation, a mixture of equimolar amounts of $NaN(Si(CH_3)_3)_2$ and $Be(C_5H_5)X$ (X = Cl, Br) in C_6H_6 is stirred for 3 to 4 h. The solvent is removed under vacuum, and the residue is sublimed at room temperature and 10^{-3} Torr over a period of several days. The compound could not be prepared by reaction of $Be(C_5H_5)_2$ with $Be(N(Si(CH_3)_3)_2)_2$.

The colorless crystals melt at 33 to 34°C. The vapor pressure is 0.04 Torr at 20°C. The ^1H NMR spectrum has two sharp singlets at $\delta = 0.14$ (CH_3) and 5.88 ppm (C_5H_5) in C_6H_6 and at $\delta = -1.2$ (CH_3) and 6.08 ppm (C_5H_5) in c-C_6H_{12}. The ^9Be NMR spectrum has a single broad peak at $\delta = -17.23$ ppm in C_6H_6 with a half-height width of 44 Hz.

The vapor phase and Nujol mull IR spectra and the Raman spectrum of the pure solid were obtained. The observed vibrations are given in Table 31. No vibrational analysis was attempted [1].

Table 31
IR and Raman Spectra (in cm^{-1}) of $Be(C_5H_5)N(Si(CH_3)_3)_2$ [1].

IR (vapor)	IR (Nujol)	Raman (solid)
		3126 (vs)
3100 (m)	3100 (w)	3104 (vs)
		3070 (sh)
		3014 (m)
2970 (s)	under Nujol	
2920 (m)	under Nujol	
2910 (w)	under Nujol	
	1720 (w)	
	1630 (w)	
	1580 (w)	
		1407 (m)
1380 to 1350 (w, br)		1355 (s)
1260 (s)	1260 (m, sh)	

Table 31 (continued)

IR (vapor)	IR (Nujol)	Raman (solid)
	1252 (s, sh)	
	1245 (vs, sh)	1246 (w)
1187 (s)	1182 (w)	
	1130 (vs, sh)	1130 (vs)
1085 to 1070 (w, br)	1080 (vs)	1066 (s)
	1018 (s, sh)	
935 (vs, sh)	958 (vs)	
	888 (m, sh)	
845 (s, br)	832 (vs)	823 (m)
	816 (vs)	790 (m)
765 to 750 (m, br)	752 (s)	745 (m)
		727 (sh)
680 (m)	672 (m)	693 (m)
		675 (m)
	612 (m)	620 (w)
		591 (m)
	473 (w, sh)	470 (m)
		455 (m)
		434 (s)
		370 (vs)
		190 (m)

The compound is monomeric in C_6H_6 and c-C_6H_{12}. Mass spectra were obtained at ionizing voltages of 10, 20, 30, and 70 eV and ambient source temperature. The spectrum at 10 eV consists only of the parent ion. From 10 to 30 eV the base peak is $[Be(C_5H_5)NSi_2(CH_3)_3]^+$. Fragmentation of the amino group and retention of the C_5H_5 group is predominant. The ionization potential is 8.57 eV; the BeN bond dissociation energy is 4.98 eV (= 114.84 kcal/mol) [1].

References:

[1] Bartke, T. C. (Diss. Univ. Wyoming 1975; Diss. Abstr. Intern. B **36** [1976] 6141).
[2] Gaines, D. F., Coleson, K. M., Calabrese, J. C. (Inorg. Chem. **20** [1981] 2185/8).

3.4 Cyclopentadienylberyllium Derivatives with Organic Ligands

$Be(C_5H_5)CH_3$

$Be(C_5H_5)_2$ is reacted with an excess of $Be(CH_3)_2$ without a solvent. Warming the mixture in a 50°C warm bath causes the solid components to melt to a colorless liquid. The reaction vessel is frozen with liquid N_2 and evacuated. Upon warming, the product is vapor-transferred into a storage vessel [2, 6]. If the reaction is performed in C_6H_6, the high product volatility prevents separation from the solvent, but the presence of $Be(C_5H_5)CH_3$ is confirmed by IR and NMR spectra of the solution [6].

References on pp. 182/3

$Be(C_5H_5)CH_3$ melts at −36 to −34°C (−35°C [2]) from long colorless needles to form a clear liquid. The vapor pressure is 24 Torr at 27°C [6] or 20 Torr at 23°C [2].

The ^1H NMR spectra of $Be(C_5H_5)CH_3$ consist of two sharp singlets at $\delta = 5.89$ (5.95 [2]) (C_5H_5) and −1.74 (−1.75 [2]) (CH_3) ppm in c-C_6H_{12}, at $\delta = 5.85$ (5.75 [2]) (C_5H_5), and −1.20 (−1.22 [2]) (CH_3) ppm in C_6H_{12}, and at $\delta = 5.82$ (C_5H_5) and −1.18 (CH_3) ppm in $C_6H_5CH_3$. The C_5H_5 protons shift in toluene from $\delta = 5.82$ ppm at 23°C to 5.67 ppm at −40°C and 5.59 ppm at −80°C. A slight shift downfield is observed for the CH_3 protons: $\delta = -1.22$ ppm at 23°C, −1.02 ppm at −40°C, and −0.85 ppm at −80°C. No splitting of the ring resonance is observed [6].

The ^9Be NMR signals are singlets at $\delta = -20.6$ in C_6H_6, −20.1 in c-C_6H_{12}, −20.5 in $CH_3C_6H_5$ [6], and −19.73 ppm for the neat compound [2]. The peaks have half-height widths of 10, 7, 9 [6], and 15 Hz [2], respectively. In C_5H_{12}/CF_3Br solution, $\delta = -20.4$ ppm (half-height width of 8 Hz) is observed [9].

No ESR signals are obtained for C_6H_6 or $C_6H_5CH_3$ solutions of the complex at temperatures as low as −150°C [6].

The IR spectrum of $Be(C_5H_5)CH_3$ has been recorded in the vapor phase [2,6] and in C_6H_6 solution [6]. The observed vibrations and their assignments are shown in Table 32. Assignments are made by analogy with the spectra of $Be(CH_3)_2$, $Be(CH_3)BH_4$, and $Be(C_5H_5)_2$. The ring fundamentals are assigned under C_{5v} symmetry, and the $BeCH_3$ fundamentals under local C_{3v} symmetry. The Be–C(CH_3) stretch is tentatively assigned to a weak band at 1080 cm^{-1}. The IR data show that there is no significant difference between the species in solution and in the vapor phase [6].

Table 32
IR Vibrations (in cm^{-1}) of $Be(C_5H_5)CH_3$ [6].

vapor at 25°C	C_6H_6	assignment
3123 (w)	3114 (w)	A_1, ν(CH ring)
2938 (m)	2920 (s)	E_1, ν_{as}(CH)
2868 (m)	2847 (m)	A_1, ν_s(CH)
2426 (w)		1215 + 1215 = 2430
2222 (w)		1215 + 1014 = 2229
2082 (w)		
1809 (m)		1014 + 789 = 1803
1724 (m)	1725 (w)	
1608 (w)	1591 (w)	
1445 (w)		E_1, ν(CC ring)
1215 (vs, PQR)	1193 (s)	A_1, $\delta_s(CH_3)$
1127 (m, PQR)	1115 (m)	A_1, ν(CC ring)
1080 (m, sh)		A_1, ν(BeC)
1014 (vs)	995 (vs)	E_1, δ(CH ring) in-plane
828 (s, PQR)	820 (s)	E_1, δ(CH ring) out-of-plane
789 (s, PQR)	780 (s)	A_1, δ(CH ring) out-of-plane
684 (m)	727 (s)	E_1, $\varrho(CH_3)$

The He(I) photoelectron spectrum was measured. Assignments are made with the aid of calculated molecular orbitals (ε_J) using a modified CNDO procedure. The measured vertical ionization energies (ε_J in parentheses) of 9.43 and ~9.8 eV (10.14 eV) result from ionization of the $e_\pi(C_5H_5)$ orbitals. The band at ~10.3 eV (10.03 eV) results from the δ(BeC) orbital. A shoulder at 12.40 eV could not be assigned [8]. The ionization potential, determined by mass spectroscopy, is 9.83 eV [6] or 9.93 eV [2]. The calculated net atomic charges are −0.082 for C in C_5H_5, 0.031 for Be, and −0.282 for C in CH_3 [8].

The gas phase molecular structure has been determined by electron diffraction. Refinements were carried out on a model with total C_s symmetry in which one hydrogen atom of the CH_3 group and one hydrogen atom of the C_5H_5 group are eclipsed. The barrier of rotation of the CH_3 group relative to C_5H_5 is assumed to be low. It is assumed that the C_5H_5 group has D_{5h} symmetry; e.g., the hydrogen atoms lie in the plane of the C_5 ring, and the CH_3 group has local C_{3v} symmetry with the threefold axis coincident with the fivefold axis of the rest of the molecule. This model is shown in **Fig. 16** together with the determined parameters. It is proposed that the Be is sp hybridized and uses one hybrid orbital for the bond with the methyl C and the other hybrid for combination with the a_1 π orbital of C_5H_5. Two more degenerate bonding molecular orbitals of higher energy are formed by combination of the unhybridized Be(2p) orbitals with the two e_1 π orbitals of the ring. Be is therefore surrounded by an octet of electrons [1, 3].

It follows from PRDDO calculations that the monomer with an η^5-bonded C_5H_5 ring is energetically favored (ΔH = −246.2340 a.u.) over the σ-bonded isomer by 40 kcal/mol. Dimers of the type $C_5H_5Be(CH_3)_2BeC_5H_5$ with π- or σ-bonded C_5H_5 groups are also energetically unfavorable [11].

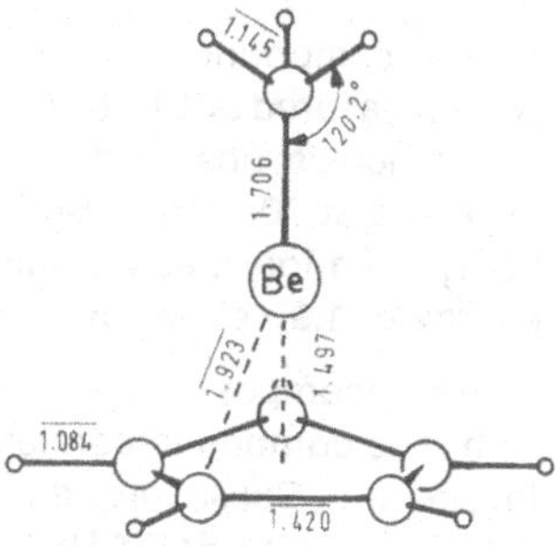

Fig. 16. Gas phase molecular structure of $Be(C_5H_5)CH_3$ [1, 3].

The liquid compound is miscible with C_6H_6 [2, 6] and c-C_6H_{12} [2]. It is monomeric in these solvents, as determined by cryoscopy [2, 6]

The mass spectrum at 12 eV/90°C shows the parent ion as the base peak and, additionally, $[BeC_5H_5]^+$. At 20 eV ionizing voltage, $[BeC_5H_5]^+$ accounts for the base peak. Other observed peaks are $[Be(C_5H_5)CH_3]^+$, $[C_5H_5]^+$, and $[CH_3]^+$. Both spectra indicate the presence of C_5H_6 and CH_4 that may have formed from thermal decomposition at 90°C. In neither spectrum is $[BeCH_3]^+$ observed, whereas the $[BeC_5H_5]^+$ species is quite stable, which is consistent with a strong covalent interaction. The mass spectrum was also recorded at 70 eV, but the data were not completely analyzed. The spectrum is dominated by fragmentation of the C_5H_5 ring [6]; see also [2]. The dissociation energy of the $BeC(CH_3)$ bond in $[Be(C_5H_5)CH_3]^+$ is determined to be 51 kcal/mol [6] or 50.27 kcal/mol [2].

In attempts to synthesize $Be(C_5H_5)H$, $Be(C_5H_5)CH_3$ was treated with $PH(CH_3)_2$, $PH(C_6H_5)_2$, $AsH(CH_3)_2$, $AsH(C_6H_5)_2$, and $SnH(CH_3)_3$, but no reactions were observed in C_6H_6 at room temperature [2].

References on pp. 182/3

$Be(C_5H_5)C_2H_5$

The compound is prepared by addition, in small aliquots, of an equimolar amount of LiC_2H_5 in C_6H_6 to a solution of $Be(C_5H_5)X$ (X = Cl, Br) in C_6H_6. After stirring for 2 h, the mixture is filtered. The solvent and $Be(C_5H_5)C_2H_5$ are vapor-transferred away from the side product $Be(C_2H_5)_2$. Trap-to-trap fractionation separates most of the compound from C_6H_6. The compound is also prepared by addition of $Mg(C_2H_5)Br$ in ether to $Be(C_5H_5)X$ (X = Cl, Br) (mole ratio 1:1). After stirring for 1 h and standing overnight, the mixture is filtered. The compound is separated from most of the ether by fractionation, using dry ice/acetone and liquid N_2 traps. According to NMR spectra it is 90 to 95% free of ether. A third method of preparation is the reaction of equimolar amounts of $Be(C_5H_5)_2$ and $Be(C_2H_5)_2$ as described for the previous compound. The compound is vapor-transferred at room temperature [2].

The compound is a colorless volatile liquid, m.p. −18°C, vapor pressure 8 Torr/23°C. ^1H NMR (C_6H_6): δ = −0.54 to −0.30 (q, CH_2), 1.06 to 1.22 (t, CH_3), and 5.75 ppm (s, C_5H_5). ^1H NMR (c-C_6H_{12}): δ = −0.32 to −0.08 (q, CH_2) and 5.87 ppm (s, C_5H_5). The signals for CH_3 are hidden under the solvent peak. The coupling constant between the CH_3 and CH_2 protons is ~8 Hz. The ^9Be NMR spectrum consists of a single peak at δ = −20.61 ppm with a half-height peak width of 27 Hz in C_6H_6; at δ = −18.5 ppm with a half-height width of 8 Hz in $C_6H_5CH_3$; at δ = −18.3 ppm in c-C_6H_{12} [2]; and at δ = −20.4 ppm in C_5H_{12}/CF_3Br with a half-height width of 8 Hz [9].

The vapor phase IR spectrum was measured and assigned. The observed vibrations (in cm^{-1}) are 3120 (w), 3100 (vw), 2943 (m), 2910 (m), 2870 (m), 2705 (w), 2420 (w), 2280 (w), 2082 (w), 1810 (w-m), 1720 (w-m), 1600 (w-m), 1460 (w, br), 1440 (w), 1418 (w), 1380 (w), 1210 (s), 1125 (m), 1024 (vs), 965 (w), 938 (s), 830 (s), 781 (s), and 630 (w-m). The band at 630 cm^{-1} is assigned to ν(Be–C(ethyl)) in contrast to the assignment made in [6] for $Be(C_5H_5)CH_3$ (see above) [2].

The compound dissolves as a monomer in C_6H_6 and c-C_6H_{12} (by cryoscopy). Mass spectra were measured at 10, 20, 30, and 70 eV ionizing voltages and ambient source temperature. The parent ion is observed in all spectra and is the only peak at 10 eV. The other fragments observed at 70 eV are $[BeC_7H_9]^+$, $[BeC_6H_n]^+$ (n = 7, 6), $[BeC_5H_5]^+$ (base peak), $[BeC_3H_3]^+$, and $[C_3H_3]^+$. An ionization potential of 9.18 eV is obtained, and a Be–C(ethyl) bond dissociation energy of 1.90 eV is found (= 43.81 kcal/mol) [2].

No decomposition is observed upon pyrolysis of the compound at 110 to 120°C for up to 48 h. The compound condenses off the vessel walls at ~65°C under 1 atm of N_2. In an attempt to prepare $Be(C_5H_5)H$, the compound was treated with $PH(CH_3)_2$, $PH(C_6H_5)_2$, $AsH(CH_3)_2$, $AsH(C_6H_5)_2$, and $SnH(CH_3)_3$. No reaction occurred in C_6H_6 at room temperature [2].

$Be(C_5H_5)C_4H_9$

The compound is prepared by dropwise addition of LiC_4H_9 to $Be(C_5H_5)X$ (X = Cl, Br) (mole ratio 1:1) in C_6H_6. After stirring for 2 h, the solution is filtered, and the solvent is removed under vacuum. The compound is isolated by vacuum distillation at 50°C/10^{-3} Torr. It is also prepared by stirring an equimolar mixture of $Be(C_5H_5)_2$ and $Be(C_4H_9)_2$ for 1 h. The reaction vessel is cooled with liquid N_2 and evaporated. The compound is then vapor-transferred upon heating with warm water [2].

The compound is a colorless liquid, m.p. −12°C, vapor pressure 4 to 5 Torr/23°C. ^1H NMR (C_6H_6): δ = −0.59 to −0.45 (br t, CH_2Be; J = 7 Hz), 1.04 to 0.92 (br t, CH_3; J = 6 Hz), 1.21 to 136 (br m, CH_2CH_2), and 5.80 ppm (s, C_5H_5). The signals appear in c-C_6H_{12} at δ = −0.84 to −0.69 (br t, CH_2Be; J = 7.5 Hz) and 5.90 ppm (s, C_5H_5). The signals for CH_3 and CH_2CH_2 are hidden under the solvent resonance.

The IR spectrum was measured for the vapor phase, but no assignment of the complex spectrum was attempted. Bands are observed at 3120(w), 3080(vw), 2990(m,sh), 2960(s), 2920(s), 2880(s), 2840(m), 2790(w-m), 2705(vw), 2420(vw), 1820(w), 1728(w), 1610(w), 1464(m), 1409(w), 1372(w), 1355(w,br), 1282(w), 1252(w), 1175(m,sh), 1160(m-s), 1125(m-s,sh), 1070(vs), 1020(vs), 962(s), 942(w), 920(w), 870(m,br), 820(w), 792(s,sh), 702(w), 655(m), 635(m), 545(w), and 509(w) cm^{-1} [2].

Monomers are present in C_6H_6 and c-C_6H_{12} (by cryoscopy). The mass spectra at 10, 20, 30, and 70 eV ionizing voltages and ambient source temperature all show a peak for the parent ion. The ion $[BeC_7H_9]^+$, obtained by loss of C_2H_5, is the base peak at 10 and 20 eV; whereas at 30 and 70 eV, the base peak is due to $[BeC_5H_5]^+$. The other fragments observed at 70 eV are $[BeC_9H_{13}]^+$, $[BeC_8H_{11}]^+$, $[BeC_7H_9]^+$, $[BeC_6H_7]^+$, $[BeC_3H_3]^+$, $[C_3H_3]^+$, $[BeH]^+$, and $[Be]^+$. The ionization potential is 8.95 eV, the Be–C(butyl) dissociation energy is 1.87 eV (= 43.12 kcal/mol) [2].

No decomposition is observed upon pyrolysis at 110 to 120°C up to 48 h. The compound condenses in the flask at ~95°C and 1 atm N_2. No reaction is observed with $SnH(CH_3)_3$ in C_6H_6 at room temperature [2].

$Be(C_5H_5)C_4H_9$-t

For the preparation, LiC_4H_9-t in C_6H_6 is added in small portions to $Be(C_5H_5)X$ (X = Cl, Br) (mole ratio 1:1). The solvent is carefully removed under vacuum. The wet white residue is washed with C_6H_{14} and filtered. The compound is obtained in pure form by sublimation at ~70°C/10^{-3} Torr [2]. The compound is also prepared by mixing equimolar amounts of $Be(C_5H_5)_2$ and $Be(C_4H_9\text{-t})_2$. After standing overnight [13], or for 2 h and subsequent heating with hot water for 0.5 h [2], the solvent is removed [13] by freezing the flask with liquid N_2 and evacuation [2]. The crystalline product is left [13]. It is vapor-transferred to a storage vessel [2].

The colorless crystals melt at 38 to 39°C [2], 39 to 40°C [13], or 40 to 41°C. The vapor pressure is 3 to 4 Torr at 23°C [2]. The compound is so volatile that some of it is lost upon evaporation of the solvent C_6H_6 [13].

The ^1H NMR spectrum consists of two sharp singlets at $\delta = 0.95$ (CH_3) and 5.70 ppm (C_5H_5) in C_6H_6. In c-C_6H_{12}, the signal for C_5H_5 is found at 5.72 ppm. The other signal is hidden under the solvent resonance. The IR spectrum was measured in Nujol, but no vibrational analysis was attempted. The observed bands (in cm^{-1}) are 3120(w), 2820(m), 2465(w), 1732(w), 1615(w), 1234(w-m), 1207(w-m), 1180(w-m), 1037(m), 1018(m), 1004(m), 925(s,br), 880(s,br), 835(s,br), 810(s,br), 786(s,br), 750(s,br), 720(s,br), 570(w-m), and 495(w-m) [2].

It follows from cryoscopic measurements that the compound dissolves as a monomer in C_6H_6 [2, 13] and c-C_6H_{12} [2]. The parent ion is observed in the mass spectra taken at 10, 20, 30, and 70 eV ionizing voltages and ambient source temperature. It is the only peak observed at 10 eV. At the other voltages the base peak corresponds to $[BeC_5H_5]^+$. The other fragments observed at 70 eV are $[BeC_9H_{12}]^+$, $[BeC_8H_{11}]^+$, $[BeC_7H_n]^+$ (n = 7,5), $[BeC_6H_n]^+$ (n = 8,7), $[BeC_5H_n]^+$ (n = 5,4), $[BeC_3H_n]^+$ (n = 3 to 1), $[C_3H_3]^+$, and $[C_2H_2]^+$. The measured ionization potential is 8.60 eV; the Be–C(butyl) bond dissociation energy is 1.60 eV (= 36.90 kcal/mol) [2].

Pyrolysis of the compound at 110 to 120°C gives no decomposition products up to 48 h. At ~90°C and 1 atm N_2, or ~75°C and 200 Torr N_2, the compound condenses on the walls of the flask [2]. It is sensitive to protic acids and gives, with an excess or an equimolar amount of t-C_4H_9OH, $(Be(OC_4H_9\text{-t})_2)_3$ [13]. No reaction is observed with $SnH(CH_3)_3$ in C_6H_6 at room temperature [2]. No adducts are formed with $N(CH_3)_2CH_2CH_2N(CH_3)_2$ and quinuclidine (1-azabicyclo[2.2.2]octane) [13].

References on pp. 182/3

$Be(C_5H_5)C_6H_5$

The compound is prepared by mixing equimolar amounts of LiC_6H_5 and $Be(C_5H_5)X$ (X = Cl, Br) in C_6H_6. After stirring for 2 h, the solution is filtered, and the solvent is removed under vacuum. The residue is sublimed at room temperature/10^{-3} Torr in the course of several days. The compound can also be vacuum-distilled. It could not be prepared from $Be(C_5H_5)_2$ and $Be(C_6H_5)_2$ [2].

The colorless crystals melt at 35 to 35.5°C. The vapor pressure is 0.03 Torr/18.5°C. The ^1H NMR spectrum shows a sharp singlet for the protons in C_5H_5 at $\delta = 5.81$ ppm in C_6H_6 and at 6.04 ppm in c-C_6H_{12}. The signals of the protons of the phenyl ring are partially lost under the solvent resonances. They are found in two multiplets at 6.28 to 6.92 ppm (meta and para positions) and 7.00 to 7.14 ppm (ortho positions). The ^9Be NMR spectrum in C_6H_6 consists of a single peak at $\delta = -21.13$ ppm with a half-height width of 32.5 Hz [2]. The fully coupled ^{13}C NMR spectrum in c-C_6H_{12} or $CDCl_3$ shows a complex multiplet with five prominent lines for the C_5H_5 protons at $\delta = 104.68$ ppm and splittings of $^1J(C,H) = 177.4$ Hz, $^{2,3}J(C,H) = 6.7$, and $^1J(C,Be) = 1.1$ Hz. The C_6H_5 resonances occur at $\delta = 127.24$(C-3,5), 127.35(C-4), and 140.00 ppm (C-2,6). The resonance of the carbon atom C-1 bonded to Be could not be located [5].

A good IR spectrum of the vapor phase could not be obtained due to the low volatility of the compound. The IR spectrum in Nujol and the Raman spectrum of the solid were measured. IR (Nujol): 3110(w), 3070 to 3000(w-m), 2720(w, br), 2415(w), 2275(w), 2080(w), 1950(w), 1880(w), 1830(w-m), 1745(br, m), 1610 to 1600 (br, w), 1575, 1418 (s, sh), 1329(w), 1290(w), 1240(w), 1209(vw), 1190(s, sh), 1165(br, vs), 1114(s, sh), 1053(w, sh), 1020(w, sh), 1015(vs, sh), 995(m, sh), 930(sh), 905(vs), 855(m), 830(m), 777(m, sh), 701 (vs, sh), 600(s, sh), 498(s, sh), and 428(m, sh) cm^{-1}. Raman (solid): 3120(vs), 3102(vs), 3096(vs), 3070(s, sh), 3047(w), 3032(w), 3014(m), 2994 to 2854, 1581, 1563, 1477, 1415(m), 1357(s), 1246(w), 1194, 1173, 1160, 1143, 1119(vs, sh), 1068(s), 1037, 1001 (sh), 937(w), 863(w), 839(m), 787(m), 740(sh), 727(w), 624(w), 609(m), 508, and 435(s) cm^{-1}. Some of the bands were assigned to ν(CH), ν(CC), and δ(CH) [2].

It follows from PRDDO calculations that the monomer with an η^5-bonded C_5H_5 group ($\Delta H = -436.5371$ a.u.) is more stable by 45.5 kcal/mol than the isomer with a σ-bonded C_5H_5 group. Dimers of the type $C_5H_5Be(C_6H_5)_2BeC_5H_5$, with π- or σ-bonded C_5H_5 rings, are energetically unfavorable [11].

The compound is monomeric in C_6H_6 and c-C_6H_{12} (by cryoscopy). Mass spectra are obtained at ionizing voltages of 70, 30, and 20 eV and ambient source temperature. The parent ion is observed at all voltages and is the major peak at 20 eV. At 30 and 70 eV, the major peak corresponds to $[BeC_5H_5]^+$. Other fragments observed at 70 eV are $[BeC_{11}H_9]^+$, $[BeC_{11}H_n]^+$ (n = 10,9), $[BeC_9H_n]^+$ (n = 8,7), $[C_9H_n]^+$ (n = 8,7), $[BeC_7H_n]^+$ (n = 6,5), $[C_7H_n]^+$ (n = 6,5), $[BeC_6H_5]^+$, $[C_6H_5]^+$, $[BeC_5H_n]^+$ (n = 5 to 3), $[BeC_4H_n]^+$ (n = 4, 3), $[BeC_3H_n]^+$ (n = 3 to 1), $[C_3H_3]^+$ [2]. The ionization potential is 11.38 eV; the Be–C(phenyl) bond dissociation energy is 3.96 eV (= 91.32 kcal/mol) [2].

$Be(C_5H_5)C_6F_5$

The compound is misprinted in the experimental part in [13] as $Be(C_5F_6)C_4H_9$-t. It is prepared from equimolar amounts of $Be(C_5H_5)_2$ and $Be(C_6F_5)_2 \cdot 2THF$ in C_6H_6. After standing for 30 h, the volatile material is removed by evaporation, and the residue is sublimed at 100°C/10^{-4} Torr.

The compound melts at 108 to 109°C with decomposition. It is monomeric in C_6H_6 (by cryoscopy) [13].

$Be(C_5H_5)C{\equiv}CH$

The preparation of the compound has not been published. It is only referred to in [10].

The He(I) photoelectron spectrum shows two peaks, corresponding to four bands, which are well separated from others. To interpret the spectrum, the molecular orbital energies (ε_J) were calculated using a modified CNDO. A comparison of calculated and experimental values suggests that the first two peaks are to be assigned to four ionization events from two e type orbitals with a Jahn-Teller splitting of 0.1 to 0.2 eV:

band	I.P. (in eV)	assignment	$-\varepsilon_J$ (in eV)
1	9.40	$e_\pi(C{\equiv}C)$	9.69
2	~9.6		
3	10.30	$e_\pi(C_5H_5)$	11.21
4	~10.5		
5	12.36	$\sigma(BeC)$	12.12

The calculated net atomic charges are −0.235 (α-C of C_2H), −0.170 (β-C of C_2H), −0.072 (C of C_5H_5), and 0.068 (Be) [8].

The gas phase structure was determined by electron diffraction. The molecular model, with the measured parameters is shown in **Fig. 17**. It is assumed that the compound has C_{5v} symmetry and that the symmetrically bonded C_5H_5 ring is completely planar. The bond distances, except for the assumed CH bond length of the acetylenic group, were refined to R = 14.8% [4].

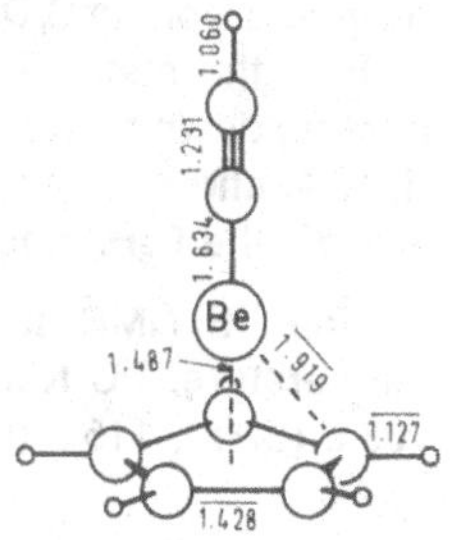

Fig. 17. Gas phase molecular structure of $Be(C_5H_5)C{\equiv}CH$ [4].

It follows from MNDO calculations that the C_{5v} structure with pentahapto C_5H_5 is, with $\Delta H_f = 33.9$ kcal/mol, more stable by 7.6 kcal/mol than the C_s model with a monohapto C_5H_5 group. The calculated ionization potentials for these two models are 9.45 and 8.66 eV; the dipole moments are 1.27 and 1.52 D, respectively. The HOMO is ascribed to a degenerate $e_\pi(C_5H_5)$ orbital, and the next lower orbital to $e_\pi(C{\equiv}C)$, which contrasts with the assignment made in [8] (see above). The distances obtained for the C_{5v} structure are (in Å) 1.209 for C≡C, 1.457 for CC in C_5H_5, 1.609 for BeC_2, and 1.993 for BeC_5 [7].

$Be(C_5H_5)C{\equiv}CCH_3$

The preparation of this compound has not been published. It is only referred to in [10].

The He(I) photoelectron spectrum has been recorded. See the preceding compound for details. The experimental vertical ionization potentials (in eV) are compared with the calculat-

References on pp. 182/3

ed orbital energies ($-\varepsilon_J$ in parentheses, also in eV), and the assignments are e_π(C≡C) at 8.82 and ~9.1 (9.33), $e_\pi(C_5H_5)$ at 9.85 and ~10.15 (10.98), and σ(BeC) at 11.95 (12.11).

The calculated net atomic charges are −0.235 (α-C of C_3H_3), −0.107 (β-C of C_3H_3), −0.073 (C_5H_5), −0.048 (CH_3), and 0.075 (Be) [8].

$Be(C_5H_5)C_8H_{15}BN$ ($C_8H_{15}BN$ = 1-t-butyldihydro-2-methyl-1 H-azaborolyl)

Equimolar quantities of $Be(C_5H_5)_2$ and $Be(C_8H_{15}BN)_2$ ($C_8H_{15}BN$ = 1-t-butyldihydro-2-methyl-1 H-azaborolyl) are stirred for 30 min in ether at room temperature. Evaporation of the solvent leaves the product as a yellow oil in quantitative yield. It could not be crystallized.

The parent ion is observed in the mass spectrum. A 0.004 M solution in THF shows an electrical conductivity of 0.077×10^{-6} S.

The ^{1}H NMR spectrum in $CDCl_3$ (see Formula I for the numbering) shows signals at δ = 0.52 (s, CH_3B), 1.41 (s, C_4H_9), 2.04 (s, H-3), 5.37 (dd, H-4; J(H-3,5) = J(H-4,5) = 3.5 Hz), 6.01 (s, C_5H_5), and 6.40 ppm (d, H-5) at room temperature. It is proposed that C-3 is σ-bonded to Be, whereas the C_5H_5 ring is η^5-coordinated (Formula I). Fluxional properties are revealed by temperature-dependent ^{1}H NMR studies. In $C_6D_5CD_3$ at −50°C, the spectrum corresponds to that in $CDCl_3$ at room temperature. The signal at 5.67 ppm (H-4) consists of a doublet of doublets (pseudotriplet), due to coupling with H-3 and H-5, with identical coupling constants. At −20°C, broad resonances at 6.5 and 3.8 ppm appear, whereas the H-4 signal has shifted from 5.67 to 5.59 ppm. This shifting continues on warming to 20°C, and the H-4 signal now appears at higher field (5.49 ppm) than the C_5H_5 signal (5.61 ppm). At 90°C, the intensities of the new and the old signals are equal. The new signals, especially the one at 3.8 ppm, may indicate the presence of η^5-coordinated $C_8H_{15}BN$ rings. The ^{1}H NMR spectrum in THF at room temperature shows signals for the $C_8H_{15}BN$ ring protons that agree with those in other solvents at low temperature. Thus, this ring is essentially η^1-bonded. However, the C_5H_5 ring signal is shifted from 5.61 ppm in $C_6D_5CD_3$ to 6.00 ppm, suggesting dissociation of $[C_5H_5]^-$ and coordination of THF to the metal. This could explain the observed electrical conductivity (see above). On cooling down to −50°C, the THF solution shows new signals at 1.93, 5.40, and 6.36 ppm for H-3, H-4, and H-5; however, the C_5H_5 signal is shifted to 5.44 ppm. This may suggest that at −50°C, the C_5H_5 ring is coordinated to Be.

The ^{11}B NMR spectrum shows a signal at δ = 40.5 ppm in $CDCl_3$ or $C_6D_5CD_3$ at room temperature. ^{13}C NMR (C_6D_6): δ = 2.8 (CH_3B), 32.1 (CH_3 in C_4H_9), 37.2 (C-3), 53.3 (C in C_4H_9), 103.6 (C_5H_5), 115.3 (C-4), and 128.3 ppm (C-5) [12].

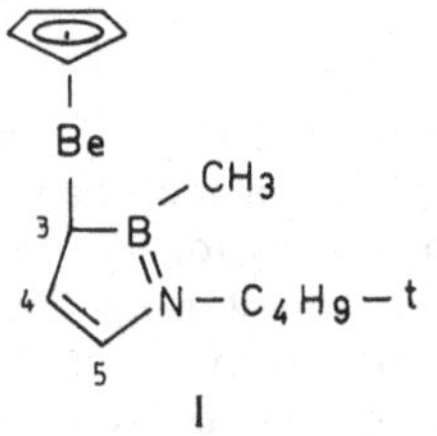

I

References:

[1] Drew, D. A., Haaland, A. (J. Chem. Soc. Chem. Commun. **1971** 1551).
[2] Bartke, T. C. (Diss. Univ. Wyoming 1975; Diss. Abstr. Intern. B **36** [1976] 6141).
[3] Drew, D. A., Haaland, A. (Acta Chem. Scand. **26** [1972] 3079/84).
[4] Haaland, A., Novak, D. P. (Acta Chem. Scand. A **28** [1974] 153/6).

[5] Fischer, P., Stadelhofer, J., Weidlein, J. (Organometal. Chem. **116** [1978] 65/73).
[6] Drew, D. A., Morgan, G. L. (Inorg. Chem. **16** [1977] 1704/8).
[7] Dewar, M. J. S., Rzepa, H. S. (J. Am. Chem. Soc. **100** [1978] 777/84).
[8] Böhm, M. C., Gleiter, R., Morgan, G. L., Lusztyk, J., Starowieyski, K. B. (J. Organometal. Chem. **194** [1980] 257/63).
[9] Gaines, D. F., Coleson, K. M., Hillenbrand, D. F. (J. Magn. Resonance **44** [1981] 84/8).
[10] Morgan, G. L., Starowieyski, K. B. (unpublished results from [4, 8]).

[11] Marynick, D. S. (J. Am. Chem. Soc. **103** [1981] 1328/33).
[12] Schmid, G., Boltsch, O., Boese, R. (Organometallics **6** [1987] 435/9).
[13] Coates, G. E., Smith, D. L., Srivastava, R. C. (J. Chem. Soc. Dalton Trans. **1973** 618/22).

3.5 Bis(cyclopentadienyl)beryllium

3.5.1 Experimental Studies

Preparation

$Be(C_5H_5)_2$ is prepared from $BeCl_2$ and MC_5H_5 (M = Na, K). Anhydrous $BeCl_2 \cdot 2O(C_2H_5)_2$ (0.5 equivalents) in ether is added dropwise to freshly prepared NaC_5H_5 (from Na and C_5H_6) in ether [1]. The temperature is kept at ~35°C [33]. After stirring for 3 to 4 h at room temperature [1] or ~35°C [33], the ether is evaporated, and the residue is sublimed in a high vacuum between 25 and 120°C [1] (45°C/10^{-3} Torr [33]). Several resublimations between 25 and 45°C give a pure compound. The yield is 74% [1]. Yields of up to 90% can be obtained with a 2-fold excess of NaC_5H_5 [30]. A reaction of $BeCl_2$ with KC_5H_5 in C_6H_6 gives a much lower yield, and the compound is not obtained with THF as solvent [1]. The statement that the reaction of NaC_5H_5 with $BeCl_2$ is carried out in THF [23] is probably erroneous.

Attempts to synthesize the compound from Be metal and C_5H_6 in the temperature range 150 to 550°C under ordinary or reduced pressure were unsuccessful [1].

Molecular Parameters and Physical Properties

$Be(C_5H_5)_2$ does not have a simple standard structure. Therefore, most of the physical and spectroscopic properties reported have been studied in great detail with the aim to obtain unequivocal information about the structure of the molecule. From all these efforts, it emerges that in the solid state the molecule is a slip-sandwich, with the Be centrally bonded to one ring and peripherally bonded to the other. In addition, the Be atoms are disordered over two equivalent positions within the framework of the two nearly planar C_5H_5 groups (Formula I). The slip-sandwich also seems to be the equilibrium structure in solution and in the gas phase. The Be oscillates between two equivalent positions, and the rings themselves interchange and are fluxional (Formula II).

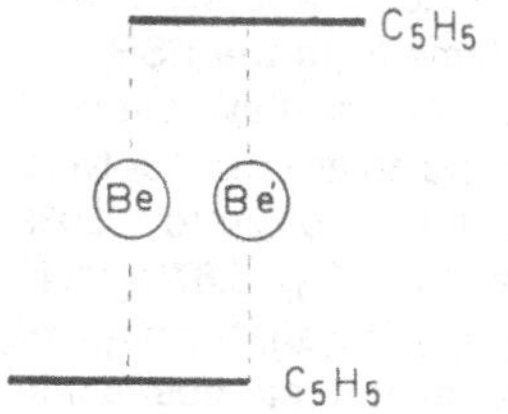

I

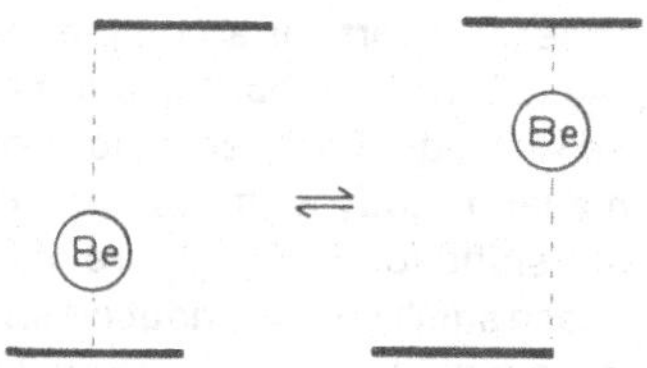

II

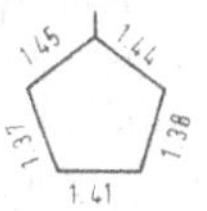

III

References on pp. 193/4

The compound crystallizes from petroleum ether at −60°C as white needles. Sublimation gives hard [33] colorless crystals [1, 33]. The melting point is 59 to 60°C [1, 27, 30, 33] or 59°C [23, 25, 31]. The extrapolated boiling point is 233°C. The vapor pressure in the temperature range between 110 and 190°C can be expressed by the equation $\log p = 8.2242 - 2671.9\,T^{-1}$ (p in Torr, T in K) [1]. The heat of vaporization is 12.2 kcal/mol [18].

$Be(C_5H_5)_2$ is diamagnetic. Molar susceptibilities of $\chi_{mol} = (-86 \pm 4) \times 10^{-6}$ (77 K), $(-99 \pm 4) \times 10^{-6}$ (190 K), and $(-107 \pm 4) \times 10^{-6}$ (291 K) cm^3/mol were determined [1].

Measured dipole moments (refractivity method) of $\mu = 2.46 \pm 0.06$ D in C_6H_6 at 25°C, 2.24 ± 0.09 D in c-C_6H_{12} at 25°C [1, 2], and 2.6 D in dioxane at 20°C [4] are strikingly different from the known values for $Mg(C_5H_5)_2$, and for other $M(C_5H_5)_2$ compounds of divalent metals, and are indicative of an asymmetric structure of the molecule [1, 2]. But it is pointed out that the refractivity method is not suited for flexible metal complexes because of a quite arbitrary estimate of the solute atomic polarization. Therefore, the dielectric properties of $Be(C_5H_5)_2$ were studied by microwave dielectric loss measurements in the region 1 to 9 GHz. The experimental absorption factors ψ lie close to a simple Debye curve (ψ vs. frequency). From the Debye curves, the dipole moments μ and the dielectric relaxation times τ are obtained. The dipole moments are significantly lower than those obtained by the refractivity method: $\mu = 1.61 \pm 0.13$ D in c-C_6H_{12}, 1.69 ± 0.09 D in decalin, 2.03 ± 0.09 D in C_6H_6, and 2.27 ± 0.08 D in dioxane. The increase in solvent viscosity in going from c-C_6H_{12} to decalin does not affect the dielectric relaxation time: $\tau = 11.8 \pm 2.2$ ps in c-C_6H_{12}, 12.5 ± 1.6 ps in decalin, 12.7 ± 1.5 in C_6H_6, and 15.9 ± 1.9 ps in dioxane. The larger value in dioxane reflects electrostatic interaction between $Be(C_5H_5)_2$ and the C–O dipoles of the solvent. An unusually large molar atomic polarization of ~57 cm^3 is calculated which is attributed to field-induced bending deformations of dipolar metal-ligand groups. The microwave study clearly confirms the dipolar nature of $Be(C_5H_5)_2$ and is in agreement with a dynamic structure with η^5- and σ-bonded nonparallel C_5H_5 groups [31].

It is mentioned that $Be(C_5H_5)_2$ shows a small electrical conductivity in ether, but no details are given [4].

^1H NMR spectra were measured and discussed for several solvents. In CS_2 and CCl_4, single sharp peaks are observed at $\delta = 4.70$ and 4.77 ppm, respectively, which are temperature-independent between −100 and 50°C [21]. In CF_2Cl_2, a single signal is also observed at $\delta = 5.60$ ppm from −50 to −135°C [22]. This result indicates that the C_5H_5 rings rotate freely relative to each other, and that the Be oscillates between its two alternate positions in the C_{5v} or slip-sandwich structure such that all hydrogens appear equivalent on the ^1H NMR time scale [21,25]. The signals at $\delta = 5.75$, 5.74, and 5.57 ppm, observed in TMS, c-C_6H_{12}, and c-$C_6H_{11}CH_3$, respectively, are also unchanged over a wide range of temperatures (range is not given). It was found that in C_6H_6 ($\delta = 5.81$ ppm) and $C_6H_5CH_3$ ($\delta = 5.88$ ppm), the signals observed for the solvents are split. From the intensity of the signals, a 2:1 complex formation to **$(Be(C_5H_5)_2 \cdot 0.5\,C_6H_6)_2$** and **$(Be(C_5H_5)_2 \cdot 0.5\,C_6H_5CH_3)_2$** is assumed. The same, but weaker, effect is observed in C_6H_5F. The signals attributed to the complexes move to higher field with decreasing temperature. Further support for a complex formation stems from the ESR and phosphorescence spectra (see below) [12], but the additional signals may also have been an artifact. They are not observed with equal concentrations of $Be(C_5H_5)_2$ and an aromatic solvent (~2%) in c-C_6H_{12} [12], nor in a later study with solutions of $Be(C_5H_5)_2$ in C_6H_6 of up to ~20% concentration [31]. The proton shifts for $Be(C_5H_5)_2$ are at $\delta = 5.73$ ppm in c-C_6D_{12}, 5.67 ppm in C_6D_6, and 5.69 ppm in C_6F_6. These small solvent induced shifts are explained by another type of solvent–$Be(C_5H_5)_2$ interaction, in which the aromatic solvents assume favored stereospecific orientations around $Be(C_5H_5)_2$ [31].

The ^9Be NMR spectra of $Be(C_5H_5)_2$ show singlets at $\delta = -18.5$ ppm in $C_6H_5CH_3$ [12, 28] and at $\delta = -18.3$ ppm in c-$C_6H_{11}CH_3$ [12]. The extreme up-field shift is explained with the ring-current effect of the η^5-C_5H_5 group [12, 28].

The ^{13}C NMR spectrum shows a single resonance at $\delta = 106.3$ ppm in ether and at 107.5 ppm in c-C_6H_{12} at room temperature. In the proton coupled spectrum, this resonance is split into the expected doublet with a C–H coupling of 168.5 Hz. Each component of the doublet is further split into a quintet (6.6 Hz) by coupling with the four remaining protons. There is additional fine structure with each line further split into an apparent doublet. This additional structure persists in the ^1H decoupled spectrum, indicating that it arises from ^{13}C–Be coupling. As the temperature is lowered, the ^{13}C line width decreases, and a sharp singlet is observed at 200 K. The line shape and the temperature dependence of the line width is similar in both solvents. For an interpretation of these results it is assumed that the quadrupolar nucleus is relaxed by inversion of the molecular dipole caused by exchange of Be between the two C_5H_5 rings. The correlation time and the activation barrier are estimated (see pp. 191/2) [24].

Solid $Be(C_5H_5)_2$ shows no ESR signal in the temperature range from −100 to 60°C. Also, no signal is observed in a solution in c-C_6H_{12}. Two very weak, superimposed signals are reported for toluene solutions in the temperature range from −50 to 60°C, a very broad signal and a somewhat sharper one at g = 2.00. Below −100°C, fine structure emerging from the broad signal is also observed. It is suggested that the spectrum arises from toluene radicals within the complex $(Be(C_5H_5)_2 \cdot 0.5\ C_6H_5CH_3)_2$, which is suggested to be formed according to the ^1H NMR spectra (see p. 184) [12]. This proposal has, however, been questioned in [31].

IR spectra are measured in the solid state, in solution, and in the gas phase. The spectrum in KBr contains bands which are assigned to the pentahapto C_5H_5 ring in analogy to $[C_5H_5]^-$: $\tilde{\nu}$ = 3067 (ν(CH), A_{2u}), 2941 (ν(CH), E_{1u}); 1428, 1400 (ν_{as}(CC), E_{1u}), 1121 (ν_{as}(CC), A_{2u}); 1014, (988) (δ(CH) in-plane, E_{1u}); 744, 738 (δ(CH) out-of-plane, A_{2u}) cm^{-1}. Bands at 1363, 956, (890), and 661 cm^{-1} are assigned to a monohapto C_5H_5 ring in analogy to C_5H_6 [3], briefly mentioned in [1]. The spectrum measured in Nujol and Hostaflon mull between 4000 and 33 cm^{-1} was first assigned on the basis of a sandwich structure with C_{5v} symmetry with the Be atom being closer to one C_5H_5 ring than to the other ring [11]. In a later publication several of these band assignments have been changed, but the C_{5v} symmetry is retained. The spectrum in Nujol between 4000 and 200 cm^{-1} is also shown in a figure in this publication [17]. Based upon the low number of bands that can be attributed to ν(BeC) in the far IR spectrum in Nujol ($\tilde{\nu}$ (in cm^{-1}) = 662 (w), 587 (vs), and 414 (m), k = 0.8 mdyn/Å for the metal-ligand vibration), the two C_5H_5 rings may be equally bonded to Be [5].

The IR vibrations in CS_2 and $CDCl_3$ between 4000 and 250 cm^{-1} are given in Table 33, p. 186 [30]. The spectrum agrees essentially with those obtained in C_6H_6 and c-C_6H_{12} [11]. The solution spectrum contains all the bands that are present in the spectra of $Be(C_5H_5)Cl$ and $Be(C_5H_5)Br$. The bands are thus due to the pentahapto C_5H_5 ring. The remaining absorptions compare well with the bands previously assigned to C_5H_5 ring vibrational modes in the spectra of $Hg(C_5H_5)Cl$ and $Hg(C_5H_5)_2$, for which the C_5H_5 rings are known to be monohapto and fluxional. Therefore, it is concluded that in the solution state Be is pentahapto-bonded to one C_5H_5 ring and monohapto to the other. The highly intense bands at 965 and 740 cm^{-1} (Table 33, p. 186) are assigned to the Be–η^5-C_5H_5 stretch and the Be–η^1-C_5H_5 stretch, respectively [30].

The gas phase spectrum recorded at 50°C in a cell with 6.2 cm pathlength shows, between 4000 and 200 cm^{-1}, absorptions at 3084 (m, br), 1740 (vs), 1598 (vw), 1434 (w), 969 (vs), 862 (vw), 828 (vvw), 772 (vw), and 736 (vs) cm^{-1} [17]; see also [33]. The bands are assigned on the basis of a C_{5v} symmetry. See the original for a figure of the spectrum. The considerably smaller number of bands observed in the gas phase, compared to the solid state and solution spectra, was explained by the ionic character ($Be^{2+}[C_5H_5]_2^-$) of the compound in the vapor phase [17]. This

References on pp. 193/4

interpretation has been questioned since the gas phase spectrum resembles the spectrum obtained with low solute concentrations where only the strongest absorptions are apparent [30].

Table 33

IR Vibrations of $Be(C_5H_5)_2$ in CS_2 and $CDCl_3$ [30].

$\tilde{\nu}$ (cm^{-1})	assignment η^5-C_5H_5	η^1-C_5H_5	$\tilde{\nu}$ (cm^{-1})	assignment η^5-C_5H_5	η^1-C_5H_5
3970(w)	+		1124(s)	+	
3910(w)			1106(w)		+
3170(vw)			1090(sh)	+	
3110(sh)		+	1074(mw)		+
3080(s)	+	+	1060(sh)	+	
3050(sh)		+	1015(s)	+	
2710(w)	+		965(vs, br)	+	
2500(vw)	+		915(sh)		+
2420(w)	+		890(sh)		+
2085(w)	+		862(ms)	+	
1818(w)	+	+	829(m)	+	+
1737(m)	+		785(sh)	+	
1605(m)	+		770(ms)		+
1428(m)	+	+	740(vs)		+
1410(sh)			663(s)		+
1365(m)		+	590(vw)	+	
1305 vw)		+	425(w)	+	
1235(vw)		+	313(w)	+	

The Raman spectra of solid $Be(C_5H_5)_2$ at −165 and 25°C and of the liquid compound at 65°C were recorded between 3000 and 30 cm^{-1}. They are shown in figures in the original. The band positions are listed in Table 34. There is no significant difference in these spectra. They contain all the bands corresponding to the normal modes of one C_5H_5 ring of local C_{5v} symmetry π-bonded to Be (see Table 34). Their positions and intensities are very similar to those observed for $Be(C_5H_5)Cl$, $Be(C_5H_5)Br$, $Be(C_5H_5)CH_3$, and $Be(C_5H_5)C{\equiv}CH$ which are known to have an η^5-C_5H_5Be moiety. The character of the second C_5H_5 ring is discussed, and a polyhapto-bonded (η^2/η^3) C_5H_5 ring is assumed to be most consistent with the data [27].

Dilute solutions of $Be(C_5H_5)_2$ in c-C_6H_{12} and C_6H_6 give rise to an intense fluorescence emission at 415 nm at room temperature. At liquid N_2 temperatures the compound is found to fluoresce at 420 nm in c-C_6H_{12} and 405 nm in C_6H_6. The spectra are not further discussed [12].

No phosphorescence is observed in a solution of $Be(C_5H_5)_2$ in c-C_6H_{12} at −196°C; whereas with a small amount of compound in C_6H_6, a phosphorescence maximum is observed at 470 nm with τ=1.56s at −196°C. Diluting this solution by one-half results in a phosphorescence maximum at 465 nm and τ=3.12s at −106°C. Benzene phosphoresces at 440 nm. The excitation maximum at 250 nm corresponds to the absorption maximum of C_6H_6. The phosphorescence spectrum is taken as a confirmation that complexation of the type

$(Be(C_5H_5)_2 \cdot 0.5C_6H_6)_2$ occurs in C_6H_6 by which stabilized triplet states or diradicals of the aromatic moiety are formed [12]. The formation of a complex between $Be(C_5H_5)_2$ and C_6H_6 was later questioned [30].

Table 34
Raman Vibrations (in cm^{-1}) of Solid and Liquid $Be(C_5H_5)_2$ [27].

solid −160°C	solid 25°C	liquid 65°C	polarization ratio	symmetry	assignment for η^5-C_5H_5Be
3122(s)	3123(s)	3128(s)	0.15	A_1	ν(CH)
3116(m, sh)					
3110(m)				E_1, E_2	ν(CH)
3100(m, sh)					
3094(s)	3102(s)	3100(s)	0.25		
3087(s)					
3076(m)					
	3070(m, sh)	3074(m, sh)	0.60		
3069(m)					
3013(s)	3020(w)	3002(w, br)	0.25		
	2869(vvw)				
	2783(vvw)				
	1506(vvw)		0.10		
1440(vvw)	1440(vw, sh)			E_1	ν(CC)
1420(w)	1420(w)	1420(m)	0.19		
1390(s)	1388(m)	1395(m)	0.28		
1356(m)	1360(m)	1362(m)	0.75	E_2	ν(CC)
1325(vw)					
1267(vvw)					
	1249(vw)	1246(vw)	0.28	A_1	
1256(vvw)					
1227(vvw)	1225(vvw)				
1129(vs)	1127(vs)	1130(vs)	0.12	A_1	ν(CC)
1118(vs)	1118(s, sh)	1116(s, sh)	0.15		
1089(m)					
	1083(m)	1082(m)	0.68	E_2	β(CH)
1075(m)					
1058(m)		1055(w, sh)	0.30		
1030(vw)	1026(vw)	1030(vw, sh)		E_1	β(CH)
996(w)	1001(vw)	994(vw)	0.50		
934(vw)	930(vw)	925(vvw)	0.75	E_2	λ(CCC)
905(vvw)					
883(w)					

References on pp. 193/4

Table 34 (continued)

solid −160°C	solid 25°C	liquid 65°C	polarization ratio	symmetry	assignment for η^5-C_5H_5Be
	870(w)	873(w)	0.75	E_1	ϱ(CH)
869(w, sh)					
846(w)					
	841(w)	839(w)	0.50	E_2	ϱ(CH)
838(w, sh)					
789(w)					
	795(w)	795(w)	0.27	A_1	ϱ(CH)
780(w, sh)					
762(w)	750(vw)	745(vvw)	0.80		
736(vvw)					
630(vw)	627(w, sh)	628(vw, sh)	0.60	E_2	χ(CCC)
600(s)	598(m)	598(m)	0.37		ν(BeC_5)
461(s)					
447(vs)	430(s, br)	447(m)	0.76	E_1	ν(BeC_5)
320(s)	313(s)	313(s)	0.23		ν(BeC_5)
187(s)					
165(m)					
150(s)					
136(s)					
124(s)					
87(s)					

The He(I) photoelectron spectrum of $Be(C_5H_5)_2$ displays three nearly equidistant peaks at 7.45, 8.42, and 9.5 eV with relative intensities ~1:1:2. The first two peaks have a Gaussian shape while the third peak is somewhat unsymmetrical and appears to consist of two strongly overlapping bands at 9.5 and 9.7 eV. A complex unresolved peak starts at 11.25 eV. The first four peaks were assigned by comparing the measured vertical ionization potentials with the calculated ones obtained by ab initio molecular orbital calculations for several structures of $Be(C_5H_5)_2$ (see Section 3.5.2, pp. 194/8). A satisfactory fit between experimental results and calculations is obtained for a slip-sandwich model of C_s symmetry [26].

The molecular structure of $Be(C_5H_5)_2$ was determined several times for the crystals at low temperatures [19, 20, 29] and at room temperature [20, 21]; and also in the gas phase [7, 13, 18, 25].

The crystal structure at −120°C was determined from X-ray diffraction photographs [19, 20]. The data of a preliminary study are too incomplete for a structure elucidation [20]. A more precise investigation shows that Be is disordered between two equivalent sites. In each site it is centrally bonded to one ring and peripherally bonded to the other. The bond distances and bond angles are given in the original. The estimated dipole moment of 2.55 D for this structure is in good agreement with the experimental values on p. 184 [19]. The structure was redetermined more precisely at −145°C by means of the diffractometer method, and the slip-sandwich structure is confirmed. The crystals belong to the monoclinic space group $P2_1/n$-C_{2h}^5

(No. 14) with a = 5.993(5), b = 7.478(4), c = 8.978(5) Å, β = 85.94(6)°, d_c = 1.151 g/cm³; Z = 2. The structure calculation was performed with 357 independent reflections with $I > 2.5\,\sigma(I)$. Refinement converged with R = 0.035.

Fig. 18 a and b show the molecular structure from the lateral and plane views. The disorder of the Be atoms between two crystallographically equivalent sites means that the observed C_5H_5 ring is the mean between centrally bonded and peripherally bonded C_5H_5 rings. In order to obtain a geometry for the ordered structure, it is assumed that the centrally bonded ring with fivefold symmetry has a C–C bond length of 1.41 Å, equal to the average of the five observed C–C distances. The geometry of the peripherally bonded ring can then be calculated from twice the difference between the observed and the average C–C distances. This leads to the values given in Formula III, p. 183. The alternate arrangement in bond lengths is clearly consistent with a peripherally bonded ring with partial diene character. The alternations in bond lengths are far less, however, than in C_5H_6 itself (1.34 to 1.51 Å) or in other σ-bonded C_5H_5 complexes. This suggests that the ring retains significant π-delocalized character and is consistent with the small bond angles between Be and the ring (94.4° and 91.4°). This is close to that expected for an sp^2 hybridization of the ring carbon with the bond to Be involving a p_π orbital. It also accounts for the absence of characteristic diene features in the Raman spectrum (see pp. 186/8). **Fig. 19**, p. 190, shows a packing diagram of the contents of the unit cell [29].

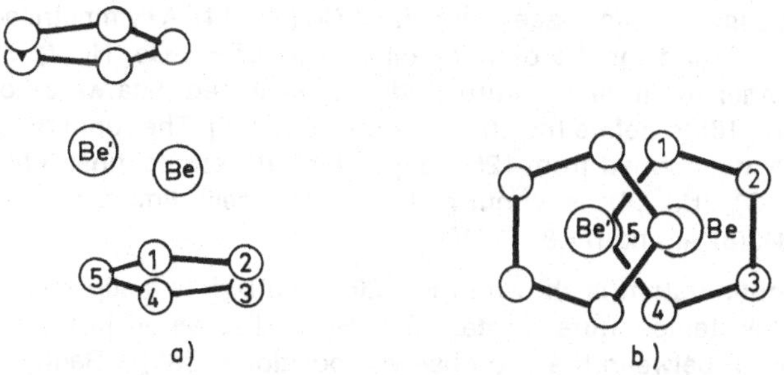

Fig. 18. Molecular structure of $Be(C_5H_5)_2$ at −145°C: a) lateral view, b) plane view [29].

Selected bond distances (in Å) and bond angles (in°):

C(1)–C(2)	1.395(4)	Be'···C(1)	2.400(6)
C(2)–C(3)	1.410(4)	Be'···C(2)	3.014(6)
C(3)–C(4)	1.386(3)	Be'···C(3)	2.984(6)
C(4)–C(5)	1.431(3)	Be'···C(4)	2.347(6)
C(5)–C(1)	1.424(3)	Be'···C(5)	1.826(6)
$\overline{\text{C–H}}$	$\overline{0.978}$	Be···Be'	1.303(12)
Be–C(1)	1.940(7)	C(1)–C(2)–C(3)	108.2(2)
Be–C(2)	1.950(6)	C(2)–C(3)–C(4)	108.7(2)
Be–C(3)	1.930(6)	C(3)–C(4)–C(5)	108.0(2)
Be–C(4)	1.913(6)	C(4)–C(5)–C(1)	108.2(2)
Be–C(5)	1.891(6)	C(5)–C(1)–C(2)	106.7(2)
		$\overline{\text{C–C–H}}$	$\overline{125.9}$
		Be–C(5)–C(1)	94.4
		Be–C(5)–C(4)	91.4

References on pp. 193/4

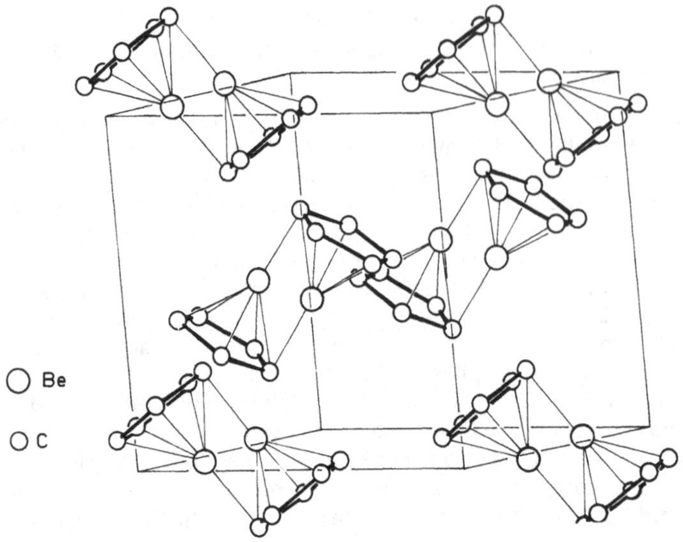

Fig. 19. Packing diagram of the unit cell of $Be(C_5H_5)_2$ [29].

An early investigation of the crystal structure at room temperature by Debye-Scherrer diagrams indicated a monoclinic space group $P2_1/c$-C^5_{2h} (No. 14). A symmetrical bonding of the two C_5H_5 groups, as found for the dicyclopentadienyls of Fe, Co, Ni, V, Mg, and Cr, was assumed [6]. In another room temperature study, the collected data were not good enough (temperature factor ~16) to define the structure precisely [20]. The room temperature crystal structure was finally determined from 125 independent reflections (final R factor 0.053). The point group $P2_1/n$-C^5_{2h} (No. 14) is confirmed, and the cell dimensions are a = 5.893(6), b = 7.683(6), c = 9.343(6) Å, and β = 87.73(10)°.

The resulting configuration is shown in **Fig. 20** a and b. It is a slip-sandwich, as already described for the low temperature crystals. The Be is disordered in the crystal, probably moving back and forth between the two observed positions (Be(I)···Be(II) = 0.95 Å). But the two rings are neither staggered as in the low temperature case nor eclipsed relative to each other, but somewhere in-between. The ring-to-ring distance of 3.40 Å is about the same as observed in the vapor phase study (3.37 Å [13]), but significantly larger than in the crystal at −120°C (3.30 Å [19]). Be(I) is nearly on the normal to the center of C_5H_5(I) and to the C–C bond of C_5H_5(II) (Fig. 20b) with Be(I)–C_5H_5(I) = 1.41 Å and Be(II)–C_5H_5(II) = 1.99 Å. This suggests that Be is π-bonded to one ring and nearly electrostatically bonded to the other ring. The average C–C bond length (1.34 Å, ranging from 1.31 to 1.39 Å) is considerably shorter than the usually observed 1.40 to 1.42 Å [21].

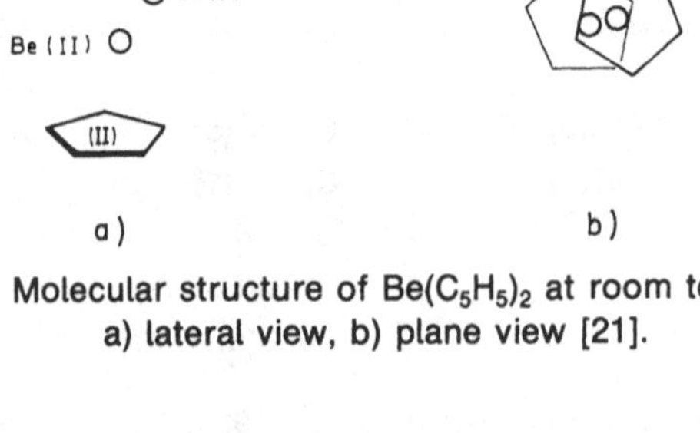

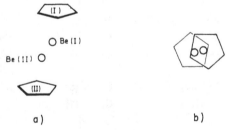

Fig. 20. Molecular structure of $Be(C_5H_5)_2$ at room temperature: a) lateral view, b) plane view [21].

The molecular geometry of gaseous $Be(C_5H_5)_2$ was determined by electron diffraction of the vapor at 70°C [7]. A later recalculation of the experimental data gave a better agreement of the experimental and theoretical radial distributions [13]. It was concluded that the molecular symmetry is C_{5v}. Two planar C_5H_5 rings lie parallel and staggered with a vertical ring-to-ring distance of 3.37 Å (3.375 Å [13]). The Be can occupy two alternate positions on the fivefold symmetry axis with 1.485 Å (1.472 Å [13]) from one ring and 1.980 Å (1.903 Å [13]) from the other ring [7]. The other molecular parameters are also given in the originals [7, 13]. It is discussed that the bonding of Be to the C_5H_5 rings may be partially ionic [7]. A refinement of the intensity data recorded in [7], and consideration of the C_{5v} model and a model with one π-bonded C_5H_5 and one σ-bonded C_5H_5 ring, showed a better agreement of experimental and calculated intensities for the C_{5v} model [25].

The electron scattering pattern of gaseous $Be(C_5H_5)_2$ was again recorded at ~120°C, and the data were considered on the basis of several molecular models. A structure with D_{5d} symmetry (symmetrical position of Be between two staggered C_5H_5 rings) or the model containing one π-bonded and one σ-bonded C_5H_5 ring were not compatible with the data. The best agreement between calculated and experimental intensities is obtained with a slip-sandwich model for a slip of 0.79(10) Å, see **Fig. 21**. The structural parameters are listed in Table 35, p.192. Since the fits obtained with the C_{5v} symmetry and with a slip of 1.21 Å (Fig. 21 and Table 35) are also satisfactory, the C_{5v} model, as well as the slip-sandwich models with slips between 0 and 1.21 Å, must be considered to be compatible with the electron diffraction data. It is therefore suggested that the potential energy of the molecule varies only slightly with the magnitude of the slip. On the potential energy surface, the minimum energy is located at a slip of ~1 Å, and the C_{5v} conformation represents an energy maximum (but the barrier is not very large). The molecule consequently undergoes large amplitude vibrations [25].

This conclusion differs from that drawn in [18]. After refinement of the electron diffraction data given in [7], it was concluded that a slip >0.6 Å could be ruled out [18].

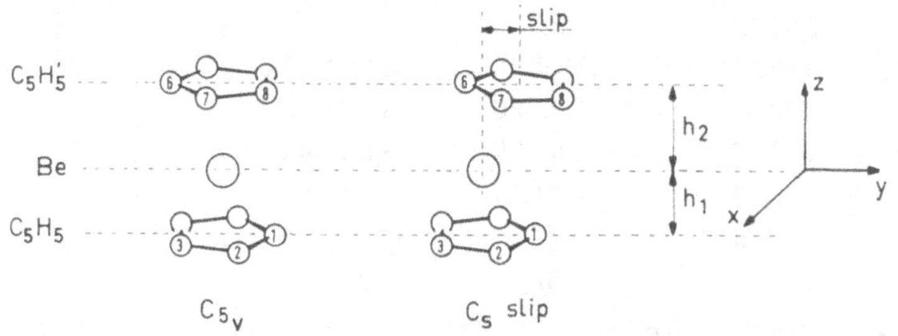

Fig. 21. Molecular models for $Be(C_5H_5)_2$ in the gas phase [25].

The movement of Be between two different energy minima in the molecule is supported by the electron density map of Be at room temperature (not given), which appears as a cigar-shaped peak [21]. Estimates have been made about the energy profile and rate of this movement. Under the assumption that the bonding between Be and the rings in the C_{5v} structure is predominantly electrostatic, a simple calculation gives an energy barrier of 2 to 5 kcal/mol [7]. The radial distribution curves of the gas phase electron diffraction indicate an energy barrier >0.8 kcal/mol [25]. Treating the Be as a one-dimensional oscillator and assuming a C_{5v} structure, the lifetime of Be in one of its two potential holes is calculated from experimental and theoretical IR data to be in the range (0.1 to 1.0) × 10^{-12} s; the energy barrier is 1.4 to 2.9 kcal/mol [16]. However, the average time required to invert the dipole moment of $Be(C_5H_5)_2$ by Be oscillation is derived in a microwave study as approximately $\geqq 60 \times 10^{-12}$s.

References on pp. 193/4

This value corresponds to the half-period of Be oscillation, and is not in accord with a C_{5v} structure [31]. From the equivalence of all H atoms in $Be(C_5H_5)_2$ even at −135°C in the ^1H NMR spectrum, it is deduced that the barrier should be <0.14 kcal/mol, and the frequency of the Be motion is estimated as $2 \times 10^{12}\ s^{-1}$ for the slip-sandwich and $4 \times 10^{-12}\ s^{-1}$ for the C_{5v} structure [22]. It follows from the ^{13}C NMR spectra that inversion of the molecular dipole caused by exchange of the Be between the two C_5H_5 rings in the slip-sandwich occurs in the order of 10^{-10} s, with an activation energy of ~5 kJ/mol in solution [24].

Table 35
Internuclear Distances (in Å) in Gaseous $Be(C_5H_5)_2$ for the C_{5v} and the Slip-Sandwich Structures [25].
See Fig. 21, p. 191, for the numbering of the atoms.

atoms	C_{5v}	slip-sandwich slip = 0.79(10) Å	slip-sandwich slip = 1.21 Å
C–C	1.423(2)	1.423(2)	1.423(2)
C–H	1.098(4)	1.104(4)	1.104(4)
C(1)···C(3)	2.303(3)	2.303(3)	2.303(3)
C(1)···H(2)	2.251(4)	2.256(4)	2.253(4)
C(1)···H(3)	3.364(5)	3.369(5)	3.366(5)
Be–C(1)	1.904(6)	1.903(8)	1.904(6)
Be–C(6)	2.27(2)	1.99(7)	1.90(4)
Be–C(7)		2.25(4)	2.33(2)
Be–C(8)		2.62(5)	2.89(6)
Be···H(1)	2.737(5)	2.741(7)	2.739(6)
Be···H(6)	3.01(2)		
C(1)···C(8)	3.48(1)	3.35 to 3.6	3.35 to 3.6
C(1)···C(7)	3.92(1)	3.6 to 4.0	3.6 to 4.0
C(1)···C(6)	4.17(1)	4.0 to 4.5	4.0 to 4.7
h_1	1.470(7)	1.468(10)	1.470(7)
h_2	1.92(2)	1.86(2)	1.82(4)
α*)	0	−4(3)°	−4(3)°

*) Angle between C_5H_5 and C_5H_5'.

Chemical Properties

The compound is soluble in C_6H_6, ether, and petroleum ether. Decomposition is observed in CS_2, CCl_4, CH_2Cl_2, and THF. According to cryoscopic measurements, it is monomeric in C_6H_6 [1, 33] and c-C_6H_{12} [33]. Vapor pressure studies in $C_6H_5CH_3$ indicate some association, but the measured mass units are not constant [12].

The mass spectrum of $Be(C_5H_5)_2$ was obtained at ionizing voltages of 70, 30, 20, and 10 eV and ambient source temperature. The spectrum at 70 eV shows the complete fragmentation of both C_5H_5 rings and is very complicated. The molecular ion is the major peak at 10 eV; whereas at 20, 30, and 70 eV, $[BeC_5H_5]^+$ is the major species. The ionization potential for the parent radical ion is 9.27 eV; the BeC bond dissociation energy is 4.10 eV (=94.54 kcal/mol) [33].

The compound is extremely sensitive to O_2, and it is vigorously decomposed by H_2O to give $Be(OH)_2$ and C_5H_6 [1]. Reaction with an excess [23, 30, 32] or an equimolar amount [33] of $BeCl_2$ gives $Be(C_5H_5)Cl$. $BeBr_2$ and BeI_2 react analogously [23, 33], but no reaction is observed with BeF_2 [23], and $Be(C_5H_5)F$ could only be identified in the ^{1}H NMR spectrum of the reaction mixture in [33]. The reaction of $Be(C_5H_5)_2$ with $FeBr_2$ in ether gives $Be(C_5H_5)Br$ [30]. $FeCl_2$ is converted to $Fe(C_5H_5)_2$ in only 19% yield in a reaction carried out in THF at 65°C. Yields of 76% are achieved in boiling ether [1]. The reactions of $Be(C_5H_5)_2$ with $NpCl_4$, $PuCl_3$, $AmCl_3$, and $CmCl_3$ give $Np(C_5H_5)_3Cl$ [10], $Pu(C_5H_5)_3$ [8], $Am(C_5H_5)_3$ [9], and $Cm(C_5H_5)_3$ [14], respectively. With $BkCl_3$, a mixture of $Bk(C_5H_5)_3$ and $Bk(C_5H_5)_2Cl$ is obtained [15].

Reactions with $Be(NR_2)_2$ ($R = CH_3$, C_2H_5, C_6H_5) give the corresponding $Be(C_5H_5)NR_2$. No reactions are observed with $Be(N(Si(CH_3)_3)_2)_2$, BeH_2, PHR_2 ($R = CH_3$, C_6H_5), and $AsHR_2$ ($R = CH_3$, C_6H_5) [33]. Reactions with equimolar amounts [33], or an excess [23], of BeR_2 ($R = CH_3$ [23, 33], C_2H_5, C_4H_9 [33], t-C_4H_9 [33, 35], C_6F_5 [35]) give the corresponding $Be(C_5H_5)R$ [23, 33, 35]. No reactions occur with $Be(C_6H_5)_2$ in C_6H_6 at 75°C within 24 h [33]. $Be(C_5H_5)_2$ reacts with $Be(C_8H_{15}BN)_2$ (Formula IV) to give $Be(C_5H_5)C_8H_{15}BN$ (Formula V) as a yellow oil in quantitative yield [34].

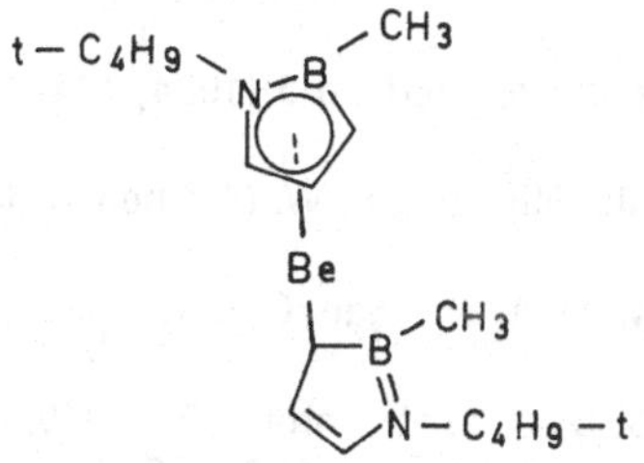

IV

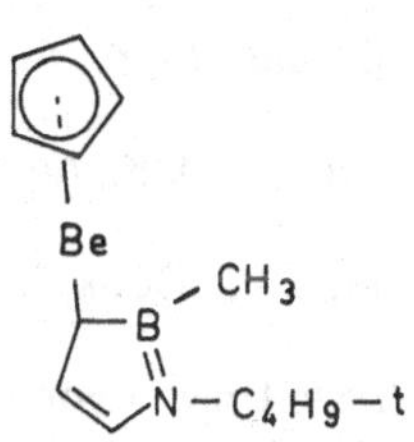

V

References:

[1] Fischer, E. O., Hofmann, H. P. (Chem. Ber. **92** [1959] 482/6).
[2] Fischer, E. O., Schreiner, S. (Chem. Ber. **92** [1959] 938/48).
[3] Fritz, H. P. (Chem. Ber. **92** [1959] 780).
[4] Strohmeier, W., von Hobe, D. (Z. Elektrochem. **64** [1960] 945/51).
[5] Fritz, H. P., Schneider, S. (Chem. Ber. **93** [1960] 1171/83).
[6] Schneider, S., Fischer, E. O. (Naturwissenschaften **50** [1963] 349).
[7] Almenningen, A., Bastiansen, O., Haaland, A. (J. Chem. Phys. **40** [1964] 3434/7).
[8] Baumgärtner, F., Fischer, E. O., Kanellakopulos, B., Laubereau, P. (Angew. Chem. **77** [1965] 866/7).
[9] Baumgärtner, F., Fischer, E. O., Kanellakopulos, B., Laubereau, P. (Angew. Chem. **78** [1966] 112/3).
[10] Fischer, E. O., Laubereau, P., Baumgärtner, F., Kanellakopulos, B. (J. Organometal. Chem. **5** [1966] 583/4).

[11] Fritz, H. P., Sellmann, D. (J. Organometal. Chem. **5** [1966] 501/5).
[12] Morgan, G. L., McVicker, G. B. (J. Am. Chem. Soc. **90** [1968] 2789/92).
[13] Haaland, A. (Acta Chem. Scand. **22** [1968] 3030/2).
[14] Baumgärtner, F., Fischer, E. O., Billich, H., Dornberger, F., Kanellakopulos, B. (J. Organometal. Chem. **22** [1970] C17/C18).
[15] Laubereau, P. G. (Inorg. Nucl. Chem. Letters **6** [1970] 611/4).
[16] Ionov, S. P., Ionova, G. V. (Izv. Akad. Nauk SSSR Ser. Khim. **1970** 2836; Bull. Acad. Sci. USSR Div. Chem. Sci. **1970** 2678/80).

[17] McVicker, G. B., Morgan, G. L. (Spectrochim. Acta A **26** [1970] 23/30).
[18] Drew, D. A., Haaland, A. (Acta Cryst. B **28** [1972] 3671).
[19] Wong, C., Lee, T., Chao, K., Lee, S. (Acta Cryst. B **28** [1972] 1662/5).
[20] Wong, C., Chao, K. J., Chih, C., Lee, T. Y. (J. Chinese Chem. Soc. [Taipei] [2] **16** [1969] 15/8).

[21] Wong, C., Lee, T. Y., Lee, T. J., Chang, T. W., Lin, C. S. (Inorg. Nucl. Chem. Letters **9** [1973] 667/73).
[22] Wong, C., Wang S.-M. (Inorg. Nucl. Chem. Letters **11** [1975] 677/8).
[23] Drew, D. A., Morgan, G. L. (Inorg. Chem. **16** [1977] 1704/8).
[24] Nugent, K. W., Beattie, J. K. (J. Chem. Soc. Chem. Commun. **1986** 186/7).
[25] Almenningen, A., Haaland, A., Lusztyk, J. (J. Organometal. Chem. **170** [1979] 271/84).
[26] Gleiter, R., Böhm, M. C., Haaland, A., Johansen, R., Lusztyk, J. (J. Organometal. Chem. **170** [1979] 285/92).
[27] Lusztyk, J., Starowieyski, K. B. (J. Organometal. Chem. **170** [1979] 293/7).
[28] Gaines, D. F., Coleson, K. M., Hillenbrand, D. F. (J. Magn. Resonance **44** [1981] 84/8).
[29] Nugent, K. W., Beattie, J. K., Hambley, T. W., Snow, M. R. (Australian J. Chem. **37** [1984] 1601/6).
[30] Pratten, S. J., Cooper, M. K., Aroney, M. J. (Polyhedron **3** [1984] 1347/50).

[31] Pratten, S. J., Cooper, M. K., Aroney, M. J., Filipczuk, S. W. (J. Chem. Soc. Dalton Trans. **1985** 1761/5).
[32] Col, D. A., Nibler, J. W., Cook, T. H., Drew, D. A., Morgan, G. L. (J. Chem. Phys. **63** [1975] 4842/53).
[33] Bartke, T. C. (Diss. Univ. Wyoming 1975; Diss. Abstr. Intern. B **36** [1976] 6141).
[34] Schmid, G., Boltsch, O., Boese, R. (Organometallics **6** [1987] 435/9).
[35] Coates, G. E., Smith, D. L., Srivastava, R. C. (J. Chem. Soc. Dalton Trans. **1973** 618/22).

3.5.2 Theoretical Studies

Molecular orbital calculations have been carried out on the various models I to VI of $Be(C_5H_5)_2$ using semiempirical [1,2] and nonempirical [5 to 7, 9 to 12] procedures.

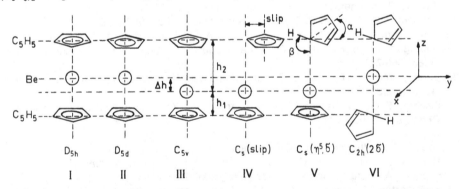

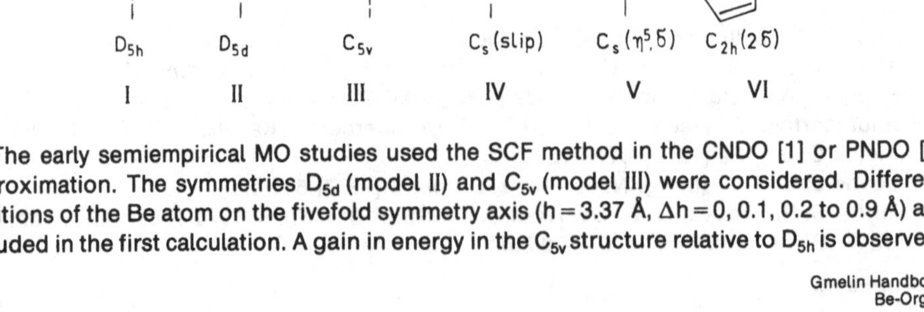

I II III IV V VI

The early semiempirical MO studies used the SCF method in the CNDO [1] or PNDO [2] approximation. The symmetries D_{5d} (model II) and C_{5v} (model III) were considered. Different positions of the Be atom on the fivefold symmetry axis (h = 3.37 Å, Δh = 0, 0.1, 0.2 to 0.9 Å) are included in the first calculation. A gain in energy in the C_{5v} structure relative to D_{5h} is observed.

The overall difference of the total energy is on the average −0.140 a.u. and is attributed mainly to a decrease in the core repulsion energy. However, this is an artifact of the method. The best fit of the calculated dipole ($\mu = 2.7$ D) with the experimental values (see Chapter 3.5.1, p. 184) is achieved at a distortion of 0.2 Å. The UV absorptions are calculated for the different distortions [1].

The calculation with the somewhat better PNDO approximation shows that the total electronic energy of the C_{5v} structure with Be displaced 0.5 Å towards one of the rings is on the average 166.0 kcal/mol lower than the D_{5d} structure with h = 3.37 Å. This is attributed mainly to the stengthening of the covalent bond. The effective positive charge on Be is lowered in the C_{5v} structure. The estimated dipole is $\mu = 3.5$ D [2].

Semiempirical calculations by the MNDO method show that the most stable form is that with one pentahapto- and one monohapto-bonded ring with C_s symmetry (model V, $\Delta H_f = 31.8$ kcal/mol). This species is calculated to be 15.8 kcal/mol lower in energy than model II (D_{5d}) and has a calculated dipole of 2.5 D. The MNDO calculations also show that the D_{5d} structure (model II) is preferred over the C_{5v} structure (model III). The rotational barrier for the C_5H_5 rings, to interconvert the models I and II, is given as 0.2 kcal/mol. The calculated ring-to-ring distance is 3.33 Å; the BeC distances are 2.07 Å. The third calculated model VI is higher in energy than model V, but slightly lower than the D_{5d} model [8].

Some important results of the nonempirical calculations (PRDDO and ab initio) are summarized in Table 36, p. 196. In no case was a full geometry optimization performed. For the η^5-C_5H_5 ring, the following bond lengths were used: CC = 1.417 [7], 1.423 [11], 1.425 [6, 9] Å, and CH = 1.100 [5, 11], 1.103 [6, 9], 1.079 [7] Å. For the σ-bonded C_5H_5 ring, the exponential geometry of C_5H_6 was used. Optimized CC and CH bonds are included in the table.

The calculations suggest that the most stable form of $Be(C_5H_5)_2$ is described either by models I/II (D_{5h}/D_{5d}) or model V (η^5, σ-C_5H_5), whereas the experimental observations reveal a slip-sandwich structure (model IV) for the solid state and a fluxional slip-sandwich structure for the gas phase (see Chapter 3.5.1, pp. 188/92). The C_{5v} symmetry (model III) appears to be energetically unfavorable [5 to 7, 9 to 12]. The model VI with two σ-bonded C_5H_5 rings is calculated to be ~45 kcal/mol less stable than model V [5].

The calculated barrier to interconvert the D_{5h} and D_{5d} structures (models I and II) is only ~0.4 kcal/mol. Several different mechanisms of intramolecular tautomerism are discussed. The tautomerism in model V is best described as an activated two-step process:

$$\sigma\text{-}\pi \xrightarrow{\sim 15\ \text{kcal/mol}} (\sigma\text{-}\pi)^{\ddagger} \xrightarrow{\sim -8.6\ \text{kcal/mol}} \pi\text{-}\pi \xrightarrow{\sim 8.6\ \text{kcal/mol}} (\pi\text{-}\sigma)^{\ddagger} \xrightarrow{\sim -15\ \text{kcal/mol}} \pi\text{-}\sigma$$

Linear synchronous transit studies show that the ring movement is energetically favored over the Be movement in the gas phase. In the solid state, with limited mobility of the C_5H_5 rings, the tautomerization by Be motion is predicted to have a barrier of ~24 to 31 kcal/mol [5].

Schematic MO interaction diagrams for the D_{5d} [4, 7 to 9] and C_{5v} [4, 9] structures reveal that the two HOMO's have antibonding character. This explains the preference of more unsymmetrical structures like models IV and V [4, 7 to 9]. Mulliken charges, overlap populations [5, 12], and the calculated energies of the HOMO's [11] are given in the originals. The C_{5v} structure is also analyzed in terms of the second-order Jahn-Teller effect [3].

From a quantitative estimation of steric hindrance, it can be concluded that $Be(C_5H_5)_2$ should have one η^5- and one η^1-bonded C_5H_5 ligand [13].

References on p. 198

Table 36
Nonempirical Molecular Orbital Calculations on $Be(C_5H_5)_2$. (Abbreviations on p. X.)

model	calculation method (basis set)	geometrical parameters distances in Å	total energy in a.u.	relative energy (lowest energy model) in kcal/mol	dipole moment μ in D	Ref.
I/II	ab initio (d-z)	h = 3.375	−398.673	0	—	[9]
	ab initio (d-z)	h = 3.375	−398.58748	0	—	[10, 12]
		h = 2.944	−398.52655	38.2 (I/II)	—	[12]
	ab initio (STO-3G)	h = 3.239, CC = 1.412, CH = 1.078	−394.278	0.7/1.1 (V)	0	[6]
II	PRDDO	h = 3.22, BeC = 2.00, CC = 1.40	−398.6472	6.4 (V)	—	[5]
	ab initio (STO-3G)	h = 3.247	−394.278	7.4 (V)	—	[7]
	ab initio (d-z)	h = 3.375	−398.6457	9.2 (V)	—	[11]
III	PRDDO	h = 3.375, Δh = 0.43	−398.6155	26.3 (V)	—	[5]
	ab initio (STO-3G)	h = 3.375, h_1 = 1.472, h_2 = 1.903	−394.250	18.8 (V)	3.06	[6]
	ab initio (STO-3G)	h = 3.375, h_1 = 1.6182	−394.274	10.2 (V)	—	[7]
		h = 3.375, h_1 = 1.3875	−394.238	32.5 (V)	—	
		h = 3.700, h_1 = 1.700	−394.232	37.6 (V)	—	
	ab initio (d-z)	h = 3.375, Δh = 0.43	−398.620	10.5 (I)	—	[9]
	ab initio (d-z)	h = 3.375, Δh = 0.2155	−398.56771	12.4 (I/II)	—	[10, 12]
	ab initio (d-z)	h = 3.375, h_1 = 1.490, h_2 = 1.885	−398.6362	16.3 (V)	2.39	[11]
IV	PRDDO	h_1 = 1.87, h_2 = 1.72	−398.6353	13.9 (V)	—	[5]
		X-ray structure at −120°C	—	21.3	—	
		X-ray structure at room temperature	—	31.0	—	

	ab initio (STO-3G)	$h = 3.239$, $h_1 = 1.343$, $h_2 = 1.896$, CC = 1.412, CH = 1.078, slip = 1.202	−394.245	21.8 (V)	5.26	[6]
	ab initio (STO-3G)	$h = 3.247$, $h_1 = 1.6235$, slip = 1.2053	−394.239	31.7 (V)	—	[7]
	ab initio (d-z)	$h_1 = 1.466$, $h_2 = 1.72$, $\alpha = 27°$, $\beta = 108.5°$	−398.625	7.5 (I)	—	[9]
		as before, but $\alpha = 90°$	−398.623	9.0 (I)	—	
	ab initio (d-z)	$h = 3.375$, $h_2 = 1.860$, slip = 0.790, $\alpha = -4°$	−398.6413	13.0 V)	—	[11]
		as before, but $h_2 = 1.821$, slip = 1.210	−398.6424	12.4	—	
V	PRDDO	$h_1 = 1.87$, $h_2 = 1.72$, $\alpha = 27°$, $\beta = 108.5°$	−398.6574	0	—	[5]
	ab initio (STO-3G)	$h_1 = 1.452$, $h_2 = 1.823$, $\alpha = 36.4°$, slip = 0.963	−394.280	0	4.69	[6]
		as before, but slip = 1.202	−394.276	2.2 (V)	4.80	
	ab initio (STO-3G)	$h_1 = 1.4655$, $h_2 = 1.69$, $\beta = 109.5°$	−394.290	0	—	[7]
	ab initio (d-z)	$h_1 = 1.466$, $h_2 = 1.72$, $\alpha = 35°$, $\beta = 108.5°$	−398.610	17.0 (I/II)	—	[9]
		as before, but $\beta = 109.5°$	−398.628	5.5 (II)	—	
	ab initio (d-z)	$h = 3.375$, $h_2 = 1.698$, $\alpha = 36.9°$	−398.6621	0	2.87	[11]

References on p. 198

References:

[1] Sundbom, M. (Acta Chem. Scand. **20** 1608/20).
[2] Lopatko, O. Ya., Klimenko, N. M., Dyatkina, M. E. (Zh. Strukt. Khim. **13** [1972] 1128/30; J. Struct. Chem. [USSR] **13** [1972] 1044/9).
[3] Glidewell, C. (J. Organometal. Chem. **102** [1975] 339/43).
[4] Collins, J. B., Schleyer, P. von R. (Inorg. Chem. **16** [1977] 152/5).
[5] Marynick, D. S. (J. Am. Chem. Soc. **99** [1977] 1436/41).
[6] Chiu, N. S., Schäfer, L. (J. Am. Chem. Soc. **100** [1978] 2604/7).
[7] Jemmis, E. D., Alexandratos, S., Schleyer, P. von R., Streitwieser, A., Schaefer III, H. F. (J. Am. Chem. Soc. **100** [1978] 5695/5700).
[8] Dewar, M. J. S., Rzepa, H. S. (J. Am. Chem. Soc. **100** [1978] 777/84).
[9] Demuynck, J., Rohmer, M. M. (Chem. Phys. Letters **54** [1978] 567/70).
[10] Boldyrev, A. I., Charkin, O. P. (Zh. Strukt. Khim. **18** [1977] 783/94; J. Struct. Chem. [USSR] **18** [1977] 623/31).

[11] Gleiter, R., Böhm, M. C., Haaland, A., Johansen, R., Lusztyk, J. (J. Organometal. Chem. **170** [1979] 285/92).
[12] Charkin, O. P., Veillard, A., Demuynck, J., Rohmer, M. (Koord. Khim. **5** [1979] 501/6; Soviet J. Coord. Chem. **5** [1979] 383/8).
[13] Lobkovsky, E. B. (J. Organometal. Chem. **277** [1984] 53/9).

4 Compounds of Unknown Structure

4.1 Hydrides

General Remarks. Synthetic work directed towards the preparation of organoberyllium hydrides has afforded in a few instances products of the composition $Be_3R_4H_2$ (R = CH_3, C_2H_5) or $Be_2(C_2H_5)_3H$ [1 to 5]. There is little evidence about their constitution. Whereas the latter example is discussed as a single compound in the literature [4, 5], solutions of $Be_3R_4H_2$ may contain many species in complex equilibria [2]. Mainly on the basis of their reaction with tertiary amines, it is suggested that the main solution species are solvated BeR_2 and solvated BeRH, but the existence of trinuclear species cannot be excluded [1, 3].

$Be_3(CH_3)_4H_2$ and $Be_3(CH_3)_4H_2 \cdot O(C_2H_5)_2$

A material of this composition is formed in the reaction of $Na_2[Be(CH_3)_2H]_2$ (prepared from excess NaH and ethereal $Be(CH_3)_2$ [3]) and $BeCl_2$ in the mole ratio 1:1 in ether. The resulting solution contains Be:CH_3:H in the ratio 3:4:2 after filtration from the NaCl precipitate. The constitution of the species is unknown [2, 3]. Evaporation of the solvent in a vacuum at room temperature leads to a colorless viscous oil of the approximate composition $Be_3(CH_3)_4H_2 \cdot O(C_2H_5)_2$. Pumping at ~50°C in a vacuum gives a white amorphous residue of $Be_3(CH_3)_4H_2$. Further heating in a sublimation apparatus with continuous pumping for 10 days results in a condensate of $Be(CH_3)_2$ at 60 to 120°C. The residue is a mixed methylberyllium hydride with an H:CH_3 ratio of 2.16:1 [3]. This ratio increases to values higher than 10:1 as the temperature rises to 170 to 180°C [2, 3]. At still higher temperatures, the evolution of CH_4 and H_2 becomes noticeable [3].

Addition of $N(CH_3)_3$ to $Be_3(CH_3)_4H_2$ in ether yields a mixture of $Be(CH_3)_2 \cdot N(CH_3)_3$ and $(Be(CH_3)H \cdot N(CH_3)_3)_2$. $N(CH_3)_2CH_2CH_2N(CH_3)_2$ and $CH_3OCH_2CH_2OCH_3$ afford the corresponding adducts $Be(CH_3)_2 \cdot {}^2D\text{-}{}^2D$ and $(Be(CH_3)H \cdot 0.5{}^2D\text{-}{}^2D)_2$ (D refers to donor) [2, 3]. With 2,2′-bipyridine, the yellow complex $Be(CH_3)_2 \cdot N_2C_{10}H_8$ is formed [3].

$Be_3(C_2H_5)_4H_2$

Treatment of an ether solution of $M_2[Be(C_2H_5)_2H]_2$ (M = Li [1], Na [3]) with an equivalent of $BeCl_2$ (M:Be = 2:1) results in the immediate precipitation of MCl. Evaporation of the supernatant liquid leaves a viscous oil, the hydrolysis of which indicates a Be:C_2H_5:H ratio of 3:4:2 [1, 3], mentioned in [2].

Heating of the residue to 70 to 80°C for 8 h leaves a glassy product consisting mainly of $Be(C_2H_5)H$ [3]. The course of further reactions at higher temperatures is affected by the composition of the mixture. Pyrolysis at 180°C/10^{-3} Torr for ~8 days in the presence of excess NaH gives complex sodium beryllium hydrides with an increased H:C_2H_5 ratio; ether, $Be(C_2H_5)_2$, and C_2H_4 are condensed [3], mentioned in [1]. In the absence of NaH, pyrolysis at 180°C/10^{-3} Torr gives a product with an H:C_2H_5 ratio of 1:2.55 [3]. Ready loss of $Be(C_2H_5)_2$ upon heating is also mentioned in [2].

$Be_2(C_2H_5)_3H$

$Be(C_2H_5)_2$ and $B(C_4H_9\text{-s})_3$ are mixed and set aside for 25 days during which a white solid forms [4]. The same compound is formed from $Be(C_2H_5)_2$ and $Be(C_2H_5)C_4H_9$-t when the mixture is first kept at room temperature for 1 week and then heated to 140°C for 1 h [5]. The solid is washed with C_6H_6 [4, 5].

References on p. 200

Melting points are given as 132 to 134°C [5] and 133 to 136°C [4]. The IR spectrum in Nujol has vibrations at 1730(vw, br), 1565(s), 1495(s), 1405(m), 1204(m), 1000(s), 992(s), 910(w), 862(s), 654(m), 604(m), 536(s), and 511 (s, br) cm^{-1}. The band at 1565 cm^{-1} may be associated with the BeHBe bridges. The solid yields an X-ray powder diffraction pattern: 6.8(s), 5.4(w), 4.05(w), 3.62(m), 3.27(m), 2.80(vw), and 2.75(vw) Å. The structure given in Formula I is tentatively proposed [4].

The hydride is insoluble in C_6H_6 and in $C_6H_5CH_3$ at 70 to 80°C. It is less vigorously hydrolyzed or alcoholyzed than other organoberyllium hydrides. Reaction with $N(CH_3)_3$ gives $(Be(C_2H_5)H \cdot N(CH_3)_3)_2$ and presumably also $Be(C_2H_5)_2 \cdot N(CH_3)_3$ [4].

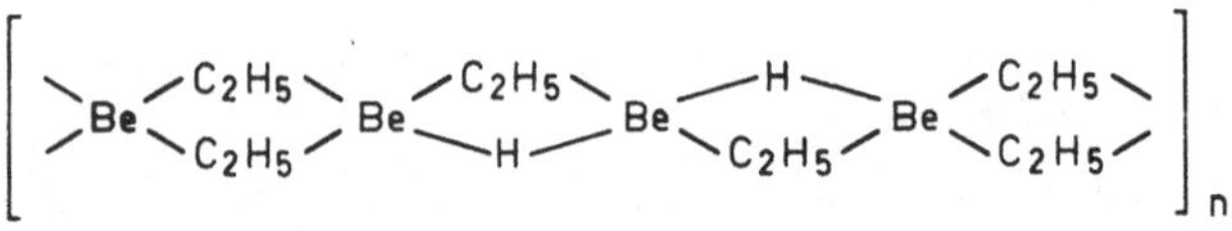

I

$Al_2Be(C_2H_5)_6H_2$

A compound of this stoichiometry is claimed to be obtained by slow addition of ether-free $Be(C_2H_5)_2$ to two equivalents of $Al(C_2H_5)_2H$ with stirring at room temperature. The exothermic reaction gives a clear colorless liquid which reacts with three equivalents of $N(CH_3)_3$ to give $BeH_2 \cdot N(CH_3)_3$ and $Al(C_2H_5)_3 \cdot N(CH_3)_3$. The compound is formulated in the original as an adduct, $Be(C_2H_5)_2 \cdot 2Al(C_2H_5)_2H$, but no further support for this is given [6].

References:

[1] Coates, G. E., Cox, G. F. (Chem. Ind. [London] **1962** 269).
[2] Bell, N. A., Coates, G. E. (Proc. Chem. Soc. **1964** 59).
[3] Bell, N. A., Coates, G. E. (J. Chem. Soc. **1965** 692/9).
[4] Coates, G. E., Francis, B. R. (J. Chem. Soc. A **1971** 1308/10).
[5] Coates, G. E., Smith, D. L., Srivastava, R. C. (J. Chem. Soc. Dalton Trans. **1973** 618/22).
[6] Shepherd, L. H., Ter Haar, G. L., Marlett, E. M. (Inorg. Chem. **8** [1969] 976/9).

4.2 Other Compounds

$Be_2(C_2H_5)_3N_3$

$Be(C_2H_5)_2$ is reacted with an excess of $Si(CH_3)_3N_3$ for 12 h at 80°C. The resulting two phases are evaporated in a high vacuum. The remaining viscous mass analyzes correctly for the title compound.

The IR and Raman spectra contain bands which can be assigned to asymmetrical N_3 vibrations and bridging C_2H_5 vibrations (Table 37). Therefore, the structures I or II are considered.

The compound is insoluble in c-C_6H_{12} and very slightly soluble in C_6H_6 or $C_6H_5CH_3$. It is nonvolatile and decomposes in a high vacuum >80°C. When heated with a flame, it decomposes without explosion. It is very sensitive to air and moisture but does not inflame in air [2].

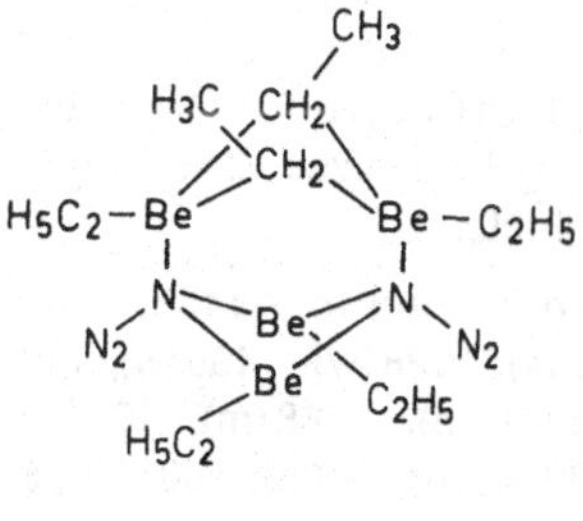

I

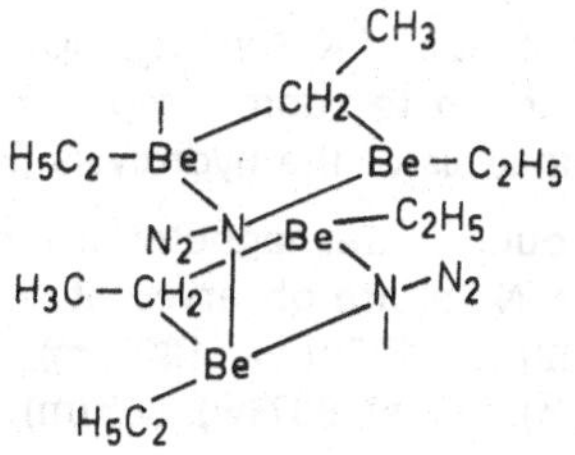

II

Table 37
IR (in Nujol or Hostaflon) and Raman (Solid) Vibrations of $Be_2(C_2H_5)_3N_3$ [2].

$\tilde{\nu}$ (cm^{-1})	intensity IR	intensity Raman*)	assignment	$\tilde{\nu}$ (cm^{-1})	intensity IR	intensity Raman*)	assignment
3420	vw, br		$\nu_{as}(N_3)+\nu_s(N_3)$	1100	w	w, br	
3000	sh			1065	sh		
2990	sh			1050	sh	vw	
2940	s	m (0.5)	$\nu(CH_3, CH_2)$	1035	sh		
2900	s	sh (0.4)		982	m, br	m, br	$\nu(CC)$
2860	s	vs (0.3)		965	sh		
2795	vw	m (0.7)		918	m	w	
2720	vw	w		875		w	$\nu(BeC)$
2235	vs			860	sh		
2180	sh	sh	$\nu_{as}(N_3)$	840	m		$\delta(CH_2)$
2160	vs	m-w (0.3)		780	w, br		
2145	sh	sh		710	m, br		$\varrho(CH_2$ bridging)
1467	m	s (0.6)	$\delta(CH_3)$	675	w, br	w	$\delta(N_3)$
1413	sh	w (1.0)	$\delta(CH_2)$	620	w	vs	
1322	sh	sh	$\delta(CH_2$ bridging)	530	w		$\nu(BeC_2Be)$
1290	m	m-w (0.2)		500	sh	vs	
1260	sh	vw (0.1)	$\nu_s(N_3)$	470		vs	
1245	m	sh		350	w	vs	δ(ring BeC_2Be, BeN_2Be)
1230	sh			315		vw	
1209	m	m (0.3)	$\tau(CH_2)$	230	w		
1175	vw						

*) Degree of depolarization in parentheses.

$Be(CH_3)NC_5H_6 \cdot 2NC_5H_5$ and $Be(C_2H_5)NC_5H_6 \cdot 2NC_5H_5$

Complexes of these compositions are obtained if pyridine in excess is added to Be(R)H ($R = CH_3$, C_2H_5) in ether at room temperature [1,5]. For $R = C_2H_5$ orange crystals deposit, and addition of C_6H_{14} causes more product to crystallize. The compound is dried under reduced pressure [1].

$Be(C_2H_5)NC_5H_6 \cdot 2NC_5H_5$ melts at 109 to 111°C and dissolves as a monomer in C_6H_6. The structure is unknown. It is only stated that it contains no Be–H bonds [1].

References on p. 202

$K_3Be_2(C_2H_5)_2(C_6H_2(CH_3)_3\text{-}2,4,6)_2(OC_4H_9\text{-}t)_2$

Boiling a solution of $K[Be(C_2H_5)_2OC_4H_9\text{-}t]$ in 1,3,5-$(CH_3)_3C_6H_3$ for 15 min yields orange-red crystals on cooling to room temperature, and C_2H_6 is evolved. The empirical formula is obtained by analysis of the hydrolysis (H_2SO_4) products.

The compound is diamagnetic. It turns black on heating to ~145°C without melting. IR absorptions in Nujol are observed at 1605(s), 1580(s), 1528(w), 1493(w), 1402(w), 1337(m), 1316(m), 1302(m), 1270(w), 1227(m), 1201(s), 1176(m), 1168(m), 1044(m,sh), 1039(s), 1016(s), 963(m), 887(s), 837(w), 805(m), 761(w), 739(w,sh), 724(m), 601(w), 561(m), 530(w), 505(w), and 479(w) cm^{-1}.

Addition of solid CO_2 to a solution of the compound in ether and subsequent hydrolysis yields 2,4,6-$(CH_3)_3C_6H_2COOH$. A detailed structure is not proposed [3].

$K_5Be_3(C_2H_5)_3(CH_2C_6H_5)_4(OC_4H_9\text{-}t)_4$

A compound that analyzes for the given empirical formula is obtained from $K[Be(C_2H_5)_2OC_4H_9\text{-}t]$ and $C_6H_5CH_3$ as described for the preceding compound.

The red prisms turn black at ~200°C without melting. IR absorptions in Nujol are given at 1590(s), 1578(s), 1498(m), 1485(m,sh), 1353(m), 1323(m), 1278(w), 1223(w), 1198(s), 1166(s), 1138(w), 1026(s), 1011(s), 954(s), 833(s), 832(w), 799(w), 758(w), 713(m), 695(m), 593(w), 563(m), 528(m), and 485(m) cm^{-1}. The reaction with CO_2 and subsequent hydrolysis (H_2SO_4) gives $C_6H_5CH_2COOH$ [3].

$Rb_5Be_3(C_2H_5)_3(C_6H_2(CH_3)_3\text{-}2,4,6)_4(OC_4H_9\text{-}t)_2$

A compound of the given composition is obtained from $Rb_2[Be_2(C_2H_5)_3(OC_4H_9\text{-}t)_3]$ and 1,3,5-$(CH_3)_3C_6H_3$ as described for the two preceding compounds.

The red crystals turn brown on heating to ~170°C without melting. IR absorptions in Nujol are observed at 1604(w), 1578(s), 1525(w), 1355(w), 1350(m), 1334(m), 1300(w), 1265(w), 1224(m), 1200(s), 1172(w), 1160(w), 1140(m,sh), 1027(s), 1013(s), 958(m), 909(w), 882(s), 838(w), 798(m), 757(w), 715(m), 610(w), 598(w), 583(w), 568(w), 557(w), 527(w), 500(w), and 472(w) cm^{-1}. Reaction with CO_2 and subsequent hydrolysis (H_2SO_4) gives 2,4,6-$(CH_3)_3C_6H_2COOH$ [3].

$Be(C_5H_4CH_3)_2$

The compound is mentioned to be liquid at room temperature, but no further information is given [4].

References:

[1] Bell, N. A., Coates, G. E. (J. Chem. Soc. A **1966** 1069/73).
[2] Atam, N., Dehnicke, K. (Z. Anorg. Allgem. Chem. **427** [1976] 193/9).
[3] Andersen, R. A., Coates, G. E. (J. Chem. Soc. Dalton Trans. **1974** 1729/36).
[4] Fischer, E. O., Hofmann, H. P. (Chem. Ber. **92** [1959] 482/6).
[5] Bell, N. A., Coates, G. E., Fishwick, A. H. (J. Organometal. Chem. **198** [1980] 113/20).

5 Theoretical Studies of Hypothetical Compounds

In this chapter, compounds are described for which only theoretical calculations exist and which are not substantiated by experiments. These hypothetical compounds are included in this volume in deviation from the Gmelin practice, since extensive theoretical calculations exist, whereas the experimental work on organoberyllium compounds has become extremely scarce because of the toxicity.

The species first described are in close analogy to the compounds included in the previous chapters. Then, the presently nonexisting compound types are mentioned.

A number of calculations were also performed with neutral and ionic organoberyllium fragments. These are only listed in the following. See the cited references for details: c-Be_2C [31], Be(CH:)H [30], Be(CH=C:)H, $(BeH)_2C{=}C$: [12], $(BeH)_2C$: [20, 21, 29], $(BeF)_2C$: [29], $[Be(CH_2)H]^-$, $[BeCH_3]^-$ [27], $[(BeH_2)_2C]^{2-}$, $[(BeH_3)_2C]^{4-}$, $[(BeF_2)_2C]^{2-}$, $[(BeF_3)_2C]^{4-}$ [29], $[Be(R)H]^+$ (R = CH_2, $CHCH_3$, CH_2CH_2, $C{=}CH_2$, CH≡CH) [6, 7], $[(BeH)_2CH_3]^+$ [17], $[(BeH)_2C{=}CH]^+$ [12], $[Be(C_6H_4)H]^+$ [11], and the fragments observed in the mass spectrum of $Be(CH_3)_2$ [28].

The calculation methods applied are MNDO, PRDDO, and the ab initio molecular orbital theory. Within the ab initio method, singlet states were computed with the spin-restricted Hartree-Fock theory, and triplet states were handled by the unrestricted version. Initially the minimal STO-3G basis set was used, usually followed by geometry optimization and extension of the basis set. The abbreviations used in the originals for the applied basis sets are given in the text together with the calculated results.

$Be(CH{=}CH_2)_2$

Divinylberyllium and its parent cation were the subject of an MNDO calculation. The molecules can exist as cis and trans isomers. The calculated parameters are as follows:

compound	point group	ΔH_f° (kJ/mol)	HOMO
$Be(CH{=}CH_2)_2$			
cis	C_{2v}	−23.5	A_2
trans	C_{2h}	−23.4	B_g
$[Be(CH{=}CH_2)_2]^+$			
cis	C_{2v}	877.2	$^2A_2$
trans	C_{2h}	876.5	^2B_g

The bond distances (in Å) and bond angles for the neutral molecule are calculated as 1.637 (BeC^1), 1.343 (C^1C^2), 1.106 (C^2H), 1.092 (C^2H_{trans}, C^2H_{cis}); 128.7° (BeC^1C^2), 123.5° ($C^1C^2H_{trans}$), and 124,0° ($C^1C^2H_{cis}$). The HOMO is a π-orbital, not involving Be, so that the principal structural change on ionization is an increase in the CC distance (in Å): 1.633 (BeC^1), 1.388 (C^1C^2), 1.108 (C^1H), 1.092 (C^2H_{trans}), 1.091 (C^2H_{cis}); 130.5° (BeC^1C^2), 113.5° (BeC^1H), 122.5° ($C^1C^2H_{trans}$), and 123.0° ($C^1C^2H_{cis}$) [24].

$Be(C{\equiv}CH)_2$

The experimentally unknown compound and its cation were investigated by an MNDO calculation. The optimized point group for the neutral molecule is $D_{\infty h}$, $\Delta H_f^\circ = 86.6$ kJ/mol, and the HOMO belongs to the E_{1u} symmetry class. The calculated bond distances are: 1.598 (BeC), 1.210 (CC), and 1.053 (CH) Å. The point group remains unchanged upon ionization; $\Delta H_f^\circ = 1172.6$ kJ/mol for the cation with $^2E_{1u}$ symmetry of the HOMO. Ionization mainly causes an increase in the CC distance: 1.601 (BeC), 1.247 (CC), and 1.060 (CH) Å [24].

References on pp. 214/5

Be(CH$_3$)C$_6$H$_5$

A detailed molecular orbital study using the PRDDO approximation was performed for this presently experimentally unknown compound. The total energy of the monomer is calculated as −284.0354 a.u.. The C_6H_5 bridged dimer $CH_3Be(C_6H_5)_2BeCH_3$, with C_6H_5 perpendicular to the BeC_2Be plane, is energetically favored over the monomer and over the CH_3 bridged dimers $C_6H_5Be(CH_3)_2BeC_6H_5$. The C_6H_5 bridged dimer with the C_6H_5 groups coplanar with the Be_2C_2 ring was not considered, since it should be too high in energy in analogy to $HBe(C_6H_5)_2BeH$. The calculated energies for the dimers are as follows [25]:

structure	total energy (a.u.)	dimerization energy (kcal/mol)
$C_6H_5Be(CH_3)_2BeC_6H_5$ (planar)	−568.0710	− 0.1
$C_6H_5Be(CH_3)_2BeC_6H_5$ (perpendicular)	−568.0623	5.3
$CH_3Be(C_6H_5)_2BeCH_3$ (perpendicular)	−568.1165	−28.7

Be(CH$_3$)F

The molecular and electronic structure of the compound was investigated by an ab initio calculation using SCF wave functions and contracted Gaussian basis sets of double zeta quality. With the assumption of a C_{3v} molecular symmetry, a CH distance of 1.09 Å, and a tetrahedral HCH angle, some of the calculated properties are: total energy −153.7777 a. u., dipole moment μ=1.75 D(Be^+F^-), distances 1.697 Å(BeC) and 1.403 Å(BeF). The predicted properties are compared with those of Mg(CH$_3$)F. The structure of the molecule in the gas phase appears to be closely related to Be(CH$_3$)$_2$ and BeF$_2$. The Mulliken population analysis shows that 1.04 less electrons are centered around Be relative to the neutral atom [2].

Be(C$_6$H$_5$)BH$_4$

The PRDDO approximation for the possible geometries of the monomer shows a preference for a double-bridged structure with the bridging hydrogens in the plane of the C_6H_5 ring (Formula I), total energy −271.4216 a. u.. The structure obtained by rotating the phenyl ring by 90° is 3.0 kcal/mol less stable, while the triple-bridged structure (Formula II) is 1.9 kcal/mol less stable. For the phenyl-bridged dimer, only the conformer with the C_6H_5 rings perpendicular to the plane defined by the two Be and bridging C was considered, since the planar conformer can be regarded as significantly less stable in analogy to $HBe(C_6H_5)_2BeH$. The dimerization is strongly favored by ΔH = −52.8 kcal/mol (total energy −542.9273 a. u.) [25].

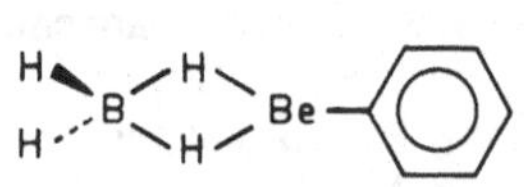

I

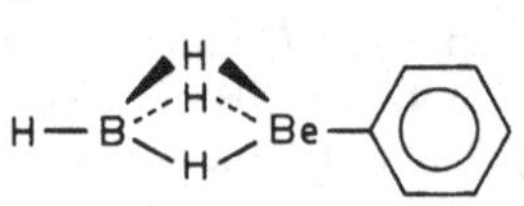

II

Be(CH$_2$X)H (X = Li, BeH, BH$_2$, NH$_2$, OH, F)

Total energies for a number of disubstituted methanes XCH_2Y (X, Y = all possible combinations of Li, BeH, BH$_2$, CH$_3$, NH$_2$, OH, F) were determined by ab initio calculations (5-21 G basis set and standard geometries) to evaluate the preferred conformations and the C–X and C–Y bond interactions. The energies of Be(CH$_2$X)H compounds are summarized in Table 38. For X = BH$_2$, NH$_2$, and OH, the following configurations of Formula III to V are considered (for NH$_2$ and OH based on the BeCNH and the BeCOH dihedral angles of 0°, 60°, 120°, and 180°):

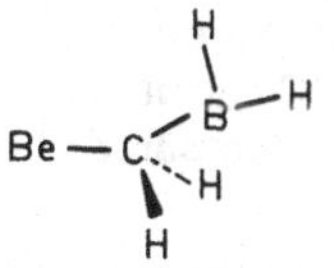
IIIa
coplanar

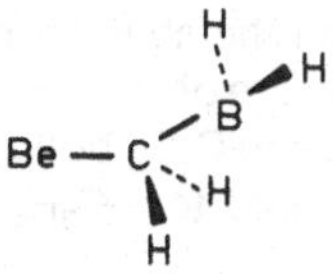
IIIb
perpendicular

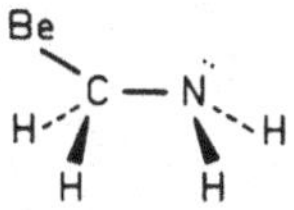
IVa
synperiplanar

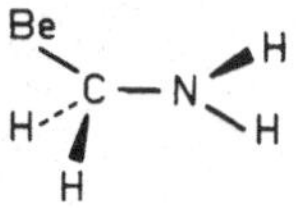
IVb
synclinal

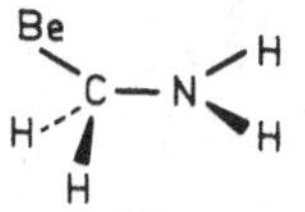
IVc
anticlinal

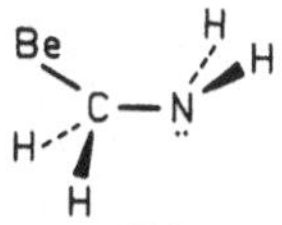
IVd
antiperiplanar

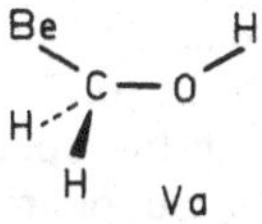
Va
synperiplanar

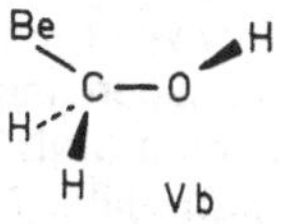
Vb
synclinal

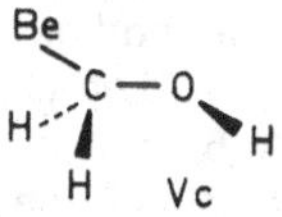
Vc
anticlinal

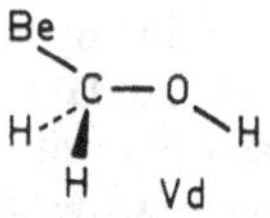
Vd
antiperiplanar

Interaction between the substituents X and BeH is revealed by the bond separation energies (the energy for the formal reaction $Be(CH_2X)H + CH_4 \rightarrow Be(CH_3)H + CH_3X$) and the calculated rotational potentials in the original [4].

Table 38
Calculated Total and Relative Energies of $Be(CH_2X)H$ Compounds [4].

No.	compound	conformation	total energy (a.u.)	relative energy (kcal/mol)
1	$Be(CH_2Li)H$		−61.56856	
2	$Be(CH_2BeH)H$		−69.33873	
3	$Be(CH_2BH_2)H$	coplanar	−79.93831	9.26
		perpendicular	−79.95306	0
4	$Be(CH_2NH_2)H$	anticlinal	−109.64687	4.69
		antiperiplanar	−109.64729	4.43
		synperiplanar	−109.65131	1.91
		synclinal	−109.65435	0
5	$Be(CH_2OH)H$	synperiplanar	−129.44765	6.9
		synclinal	−129.45127	4.63
		anticlinal	−129.45382	3.03
		antiperiplanar	−129.45865	0
6	$Be(CH_2F)H$		−153.44035	

$Be(C_3H_5\text{-c})H$ and **$Be(C_3H_7\text{-i})H$**

The geometries and energies of these compounds were determined by MNDO and ab initio calculations. The total energies obtained are −131.49861 a.u. for $Be(C_3H_5\text{-c})H$ and −132.69731 a.u. for $Be(C_3H_7\text{-i})H$ from ab initio (4-31G) calculations. The stabilization energy for the reaction $Be(C_3H_7\text{-i})H + c\text{-}C_3H_6 \rightarrow Be(C_3H_5\text{-c})H + C_3H_8$ is −6.8 kcal/mol (4-31G). The geometrical

References on pp. 214/5

parameters obtained from the ab initio method (4-31G) are presented in Formulas VI and VII. The values in parentheses are obtained from the MNDO calculations. The C_1C_2 and $C_1C_{2'}$ bonds in VI are lengthened, and the $C_2C_{2'}$ bond is shortened compared to c-C_3H_6. The CC bond in VII is also lengthened compared to c-C_3H_6 [34].

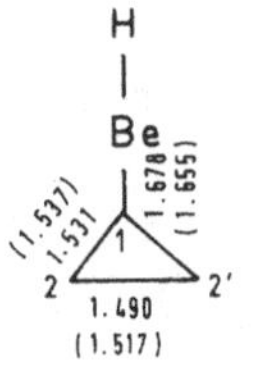

VI

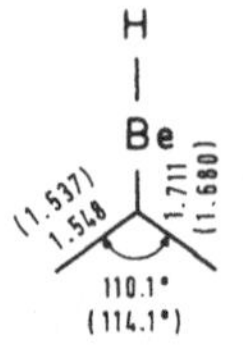

VII

Be(CH=CH$_2$)H

Calculations by the ab initio method gave total energy values of −91.50803 (STO-3G) and −92.51716 (5-21G) a.u. for a standard ethylene structure depending on the basis set. The BeC bond was optimized to 1.663 Å [6, 7]. A computation using MNDO results in the optimized point group C_s with A″ symmetry of the HOMO (π orbital). The optimized geometries are 1.277 (BeH), 1.640 (BeC^1), 1.640 (CC), 1.342 (C^{1}C^2), 1.106 (C^{1}H), 1.092 (C$^2H_{trans}$ and C$^2H_{cis}$) Å; 128.6° (BeC^{1}C^2), 114.4° (BeC^{1}H), 123.5° (C^{1}C$^2H_{trans}$), and 124.1° (C^{1}C$^2H_{cis}$). The molecular symmetry is retained in the cation. The principal structural difference is an increase in the BeC and CC distances upon ionization. The optimized geometries for the cation are 1.216 (BeH), 1.719 (BeC^1), 1.389 (C^{1}C^2), 1.111 (C^{1}H), 1.099 (C$^2H_{trans}$), 1.097 (C$^2H_{cis}$) Å; 129.8° (BeCC), 111.1° (BeC^{1}H), 122.2° (C^{1}C$^2H_{trans}$), and 123.2° (C^{1}C$^2H_{cis}$) [24].

Be(C≡CH)H

A computation using MNDO results in the optimized point group $C_{\infty v}$ with π(E_1) symmetry of the HOMO. The optimized bond distances and bond angles are 1.604 (BeC), 1.275 (BeH), 1.290 (CC), and 1.053 (CH) Å. The molecular symmetry is retained in the cation. Only an increase in the BeC and CC distances occurs upon ionization: 1.601 (BeC), 1.247 (CC), and 1.060 (CH) Å [24].

Be(C$_5$H$_7$)H

An MNDO study of the structures and stabilities of some substituted pentadienyl anions was performed. Calculations of the relative stability of $[BeH]^+$ adducts with different isomeric pentadienyl anions (Formulas VIII to XIII) show that isomer VIII is the most stable ($\Delta H_f = 2.1$ kcal/mol). The calculated bond distances for the isomers VIII to XIII are given in the original literature [19].

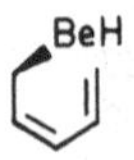

VIII

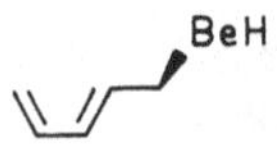

IX

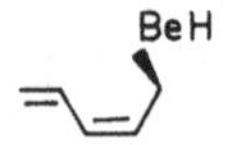

X

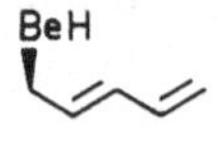

XI

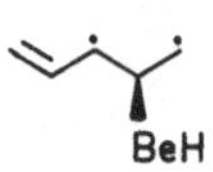

XII

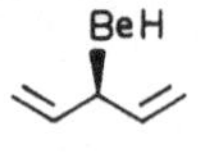

XIII

Be$_2$(CH$_3$)$_3$H and **Be$_2$(CH$_3$)H$_3$**

Two compounds of this composition are included in a detailed molecular orbital study (PRDDO) of molecules of the type $R^tBeR^b_2BeR^t$ (t = terminal, b = bridging). Total energies for $HBe(CH_3)_2BeCH_3$ and $HBeH_2BeCH_3$ are given as −148.4674 and −70.5042 a.u., respectively [25].

The structure of $Be_2(CH_3)H_3$ is optimized by the MNDO method starting from a configuration like that in $Be(CH_3)_2$. The optimized structure has C_{3v} symmetry and is shown in Formula XIV together with the calculated bond lengths and bond angles. There is no direct BeBe bond. The molecule is remarkably stable, its formation as a vapor from $Be(CH_3)H$ and monomeric BeH_2 being exothermic by 160.2 kJ/mol. The HOMO is of A_1 symmetry, concentrated in the BeC bond so that ionization effects no change in the point group, but causes a large increase in the BeC bond. The calculated parameters for the cation $[Be_2(CH_3)H_3]^+$ (symmetry class of HOMO $^2A_1$) are included in parentheses in Formula XIV [24].

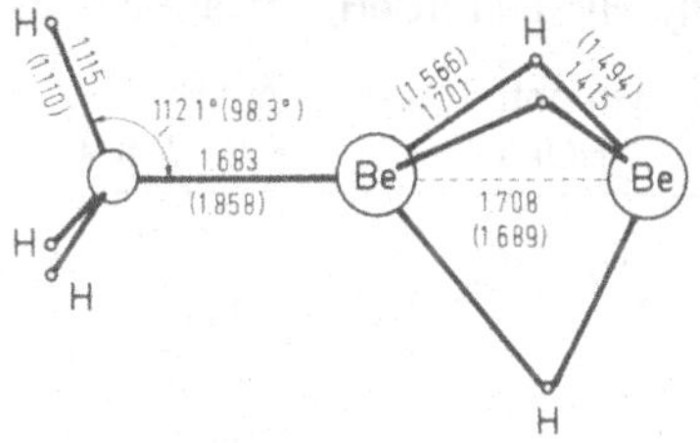

XIV

$(BeH)_{4-n}CH_n$ (n=3 to 0)

The experimentally known $Be(CH_3)H$ is included for comparison. See also Chapter 1.3.2, pp. 93/5, for this compound. The energies of planar vs. tetrahedral geometry around C for the neutral molecules have been surveyed by ab initio molecular orbital calculations using various levels of sophistication and by the MNDO method. The classical tetrahedral configuration around C is strongly favored, but a steady decrease in the planar-tetrahedral energy difference is noted on going from $BeH(CH_3)$ to $(BeH)_4C$ [3, 24]:

compound	planar-tetrahedral energy differences (kcal/mol) ab initio [3]	MNDO [24]
$BeH(CH_3)$	79 to 140	66
cis-$(BeH)_2CH_2$	45 to 72	28
trans-$(BeH)_2CH_2$	70 to 101	63.4
$(BeH)_3CH$	41 to 48	23.9
$(BeH)_4C$	32	25.5

The optimized geometrical parameters (distances in Å) from the ab initio calculations (STO-3G) are [3]:

$(BeH)_2CH_2$	C_{2v} tetrahedral:	1.666(BeC), 1.088(CH), 1.290(BeH); 107.0°(HCH), 113.9°(BeCBe), 180.0°(CBeH)
	planar:	1.557(BeC), 1.097(CH), 1.284(BeH); 105.3°(HCH), 86.4°(BeCBe), 140.1°(BeBeH)
	D_{2h}:	1.580(BeC), 1.286(CH), 1.290(BeH)
$(BeH)_3CH$	C_{3v}:	1.644(BeC), 1.093(CH); 108.7°(HCBe)
	C_{2v}:	1.616 and 1.546(BeC), 1.102(CH); 89.5°(HCBe)
$(BeH)_4C$	D_{4h}:	1.622(BeC)
	T_d:	1.630(BeC)

References on pp. 214/5

The MNDO method gives similar optimized geometries. The BeC distance steadily decreases, and the CH distance increases with increasing n [24]. Calculated ionization potentials for $(BeH)_4C$ (by Koopmans' theorem) are 11.69 eV for T_d and 10.26 eV for D_{4h} symmetry [13].

The MNDO method gives geometries for the cations which have symmetries in the precise accord with the predictions of the Jahn-Teller theorem. When n = 0 or 3, the cation optimizes to a different point group than the neutral molecules. For n = 1 or 2, no change in point group is observed on ionization. The barrier to inversion via a planar tetracoordinated transition state decreases with increasing n, and the barriers are significantly smaller for the cations than for the neutral molecules; $[(BeH)_3CH]^+$ and $[(BeH)_4C]^+$ are strictly planar at C.

compound	optimized point group	symmetry class of HOMO	$\delta(\Delta H_f^\circ)$ *) (in kJ/mol)
$[Be(CH_3)H]^+$	C_{3v}	$^2A_1$	10.4
$[(BeH)_2CH_2]^+$	C_{2v}	$^2B_1$	0.2(cis) 99.0(trans)
$[(BeH)_3CH]^+$	C_{2v}	$^2B_1$	0.0
$[(BeH)_4C]^+$	D_{4h}	$^2A_1$	0.0

*) Inversion barrier via a planar intermediate.

Throughout this series the BeC distances are increased in the cations compared to the neutral molecules [24]. The optimized bond distances (in Å) and bond angles for the cations are as follows [24]:

$[Be(CH_3)H]^+$: 1.261 (BeH), 1.786 (BeC), 1.116 (CH); 98.8° (BeCH)

$[(BeH)_2CH_2]^+$: 1.258 (BeH), 1.756 (BeC), 1.126 (CH); 104.9° (HCH), 89.0° (BeCBe), 172.7° (CBeH)

$[(BeH)_3CH]^+$: 1.262 (Be^1H), 1.260 (Be^2H, Be^3H), 1.723 (Be^1C), 1.750 (Be^2C, Be^3C), 1.149 (CH); 91.2° (BeCBe), 176.0° (CBeH)

$[(BeH)_4C]^+$: 1.266 (BeH), 1.745 (BeC)

$(BeH)_2C{=}CH_2$

The ab initio unrestricted Hartree-Fock theory predicts that the rotational barrier around the C=C bond, from a planar to a perpendicular conformation, is lowered relative to C_2H_4, and the bond population between the C atoms is decreased in the perpendicular form [14].

$Be(CB_2H_3)_2$ and **$Be(C_3BH_4)_2$**

The relative stability of the σ-bonded structures XV and XVII compared to the π-bonded structures XVI and XVIII are calculated by the ab initio method (STO-3G basis set). With optimized Be–ligand bond distances, the π-bonded structure in $Be(CB_2H_3)_2$ is calculated to be 6 kcal/mol more stable than the σ-bonded structure; whereas in $Be(C_3BH_4)_2$, the structure XV is 57 kcal/mol less stable than structure XVI. The result is explained with unfavorable antibonding interactions in the ligands (compare $Be(C_3H_3)_2$) [9]. The optimized geometries are included in the formulas.

$Be(C_3B_2H_5)_2$ and **$Be(C_3B_3H_6)_2$**

These compounds are predicted to exist as symmetrical sandwich complexes as shown in Formulas XIX and XX, since the number of interstitial electrons involved in ligand–metal bonding does not exceed eight (compare $Be(C_3H_3)_2$) [9].

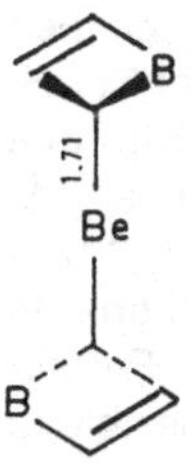

XV

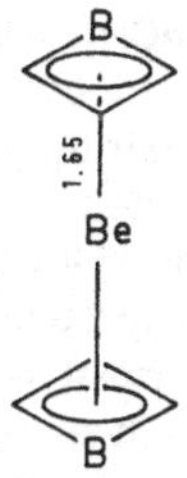

XVI

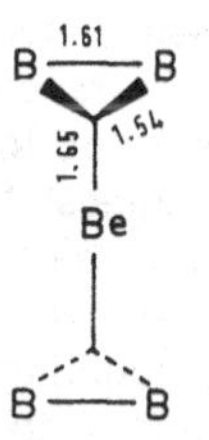

XVII

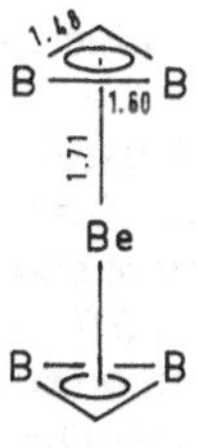

VIII

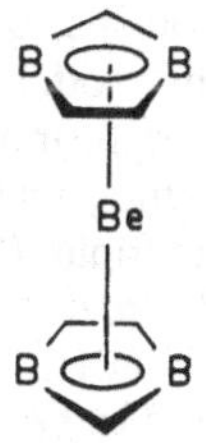

IX

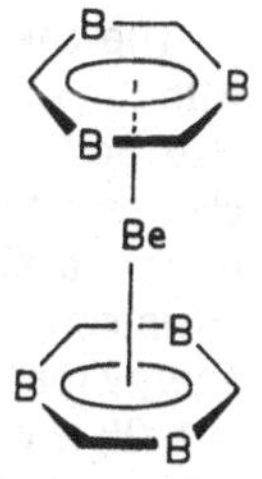

XX

$BeCH_2$

The molecule is, in the planar C_{2v} form, an important prototype of a possible BeC double bond and of a metal-carbene complex. Ab initio molecular orbital theory [1,8,10,30] and the MNDO method [28] were used to study the stability of this complex and the energies of the lowest singlet and triplet states. The ab initio calculations differ by the basis sets employed, e.g., an sp basis set [1], STO-3G [8,10], 6-31G* [8,10,30], MP2/6-31G* [10], MP3/6-31G* [10], or MP4/6-311G** [30]. It follows from all ab initio calculations that the ground state is a triplet with one π electron largely on C and the other electron on Be in a largely nonbonding σ orbital ($^3B_1$). The singlet state with two π electrons of C_{2v} structure ($^1A_1$) was first predicted to be of low stability [1,8], but the inclusion of correlation corrections show it to be also a bonding state [10,30]. The results presented in [30] suggest that the lowest energy singlet state is an open-shell singlet ($^1B_1$) with the same electronic configuration as $^3B_1$. The calculated geometry parameters (distances in Å) for the triplet ground state range from 1.65 to 1.671 for BeC, 1.08 to 1.086 for CH, and 111° to 111.2° for ∢HCH. For the $^1A_1$ singlet state the values are 1.472 to 1.524 for BeC, 1.080 to 1.083 for CH, and 111.2° to 113.2° for ∢HCH [1,8,10,30]. The parameters obtained by the MNDO method are 1.432 for BeC, 1.098 for CH, and 122.9° for ∢ HCBe [28]. The $^1B_1$ singlet state is calculated to have a BeC distance of 1.672 Å, a BeH distance of 1.083 Å, and an HCH angle of 111° (6-31G*) [30]. For the energy difference between the $^1A_1$ singlet and $^3B_1$ triplet state, values between 54.9 kcal/mol (6-31G*) [8,10] and 20.9 kcal/mol (MP4/6-311G**) [30] are calculated. A better estimate is 15.9 kcal/mol [30]. The lowest singlet state $^1B_1$ lies probably more than 2.5 kcal/mol higher than $^3B_1$ [30].

Dipole moments of 4.19 D and 0.56 D are calculated for the $^1A_1$ and $^3B_1$ states, respectively [10]. The Mulliken population analysis for the $^1A_1$ singlet state indicates that there is a drift of electrons away from Be, resulting in a charge of 0.156 [22]. The π orbital population is almost equally shared between Be(0.933) and C (1.067) [8,22]. The BeC bond order is 1.929 [30].

The BeC dissociation energy for the process $BeCH_2$ ($^3B_1$)$\rightarrow$$CH_2$($^3B_1$) + Be is >50 kcal/mol, e.g., 62.3 kcal/mol (6-31G*) in [8] and 57.0 kcal/mol (MP4/6-311G**) in [30]. The dissociation energy from the $^1A_1$ singlet state for the process $BeCH_2$($^1A_1$)$\rightarrow$$CH_2$($^3B_1$) + Be is arrived at 36.8 kcal/mol (MP4/6-311G**) in [30], but only at 1.6 to 3.2 kcal/mol (6-31G*) without the inclusion of correlation corrections [1,8,10]. The bonding energy for the $^1B_1$ state is ~54.5 kcal/mol relative to the ground states of CH_2 and Be [30]. Calculated heats of hydrogenation are $\Delta H = -135.7$ kcal/mol for $^3B_1$ and -80.9 kcal/mol for $^1A_1$ (6-31G*) [10].

Based upon MNDO calculations, the $^2B_1$ state of $[BeCH_2]^+$ is 146.1 kcal/mol (611.4 kJ/mol) higher in energy than the $^1A_1$ state of $BeCH_2$. Ionization gives mainly an increase in the BeC distance: 1.573 for BeC, 1.095 for CH, and 123.2° for ∢HCBe in the cation [30]. Calculated heats of hydrogenation for the $^3B_1$ and $^1A_1$ states in $BeCH_2$ are given as -135.7 kcal/mol and -80.9 kcal/mol, respectively (6-31G*) [10].

References on pp. 214/5

The carbene species, Be(CH:)H, which is a possible isomer of $BeCH_2$, was also examined with the ab initio method. Formula XXII represents the transition state for conversion of $BeCH_2$ into Be(CH:)H. The ground state of Be(CH:)H was found to be a linear triplet ($^3\Sigma^-$) with one electron in each orthogonal π system. The lowest singlet state is predicted to be also linear ($^1\Delta$) (Formula XXI). The triplet ($^3\Sigma^-$) is more stable than the singlet ($^1\Delta$) by 35.4 kcal/mol (6-311G** basis set) [14], 52.9 kcal/mol with an STO-3G basis set [26], and is only 2.9 kcal/mol higher in energy than triplet $BeCH_2$. The transition state connecting these two triplet molecules (Formula XXII) is calculated to be 50.2 kcal/mol higher in energy than $BeCH_2$. It should be possible to observe both triplet states as independent species. The optimized geometries (6-31G*) are given in Formulas XXI and XXII. The computed harmonic vibrations of all structures are given in the original literature [30], and for $BeCH_2$ also in [1].

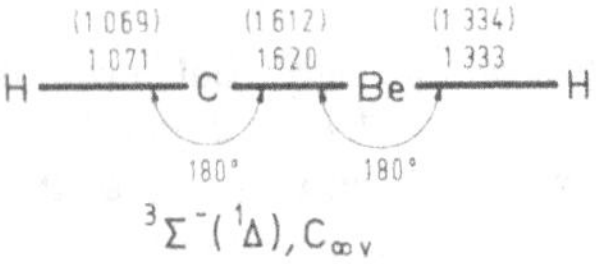

XXI

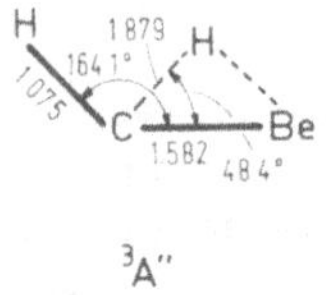

XXII

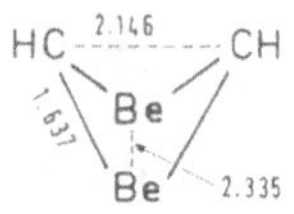

XXIII

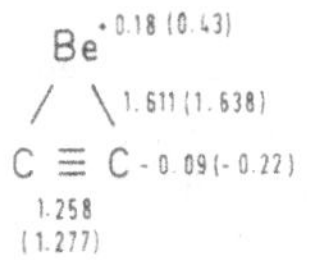

XXIV

BeC_2H_2 and BeC_2H_4

Several electronic states of the Be–C_2H_2 and Be–C_2H_4 systems have been investigated using the ab initio electronic structure theory with double ζ basis sets. For ten different electronic arrangements, potential energy curves have been predicted for the perpendicular approach of the Be nucleus to the midpoint of the CC bond. The interaction of Be^+ ion (ground state ^2S) with both CH≡CH and CH_2=CH_2 is attractive in nature, yielding equilibrium Be midpoint separations of 2.03 and 2.09 Å and binding energies of 30 and 33 kcal/mol, respectively. In contrast, the ^1S ground state of the neutral Be has an essentially repulsive interaction. The threefold degenerate ^3P excited state of Be splits into $^3A_1$, $^3B_1$, and $^3B_2$ components as it approaches C_2H_2 or C_2H_4. The $^3B_2$ state is rather strongly bound, 19 kcal/mol for BeC_2H_2, and 25 kcal/mol for BeC_2H_4. Equilibrium Be midpoint separations are 1.771 and 1.782 Å, respectively. The $^3A_1$ state is strongly repulsive, while the $^3B_1$ state assumes an intermediate role [5].

The structure of the three-membered ring c-Be$(CH)_2$ was optimized with the semiempirical MO method SINDO1. Bond orders of 1.565 for BeC and 1.836 for CC indicate aromaticity. The molecule thus represents the smallest aromatic ring (ring with the smallest number of electrons). The bond lengths are calculated as 1.602 Å for BeC and 1.376 Å for CC. The charge on Be is 1.376, and on C −0.12 [31]. The existence of such a ring is supported by ab initio calculations which gave the following ring geometry: BeC = 1.553 Å, and CC = 1.373 Å (6-31G*). Be$(CH)_2$ is 15.3 kcal/mol (6-31G*) or 22.9 kcal/mol (MP2/6-31G*) more stable than Be + C_2H_2. The repulsive interaction found in the study described above is obviously due to the frozen geometry of C_2H_2 [32].

$(BeCH)_2$

The structure of the smallest nonplanar aromatic compound (Formula XXIII) was optimized by the semiempirical MO method SINDO1. The molecule is folded with a puckering angle of 40.7°. The BeC bond order of 1.533 classifies it as moderately aromatic. The bond lengths are included in Formula XXIII. The charge on Be is 0.22 and on C −0.25 [31].

BeC_2

Berylliryne (Formula XXIV) is predicted by theory to be quite a stable molecule in the gas phase. Ab initio calculations show that it is a global minimum on the energy hypersurface at the MP2/6-31G* level, and it is very stable towards dissociation into Be and C_2. The bond lengths and charge distributions are included in Formula XXIV (6-31G*) [33]. The values in parentheses are obtained from SINDO1 calculations. The BeC bond order of 1.248 (2.439 for CC) shows that the molecule is nonaromatic [31]. There is very little delocalization of the MO including the Be atom, and the shape of the orbital suggests a T structure rather than a cyclic structure with maximum density between the Be atom and the CC midpoint. A linear isomer C=C=Be is higher in energy by 35.9 kcal/mol [33].

$Be(C_3H_3)H$

In a theoretical investigation, using MNDO, no minimum was found corresponding to a Be(η^3-C_3H_3)H derivative with C_{3v} symmetry. The Be undergoes insertion into the C_3H_3 ring to yield a completely planar beryllacyclobutadiene of overall C_{2v} symmetry. This is a system in which four molecular orbitals, normal to the molecular plane, are occupied by only two electrons. The energies and symmetries of these orbitals are given in the original literature. The geometrical parameters are included in Formula XXV. It is not possible to draw a single classical valence structure [23].

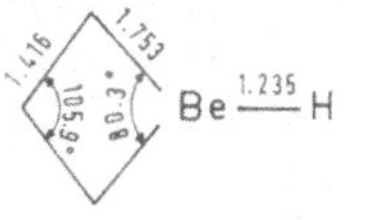

XXV

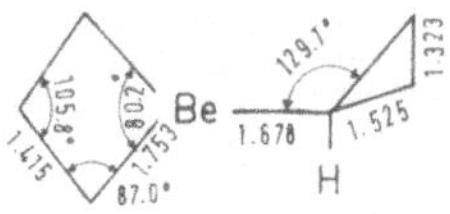

XXVI

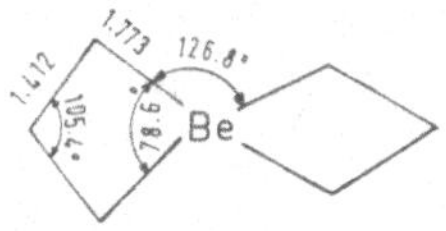

XXVII

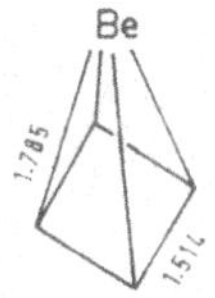

XXVIII

$Be(C_3H_3)_2$

Structure calculations with the MNDO method gave no minimum for the isomer with two trihapto-bonded C_3H_3 groups of either D_{3h} or D_{3d} symmetry. A structure is obtained in which the Be is inserted into one of the C_3 rings. The molecule has an overall symmetry of C_s and is shown in Formula XXVI, which also includes the geometrical parameters. The dihedral angle is 134.3°. The heterocycle is completely planar with two π electrons, and the molecular energy levels are very similar to those in the preceding compound, being only weakly perturbed by the orbitals of the cyclopropenyl ring. Another plausible structure is the spiro compound with D_{2d} symmetry shown in Formula XXVII, which is some 180 kJ/mol lower in energy than the spiro compound with D_{2h} symmetry [23].

It follows from ab initio calculations (STO-3G) that a structure with two σ-bonded cyclopropenyl groups (BeC = 1.70 Å) is energetically favored. The structure with one η^3-bonded C_3H_3 and one σ-bonded C_3H_3 group is 117 kcal/mol higher in energy, and the structure with two η^3-bonded C_3H_3 groups (D_{3h}) is ~139 kcal/mol higher in energy. The low stability of the sandwich-type structure is explained with the antibonding character of the HOMO (degenerate e') with respect to the C_3H_3 rings [9].

BeC_4H_4

The molecule was investigated in the singlet and triplet states by the MNDO method. The $^1A_1$ singlet state optimizes to a structure with full C_{4v} symmetry in which the C_4 ring is entirely planar, while the $^3B_2$ triplet state optimizes to a C_{2v} structure containing a puckered ring (dihedral angle 150.5°) which may be described as a dihapto ligand in contrast to the tetrahapto ligand of the singlet. The triplet lies 26.3 kJ/mol higher in energy than the singlet state. The two structures are shown in Formulas XXVIII and XXIX together with some geometrical parameters. Ionization of the singlet ($\Delta E = 706.2$ kJ/mol $= 168.8$ kcal/mol) or triplet ($\Delta E = 679.9$ kJ/mol $= 162.5$ kcal/mol) state produces a cation ($^2A_1$) which has a similar structure compared to the triplet state [24]. The interaction between Be and the C_4H_4 ring in the singlet state is largely via the π system of the ring [23].

Optimization of an input geometry, corresponding to a beryllaheterocycle, yields a structure of strict C_{2v} symmetry with a strong π bond fixation (see Formula XXX) [23].

$Be(C_4H_4)_2$

Investigation of the possible structures by the MNDO method gives the two geometries shown in Formulas XXXI and XXXII. The structure in Formula XXXI has exact D_{2h} symmetry. The rectangular rings are planar and parallel. The interaction between the Be atom and the rings occurs almost exclusively with the π orbitals of the rings. The six available electrons occupy three bonding orbitals. The second stable structure has overall C_{2h} symmetry and contains dihapto cyclobutenyl ligands. The Be atom forms four bonds to four individual C atoms, and these bonds are strictly coplanar creating an unusual stereochemistry. A stable spiro derivative in analogy to Formula XXVII is not predicted by the MNDO calculations [23].

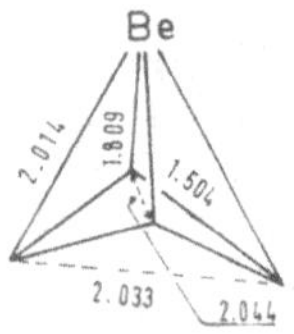

XXIX

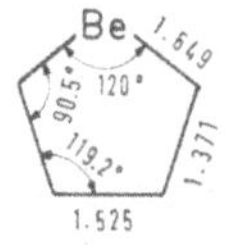

XXX

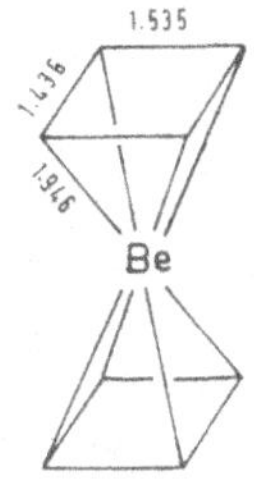

XXXI

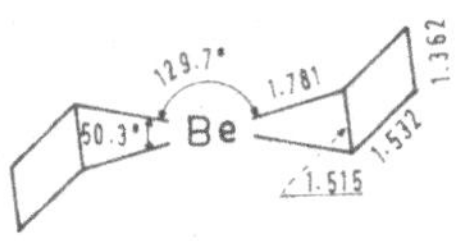

XXXII

BeC_6H_6, $Be(C_6H_6)_2$, and **$Be_2C_6H_6$**

The most stable structures were searched by the MNDO method. For BeC_6H_6, a structure of exact C_{2v} symmetry as shown in Formula XXXIII is obtained. In this dihapto complex, Be is a four-electron atom. The eight electron η^6 isomer of C_{6v} symmetry is some 212 kJ/mol higher in energy. Another stable isomer with C_{2v} symmetry is shown in Formula XXXIV. For $Be(C_6H_6)_2$, energy minima are obtained for the structures shown in Formulas XXXV (D_{2d}) and XXXVI (C_s). For $Be_2C_6H_6$, the structure shown in Formula XXXVII (C_s) is obtained [23].

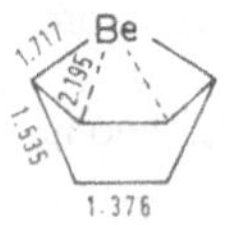

XXXIII

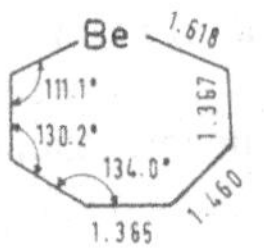

XXXIV

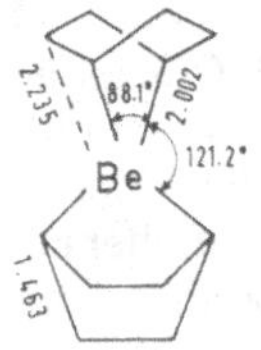

XXXV

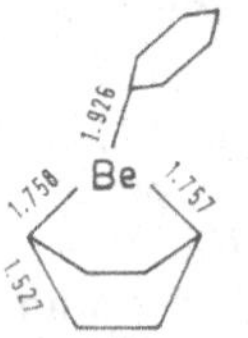

XXXVI

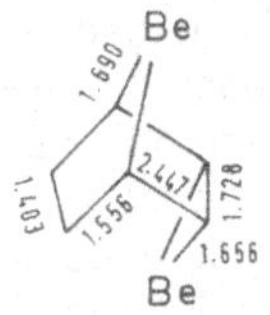

XXXVII

$Be(C_7H_7)H$

According to calculations by the MNDO method, the C_7H_7 ring is η^3-bonded with overall C_s symmetry and a strong π bond fixation in the ring as shown in Formula XXXVIII. A beryllaheterocyclic isomer is not planar (Formula XXXIX) [23].

BeC_8H_8, $(BeH)_2C_8H_8$, $Be_2C_8H_8$, and $(BeH)_2C_8H_6$

An attempted preparation of a C_8H_8 compound of Be is described in the literature. The reaction of $K_2[C_8H_8]$ and anhydrous $BeCl_2$ in THF yields a yellow, insoluble, nonvolatile powder which could not be characterized further. Upon exposure to air, a ν(OH) band is observed in the IR spectrum. It seems that THF is involved in the reaction, since hydrolysis gives a mixture of alcohols, probably 4-hydroxybutylated cyclooctapolyenes [18].

MNDO calculations predict that Be in BeC_8H_8 is η^2-bonded to C_8H_8 as shown in Formula XL (C_s). There is a strong π bond fixation in the ring. The η^8-bonded isomer with a planar C_8H_8 ring is 65.0 kcal/mol (272.2 kJ/mol) higher in energy [23]. For Be(η^8-C_8H_8), the optimum Be–ring distance is calculated by the ab initio (STO-3G) method as 1.0 Å. A planar structure, in which Be lies in the center of the ring, is highly unfavorable [15]. It follows from the MNDO calculations that a beryllaheterocycle should have the structure shown in Formula XLI (C_2). Because of the close approach of Be to the carbon atoms C(4) and C(4') (2.296 Å), the compound can be viewed as a tetrahapto species. There is a strong bond fixation along the chain. This structure is ~70 kJ/mol more stable than a planar beryllaheterocycle [23].

According to the MNDO calculations, the Be atoms in $Be_2C_8H_8$ span the 1,4- and 5,8-positions of a bis-dihapto C_8H_8 ligand (C_{2h}), in which again strong π bond fixation occurs (see Formula XLII) [23].

The structure shown in Formula XLIII for $(BeH)_2C_8H_8$ is predicted by ab initio calculations (STO-3G) to be 60 kcal/mol more stable than Be(η^8-C_8H_8). It represents a model for a $(BeC_8H_8)_n$ polymer. The structures in Formulas XLIV and XLV for $(BeH)_2C_8H_6$, in which each Be is localized on one C atom, are slightly favored relative to XLIII by 0.25 kcal/mol for the syn form XLIV and 2.4 kcal/mol for the anti form XLV [15].

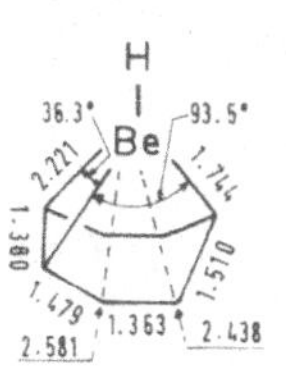

XXXVIII

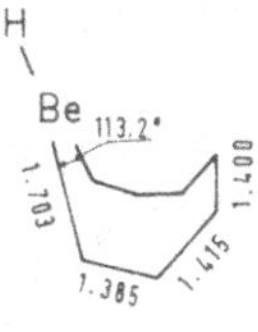

XXXIX

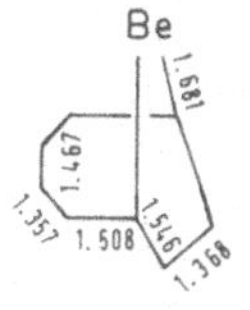

XL

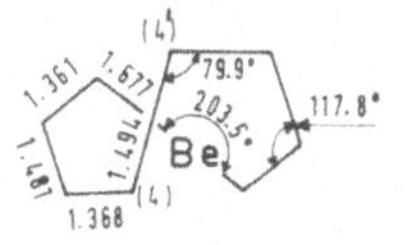

XLI

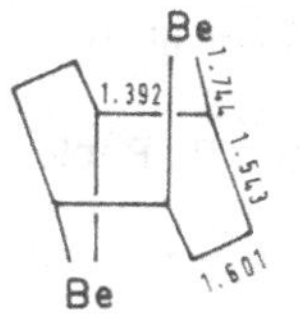

XLII

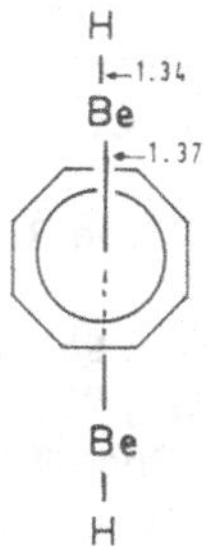

XLIII

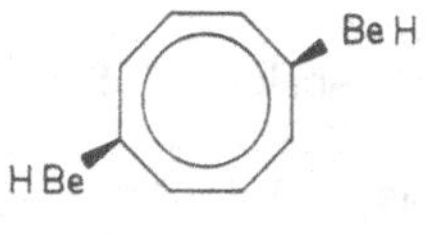

XLIV

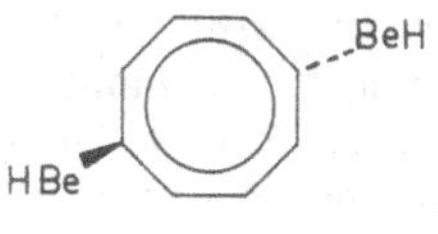

LV

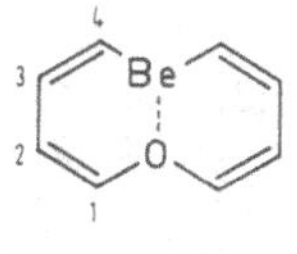

XLVI

BeC_8H_8O (Formula XLVI)

This hypothetical molecule, with the structure shown in Formula XLVI, is included in MNDO calculations. The calculated point group is C_{2v}, and the bond distances (in Å) are 1.833(BeO), 1.389(OC^1), 1.363(C^1C^2), 1.459(C^2C^3), 1.360(C^3C^4), and 1.680(C^4Be). The net atomic charge on

References on pp. 214/5

Be is 0.280 and −1.125 on O. The HOMO has appreciable electron density on Be and is bonding in the Be–O region. There is, however, no σ orbital which is appreciably bonding in the Be–O region, suggesting that the molecule may be homoaromatic [13].

$Be(C_9H_7)X$ (X = H, Cl) and $Be(C_{13}H_9)H$

MNDO calculations have been carried out for the complexes formed between indenyl and fluorenyl anions and $[BeH]^+$ or $[BeCl]^+$. The three stable isomers XLVII to XLIX were located for the indenyl complexes. The calculated energies increase from $\eta^1<\eta^5<\eta^6$ coordination. The calculated CC bond lengths for X = H are included in the Formulas. In XLVII and XLIX, the ligand is slightly nonpolar. The structures vary only slightly for X = Cl as compared to X = H. For the fluorenyl system, the η^6 isomer L was found to be stable, but no minimum corresponding to η^5 coordination could be located. All geometry optimizations produce the η^1 isomer LI. The dipole moments and the first ionization potentials for the stable isomers are given in the original literature [16].

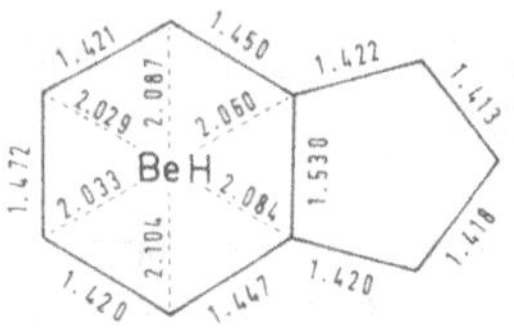
XLVII

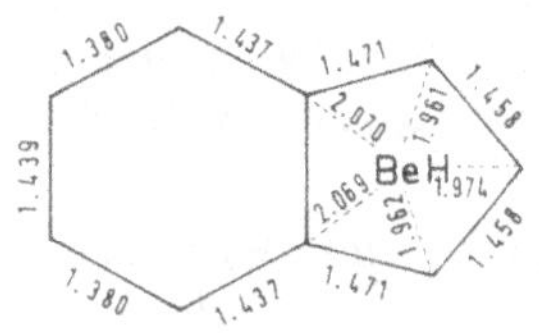
XLVIII

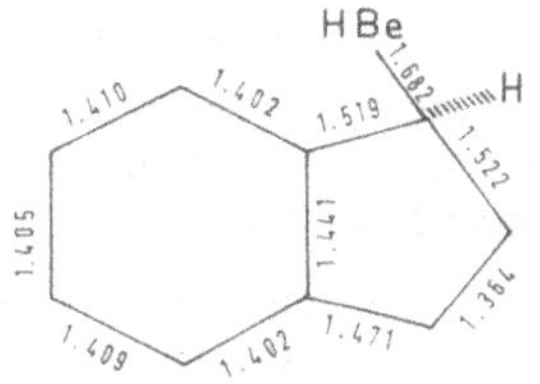
XLIX

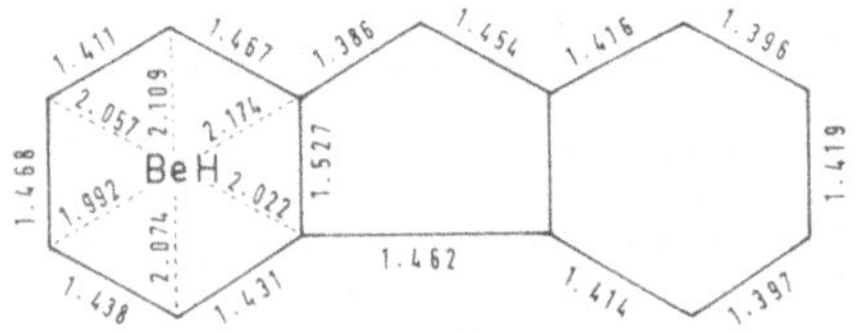
L

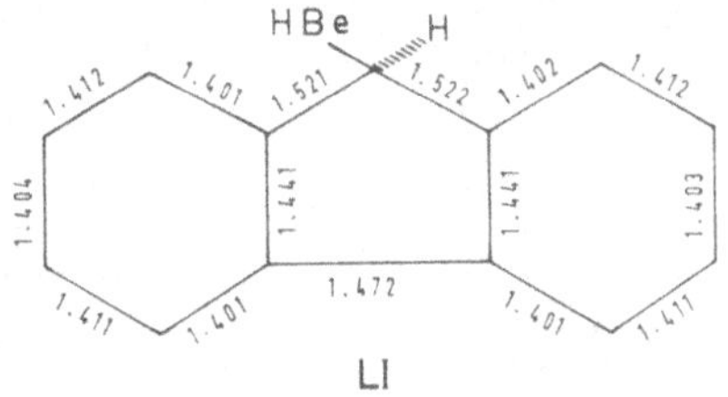
LI

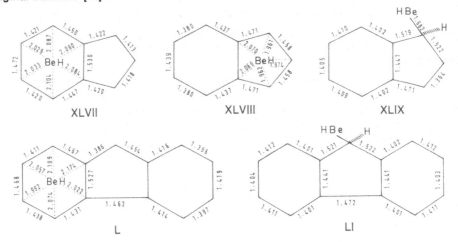

References:

[1] Lamanna, U., Maestro, M. (Theor. Chim. Acta **36** [1974/75] 103/8).
[2] Baskin, C. P., Bender, C. F., Lucchese, R. R., Bauschlicher, C. W., Schaefer III, H. F. (J. Mol. Struct. **32** [1976] 125/31).
[3] Collins, J. B., Dill, J. D., Jemmis, E. D., Apeloig, Y., Schleyer, P. von R., Seeger, R., Pople, J. A. (J. Am. Chem. Soc. **98** [1976] 5419/27).
[4] Dill, J. D., Schleyer, P. von R., Pople, J. A. (J. Am. Chem. Soc. **98** [1976] 1663/8).
[5] Swope, W. C., Schaefer III, H. F. (J. Am. Chem. Soc. **98** [1976] 7962/7).
[6] Apeloig, Y., Schleyer, P. von R., Pople, J. A. (J. Am. Chem. Soc. **99** [1977] 1291/5).
[7] Apeloig, Y., Schleyer, P. von R., Pople, J. A. (J. Am. Chem. Soc. **99** [1977] 5901/9).
[8] Binkley, J. S., Seeger, R., Pople, J. A., Dill, J. D., Schleyer, P. von R. (Theor. Chim. Acta **45** [1977] 69/72).
[9] Collins, J. B., Schleyer, P. von R. (Inorg. Chem. **16** [1977] 152/5).
[10] Dill, J. D., Schleyer, P. von R., Binkley, J. S., Pople, J. A. (J. Am. Chem. Soc. **99** [1977] 6159/73).

[11] Dill, J. D., Schleyer, P. von R., Pople, J. A. (J. Am. Chem. Soc. **99** [1977] 1/8).
[12] Apeloig, Y., Schreiber, R. (Tetrahedron Letters **1978** 4555/8).

[13] Dewar, M. J. S., Rzepa, H. S. (J. Am. Chem. Soc. **100** [1978] 777/84).
[14] Nagase, S., Morokuma, K. (J. Am. Chem. Soc. **100** [1978] 1661/6).
[15] Streitwieser, A., Williams, J. E. (J. Organometal. Chem. **156** [1978] 33/6).
[16] Dewar, M. J. S., Rzepa, H. S. (Inorg. Chem. **18** [1979] 602/5).
[17] Jemmis, E. D., Chandrasekhar, J., Schleyer, P. von R. (J. Am. Chem. Soc. **101** [1979] 527/33).
[18] Berke, C. M., Streitwieser, A. (J. Organometal. Chem. **197** [1980] 123/34).
[19] Dewar, M. J. S., Fox, M. A., Nelson, D. J. (J. Organometal. Chem. **185** [1980] 157/81).
[20] Pauling, L. (J. Chem. Soc. Chem. Commun. **1980** 688/9).

[21] Schoeller, W. W. (J. Chem. Soc. Chem. Commun. **1980** 124/5).
[22] Armstrong, D. R., Perkins, P. G. (Coord. Chem. Rev. **38** [1981] 139/275).
[23] Bews, J. R., Glidewell, C. (J. Organometal. Chem. **219** [1981] 279/93).
[24] Glidewell, C. (J. Organometal. Chem. **217** [1981] 273/80).
[25] Marynick, D. S. (J. Am. Chem. Soc. **103** [1981] 1328/33).
[26] Mueller, P. H., Rondan, N. G., Houk, K. N., Harrison, J. F., Hooper, D., Willen, B. H., Liebman, J. F. (J. Am. Chem. Soc. **103** [1981] 5049/52).
[27] Pross, A., De Frees, D. J., Levi, B. A., Pollack, S. K., Radom, L., Hehre, W. J. (J. Org. Chem. **46** [1981] 1693/9).
[28] Bews, J. R., Glidewell, C. (J. Mol. Struct. **90** [1982] 151/63).
[29] Cuthbertson, A. F., Glidewell, C. (J. Mol. Struct. **87** [1982] 71/9).
[30] Luke, B. T., Pople, J. A., Schleyer, P. von R. (Chem. Phys. Letters **97** [1983] 265/9).

[31] Jug, K. (J. Org. Chem. **49** [1984] 4475/8).
[32] Schleyer, P. von R. (private communication from [31]).
[33] Frenking, G. (Chem. Phys. Letters **111** [1984] 529/34).
[34] Clark, T., Spitznagel, G. W., Klose, R., Schleyer, P. von R. (J. Am. Chem. Soc. **106** [1984] 4412/9).

Empirical Formula Index

In the following index, the compounds are listed by their empirical formula in the order of increasing carbon content. Lewis bases present in the adducts are not included in the empirical formula; e.g., $Be(CH_3)_2 \cdot N(CH_3)_3$ is listed as $C_2H_6Be \cdot N(CH_3)_3$. Adducts are so treated due to the presence of a nonstoichiometric amount of the Lewis base found in some compounds; e.g. $Be(C{\equiv}CCH_3)_2 \cdot nO(C_2H_5)_2$ ($0<n\leq1$). Deuterated compounds are only listed when they appear in a table. Otherwise, they are handled together with the ^1H isotopes. Formulas of ionic compounds are given in square brackets; ions as well as solvates are separated by a dot.

In the second column, page references are printed in ordinary type, table numbers in boldface, and compound numbers within the table in italics.

Ligand Formula Index

The ligands containing carbon atoms can be used to locate a compound in this volume. These ligands are listed in the ligand Formula Index by the number of carbon atoms. The number of identical ligands in a compound is not taken into consideration. Thus, several compounds may be listed at one position. Compounds having two or more different carbon-containing ligands occur at more than one position. The variable C-containing ligands are placed in the first three columns, while nonorganic ligands such as OH, halogen, chalcogen, etc., appear in the fourth column.

Page references are printed in ordinary type, table numbers in boldface, and compound numbers within the tables in italics.

CH_3	CH_3O	—	—	120, **25**, *1* 128, **26**, *1* 128, **26**, *2*
CH_3	CH_3S	—	—	133, **27**, *1*
CH_3	CH_4N	—	—	137
CH_3	C_2H_5O	—	—	120, **25**, *2*
CH_3	C_2H_6N	—	—	137/8 145
CH_3	C_2H_6P	—	—	150
CH_3	C_3H_3	—	—	78, **17**, *10*
CH_3	C_3H_6NO	—	—	131
CH_3	C_3H_7O	—	—	121, **25**, *3* 121, **25**, *4*
CH_3	$C_3H_7O_2$	—	—	122, **25**, *10*
CH_3	C_3H_7S	—	—	133, **27**, *2*
CH_3	C_3H_9OSi	—	—	122, **25**, *13*
CH_3	C_4H_8NO	—	—	139
CH_3	C_4H_9	—	—	77, **17**, *1* 78, **17**, *5* 78, **17**, *6* 78, **17**, *7*
CH_3	C_4H_9O	—	—	121, **25**, *5* 128, **26**, *3* 130
CH_3	C_4H_9S	—	—	133, **27**, *3*
CH_3	$C_4H_{10}N$	—	—	138
CH_3	$C_4H_{10}NO$	—	—	122, **25**, *11*
CH_3	$C_4H_{10}NS$	—	—	133, **27**, *4*
CH_3	$C_4H_{11}N_2$	—	—	139
CH_3	C_5H_5	—	—	175/7
CH_3	C_5H_6N	—	—	201
CH_3	$C_5H_{10}N$	—	—	139 146
CH_3	C_5H_{11}	—	—	78, **17**, *8* 78, **17**, *9*
CH_3	$C_5H_{13}N_2$	—	—	139
CH_3	C_6H_5	—	—	77 204
CH_3	C_6H_5O	—	—	129, **26**, *7*

C_2H	—	—	H	206
C_2H	C_5H_5	—	—	181
C_2H_2	—	—	—	210
C_2H_2	—	—	H	208
C_2H_3	—	—	—	62 203
C_2H_3	—	—	H	206
C_2H_3O	—	—	Br	117
C_2H_3O	—	—	Cl	117
C_2H_3O	—	—	I	117
C_2H_4	—	—	—	210
C_2H_5	—	—	—	6 26/37 92
C_2H_5	—	—	Br	114
C_2H_5	—	—	CN	87/8
C_2H_5	—	—	CNS	89
C_2H_5	—	—	Cl	86/7 112/4
C_2H_5	—	—	D	81, **18**, *6*
C_2H_5	—	—	F	84/6
C_2H_5	—	—	H	81, **18**, *4* 81, **18**, *5* 82, **18**, *11* 82, **18**, *12* 82, **18**, *13* 98/9 199/200
C_2H_5	—	—	I	114
C_2H_5	—	—	NH_2	141
C_2H_5	—	—	N_3	89/90 200/1
C_2H_5	CH_3S	—	—	133, **27**, *5*
C_2H_5	CH_5N_2	—	—	149
C_2H_5	C_2H_5S	—	—	133, **27**, *6*
C_2H_5	C_2H_5Se	—	—	135, **27**, *21*
C_2H_5	C_2H_6N	—	—	141
C_2H_5	$C_2H_7N_2$	—	—	149
C_2H_5	$C_3H_7O_2$	—	—	123, **25**, *17*

C_2H_5	C_3H_7S	—	—	133, **27**, *7* 134, **27**, *12*
C_2H_5	$C_3H_9N_2$	—	—	149
C_2H_5	C_4H_9	—	—	77, **17**, *2*
C_2H_5	C_4H_9O	—	—	91/2 122, **25**, *14* 150
C_2H_5	C_4H_9O	C_7H_7	—	202
C_2H_5	C_4H_9O	C_9H_{11}	—	202
C_2H_5	C_4H_9S	—	—	133, **27**, *8* 134, **27**, *13* 134, **27**, *14* 134, **27**, *15* 134, **27**, *16*
C_2H_5	$C_4H_{10}N$	—	—	142
C_2H_5	C_5H_5	—	—	178
C_2H_5	C_5H_6N	—	—	201
C_2H_5	C_6H_5Se	—	—	135, **27**, *22*
C_2H_5	$C_7H_{15}O$	—	—	123, **25**, *15*
C_2H_5	$C_9H_{19}O$	—	—	123, **25**, *16*
C_2H_5	$C_{10}H_{14}N$	—	—	142
C_2H_5	$C_{11}H_{14}N$	—	—	79, **17**, *15*
C_2H_5	$C_{12}H_{10}N$	—	—	142/3
C_2H_5	$C_{15}H_{16}N$	—	—	142
C_2H_5O	CH_3	—	—	120, **25**, *2*
C_2H_5S	C_2H_5	—	—	133, **27**, *6*
C_2H_5S	C_3H_7	—	—	134, **27**, *17* 134, **27**, *18*
C_2H_5Se	C_2H_5	—	—	135, **27**, *21*
C_2H_6N	CH_3	—	—	137/8 145
C_2H_6N	C_2H_5	—	—	141
C_2H_6N	C_3H_7	—	—	143
C_2H_6N	C_4H_9	—	—	143 151
C_2H_6N	C_5H_5	—	—	172
C_2H_6N	C_6H_5	—	—	144
C_2H_6P	CH_3	—	—	150

$C_3H_7O_2$	CH_3	—	—	122, **25**, *10*
$C_3H_7O_2$	C_2H_5	—	—	123, **25**, *17*
C_3H_7S	CH_3	—	—	133, **27**, *2*
C_3H_7S	C_2H_5	—	—	133, **27**, *7* 134, **27**, *12*
C_3H_7S	C_3H_7	—	—	134, **27**, *9* 134, **27**, *19*
$C_3H_9N_2$	C_2H_5	—	—	149
C_3H_9OSi	CH_3	—	—	122, **25**, *13*
$\mathbf{C}_4H_4$	—	—	—	211/2
C_4H_7	—	—	—	62
C_4H_7O	—	—	Br	117
C_4H_7O	—	—	Cl	117
C_4H_7O	—	—	I	117
C_4H_8NO	CH_3	—	—	139
C_4H_9				
C_4H_9	—	—	—	40/1
C_4H_9	—	—	Br	114
C_4H_9	—	—	I	114/5
C_4H_9	C_5H_5	—	—	179
$(CH_3)_2CHCH_2$	—	—	—	41/2
$(CH_3)_2CHCH_2$	—	—	H	82, **18**, *9* 82, **18**, *10* 82, **18**, *15* 99/100
$(CH_3)_2CHCH_2$	$C_5H_{11}O$	—	—	123, **25**, *19*
$(CH_3)_2CHCH_2$	C_7H_6N	—	—	148
$(CH_3)_2CHCH_2$	$C_9H_{19}O$	—	—	123, **25**, *20*
$C_2H_5(CH_3)CH$	—	—	—	42
$C_2H_5(CH_3)CH$	—	—	H	99
$(CH_3)_3C$	—	—	—	42/57
$(CH_3)_3C$	—	—	Br	115/6
$(CH_3)_3C$	—	—	Cl	115
$(CH_3)_3C$	—	—	H	82, **18**, *16* 100
$(CH_3)_3C$	CH_3	—	—	77, **17**, *1* 78, **17**, *5* 78, **17**, *6* 78, **17**, *7*
$(CH_3)_3C$	CH_3O	—	—	123, **25**, *21*
$(CH_3)_3C$	C_2H_5	—	—	77, **17**, *2*
$(CH_3)_3C$	C_2H_6N	—	—	143 151
$(CH_3)_3C$	C_3H_7O	—	—	124, **25**, *22*
$(CH_3)_3C$	C_4H_9O	—	—	124, **25**, *23*

C_4H_9				
$(CH_3)_3C$	C_4H_9O	—	—	129, **26**, *8* 130 150/1
$(CH_3)_3C$	$C_4H_{10}NO$	—	—	124, **25**, *28*
$(CH_3)_3C$	$C_4H_{10}NS$	—	—	134, **27**, *10*
$(CH_3)_3C$	C_5H_5	—	—	178/9
$(CH_3)_3C$	$C_5H_{10}N$	—	—	148
$(CH_3)_3C$	$C_5H_{11}O$	—	—	124, **25**, *24*
$(CH_3)_3C$	$C_5H_{13}N_2$	—	—	144
$(CH_3)_3C$	C_6F_5	—	—	76 78, **17**, *14*
$(CH_3)_3C$	C_7H_6N	—	—	148
$(CH_3)_3C$	C_7H_7	—	—	77, **17**, *3* 78, **17**, *12*
$(CH_3)_3C$	$C_7H_{15}O$	—	—	151
$(CH_3)_3C$	C_8H_9	—	—	77, **17**, *4* 78, **17**, *13*
$(CH_3)_3C$	$C_8H_{10}N$	—	—	144
$(CH_3)_3C$	$C_9H_{19}O$	—	—	124, **25**, *25* 129, **26**, *9* 129, **26**, *10*
$(CH_3)_3C$	$C_{11}H_{14}N$	—	—	79, **17**, *16*
$(CH_3)_3C$	$C_{13}H_{11}O$	—	—	124, **25**, *26*
$(CH_3)_3C$	$C_{15}H_{14}N$	—	—	144
$(CH_3)_3C$	$C_{19}H_{15}O$	—	—	124, **25**, *27*
C_4H_9O	CH_3	—	—	121, **25**, *5* 128, **26**, *3* 130
C_4H_9O	C_2H_5	—	—	91/2 122, **25**, *14* 150
C_4H_9O	C_2H_5	C_7H_7	—	202
C_4H_9O	C_2H_5	C_9H_{11}	—	202
C_4H_9O	C_4H_9	—	—	124, **25**, *23* 129, **26**, *8* 130 150/1
C_4H_9O	C_6H_5	—	—	124, **25**, *30* 129, **26**, *11* 130
C_4H_9S	CH_3	—	—	133, **27**, *3*
C_4H_9S	C_2H_5	—	—	133, **27**, *8* 134, **27**, *13*

C_4H_9S	C_2H_5	—	—	134, **27**, *14* 134, **27**, *15* 134, **27**, *16*
C_4H_9S	C_6H_5	—	—	135, **27**, *20*
$C_4H_{10}N$	CH_3	—	—	138
$C_4H_{10}N$	C_2H_5	—	—	142
$C_4H_{10}N$	C_5H_5	—	—	173
$C_4H_{10}NO$	CH_3	—	—	122, **25**, *11*
$C_4H_{10}NO$	C_4H_9	—	—	124, **25**, *28*
$C_4H_{10}NS$	CH_3	—	—	133, **27**, *4*
$C_4H_{10}NS$	C_4H_9	—	—	134, **27**, *10*
$C_4H_{11}N_2$				
$NHCH_2CH_2N(CH_3)_2$	CH_3	—	—	139
$N(CH_3)CH_2CH_2N(CH_3)H$	CH_3	—	—	139
$C_4H_{11}Si$	—	—	—	60 61
$C_4H_{11}Si$	$C_{17}H_{21}OSi$	—	—	124, **25**, *29*
C_5H_5	—	—	—	183/98
C_5H_5	—	—	BH_4	157/61
C_5H_5	—	—	B_3H_8	161
C_5H_5	—	—	B_5H_8	162/3
C_5H_5	—	—	B_5H_{10}	163/4
C_5H_5	—	—	Br	170/1
C_5H_5	—	—	Cl	165/9
C_5H_5	—	—	F	165
C_5H_5	—	—	H	155/7
C_5H_5	—	—	NH_2	172
C_5H_5	—	—	OH	172
C_5H_5	—	—	I	171
C_5H_5	CH_3	—	—	175/7
C_5H_5	$CH_{10}B_5Zn$	—	—	163
C_5H_5	C_2H	—	—	181
C_5H_5	C_2H_5	—	—	178
C_5H_5	C_2H_6N	—	—	172
C_5H_5	C_3H_3	—	—	181/2
C_5H_5	C_4H_9	—	—	178/9
C_5H_5	$C_4H_{10}N$	—	—	173

C_6F_5	C_5H_5	—	—	180/1
C_6H_4Cl	—	—	—	74, **15**, *2*
C_6H_5	—	—	—	6
C_6H_5	—	—	—	69/73
C_6H_5	—	—	BH_4	204
C_6H_5	—	—	Br	116/7
C_6H_5	—	—	H	82, **18**, *17* 101/3
C_6H_5	—	—	I	117
C_6H_5	CH_3	—	—	77 204
C_6H_5	CH_3O	—	—	124, **25**, *31* 129, **26**, *13*
C_6H_5	C_2H_6N	—	—	144
C_6H_5	C_5H_5	—	—	180
C_6H_5	$C_5H_{13}N_2$	—	—	145
C_6H_5	$C_{12}H_{10}N$	—	—	144/5
C_6H_5	$C_{13}H_9$	—	—	6
C_6H_5	$C_{16}H_{13}BeN_2$	—	—	145
C_6H_5	$C_{19}H_{15}$	—	—	6
C_6H_5	$C_{19}H_{15}O$	—	—	125, **25**, *32*
C_6H_5O	CH_3	—	—	129, **26**, *7*
C_6H_5O	C_8H_5	—	—	129, **26**, *12*
C_6H_5S	CH_3	—	—	134, **27**, *11*
C_6H_5Se	C_2H_5	—	—	135, **27**, *22*
C_6H_6	—	—	—	212
C_6H_7	—	—	—	202
C_6H_9	—	—	—	64 66, **14**, *5* 67, **14**, *21*
C_6H_9	CH_3	—	—	78, **17**, *11*
$C_6H_{11}O$	—	—	Br	117
$C_6H_{11}O$	—	—	Cl	117
$C_6H_{11}O$	—	—	I	117
$C_6H_{14}N$	CH_3	—	—	138
$C_6H_{18}NSi_2$	C_5H_5	—	—	174/5

C_7H_6N	CH_3	—	—	146/7
C_7H_6N	C_4H_9	—	—	148
C_7H_7				
$C_6H_5CH_2$	—	—	—	59
$C_6H_5CH_2$	C_2H_5	C_4H_9O	—	202
2-$CH_3C_6H_4$	—	—	—	74, **15**, *3*
				75, **16**, *4*
				75, **16**, *5*
				75, **16**, *6*
				75, **16**, *7*
2-$CH_3C_6H_4$	—	—	H	103
2-$CH_3C_6H_4$	C_4H_9	—	—	77, **17**, *3*
				78, **17**, *12*
3-$CH_3C_6H_4$	—	—	—	74, **15**, *4*
				75, **16**, *8*
				75, **16**, *9*
				75, **16**, *10*
3-$CH_3C_6H_4$	—	—	H	103/4
C_7H_7	—	—	H	212
C_7H_7O	CH_3	—	—	122, **25**, *7*
$C_7H_{15}O$	C_2H_5	—	—	123, **25**, *15*
$C_7H_{15}O$	C_4H_9	—	—	151
C_8H_5	—	—	—	2
				65
				65, **14**, *2*
				65, **14**, *3*
				66, **14**, *6*
				66, **14**, *7*
				66, **14**, *8*
				66, **14**, *12*
				66, **14**, *13*
				66, **14**, *14*
				66, **14**, *15*
				67, **14**, *16*
				67, **14**, *17*
C_8H_5	C_4H_9O	—	—	124, **25**, *30*
				129, **26**, *11*
				130
C_8H_5	C_4H_9S	—	—	135, **27**, *20*
C_8H_5	C_6H_5O	—	—	129, **26**, *12*
C_8H_6	—	—	H	213
C_8H_8	—	—	—	213
C_8H_8	—	—	H	213

C_8H_8N				
$N{=}CHC_6H_4CH_3$-2	CH_3	—	—	147
$N{=}CHC_6H_4CH_3$-3	CH_3	—	—	147
C_8H_8O	—	—	—	213/4
C_8H_9	—	—	—	74, **15**, *5* 75, **16**, *11*
C_8H_9	C_4H_9	—	—	77, **17**, *4* 78, **17**, *13*
$C_8H_{10}N$	C_4H_9	—	—	144
$C_8H_{11}N_2$	CH_3	—	—	139
$C_8H_{15}BN$	—	—	—	153/4
$C_8H_{15}BN$	C_5H_5	—	—	182
C_8H_{17}	—	—	I	116
C$_9H_6NO$	CH_3	—	—	122, **25**, *12*
C_9H_7	—	—	Cl	214
C_9H_7	—	—	H	214
C_9H_{11}	—	—	—	74, **15**, *6* 75, **16**, *12*
C_9H_{11}	C_2H_5	C_4H_9O	—	202
$C_9H_{19}O$	C_2H_5	—	—	123, **25**, *16*
$C_9H_{19}O$	C_4H_9	—	—	123, **25**, *20* 124, **25**, *25* 129, **26**, *9* 129, **26**, *10*
C$_{10}H_7$	—	—	—	74, **15**, *7* 75, **16**, *13*
$C_{10}H_{14}N$	C_2H_5	—	—	142
$C_{10}H_{17}$	—	—	—	60
$C_{10}H_{17}$	—	—	Cl	116
$C_{10}H_{21}O$	CH_3	—	—	121, **25**, *6*
C$_{11}H_{14}N$	C_2H_5	—	—	79, **17**, *15*
$C_{11}H_{14}N$	C_4H_9	—	—	79, **17**, *16*
C$_{12}H_{10}N$	CH_3	—	—	140/1
$C_{12}H_{10}N$	C_2H_5	—	—	142/3

$C_{12}H_{10}N$	C_5H_5	—	—	173/4
$C_{12}H_{10}N$	C_6H_5	—	—	144/5
C$_{13}H_9$	—	—	H	214
$C_{13}H_9$	C_6H_5	—	—	6
$C_{13}H_{11}O$	CH_3	—	—	122, **25**, *8* 128, **26**, *4* 128, **26**, *5*
$C_{13}H_{11}O$	C_4H_9	—	—	124, **25**, *26*
$C_{13}H_{12}N$	CH_3	—	—	139/40
C$_{14}H_{10}$	—	—	—	59/60
$C_{14}H_{14}N$	CH_3	—	—	140
C$_{15}H_{14}N$	C_4H_9	—	—	144
$C_{15}H_{16}N$	CH_3	—	—	140
$C_{15}H_{16}N$	C_2H_5	—	—	142
C$_{16}H_{13}BeN_2$	C_6H_5	—	—	145
C$_{17}H_{21}OSi$	$C_4H_{11}Si$	—	—	124, **25**, *29*
C$_{19}H_{15}$	C_6H_5	—	—	6
$C_{19}H_{15}O$	CH_3	—	—	122, **25**, *9* 128, **26**, *6*
$C_{19}H_{15}O$	C_4H_9	—	—	124, **25**, *27*
$C_{19}H_{15}O$	C_6H_5	—	—	125, **25**, *32*
$C_{19}H_{17}P$	—	—	Cl	117/8
C$_{20}H_{19}P$	—	—	Cl	117/8

Table of Conversion Factors

Following the notation in Landolt-Börnstein [7], values which have been fixed by convention are indicated by a bold-face last digit. The conversion factor between calorie and Joule that is given here is based on the thermochemical calorie, $cal_{th\,ch}$, and is defined as 4.184**0** J/cal. However, for the conversion of the "Internationale Tafelkalorie", cal_{IT}, into Joule, the factor 4.1868 J/cal is to be used [1, p. 147]. For the conversion factor for the British thermal unit, the Steam Table Btu, BTU_{ST}, is used [1, p. 95].

Force	N	dyn	kp
1 N (Newton)	1	10^5	0.1019716
1 dyn	10^{-5}	1	1.019716×10^{-6}
1 kp	**9.80665**	**9.80665** $\times 10^5$	1

Pressure	Pa	bar	kp/m²	at	atm	Torr	lb/in²
1 Pa (Pascal) = 1 N/m²	1	10^{-5}	1.019716×10^{-1}	1.019716×10^{-5}	0.986923×10^{-5}	0.750062×10^{-2}	145.0378×10^{-6}
1 bar = 10^6 dyn/cm²	10^5	1	10.19716×10^3	1.019716	0.986923	750.062	14.50378
1 kp/m² = 1 mm H_2O	**9.80665**	**0.980665** $\times 10^{-4}$	1	10^{-4}	0.967841×10^{-4}	0.735559×10^{-1}	1.422335×10^{-3}
1 at = 1 kp/cm²	**0.980665** $\times 10^5$	**0.980665**	10^4	1	0.967841	735.559	14.22335
1 atm = 76**0** Torr	1.0132**5** $\times 10^5$	1.0132**5**	1.033227×10^4	1.033227	1	760	14.69595
1 Torr = 1 mm Hg	133.3224	1.333224×10^{-3}	13.59510	1.359510×10^{-3}	1.315789×10^{-3}	1	19.33678×10^{-3}
1 lb/in² = 1 psi	6.89476×10^3	68.9476×10^{-3}	703.069	70.3069×10^{-3}	68.0460×10^{-3}	51.7149	1

Work, Energy, Heat	J	kWh	kcal	Btu	MeV
1 J (Joule) = 1 Ws = 1 Nm = 10^7 erg	1	2.778×10^{-7}	2.39006×10^{-4}	9.4781×10^{-4}	6.242×10^{12}
1 kWh	3.6×10^6	1	860.4	3412.14	2.247×10^{19}
1 kcal	4184.0	1.1622×10^{-3}	1	3.96566	2.6117×10^{16}
1 Btu (British thermal unit)	1055.06	2.93071×10^{-4}	0.25164	1	6.5858×10^{15}
1 MeV	1.602×10^{-13}	4.450×10^{-20}	3.8289×10^{-17}	1.51840×10^{-16}	1

1 eV ≙ 23.0578 kcal/mol = 96.473 kJ/mol

Power	kW	PS	kp·m/s	kcal/s
1 kW = 10^{10} erg/s	1	1.35962	101.972	0.239006
1 PS	0.73550	1	75	0.17579
1 kp·m/s	9.80665×10^{-3}	0.01333	1	2.34384×10^{-3}
1 kcal/s	4.1840	5.6886	426.650	1

References:

[1] A. Sacklowski, Die neuen SI-Einheiten, Goldmann, München 1979. (Conversion tables in an appendix.)
[2] International Union of Pure and Applied Chemistry, Manual of Symbols and Terminology for Physicochemical Quantities and Units, Pergamon, London 1979; Pure Appl. Chem. **51** [1979] 1/41.
[3] The International System of Units (SI), National Bureau of Standards Spec. Publ. 330 [1972].
[4] H. Ebert, Physikalisches Taschenbuch, 5th Ed., Vieweg, Wiesbaden 1976.
[5] Kraftwerk Union Information, Technical and Economic Data on Power Engineering, Mülheim/Ruhr 1978.
[6] E. Padelt, H. Laporte, Einheiten und Größenarten der Naturwissenschaften, 3rd Ed., VEB Fachbuchverlag, Leipzig 1976.
[7] Landolt-Börnstein, 6th Ed., Vol. II, Pt. 1, 1971, pp. 1/14.
[8] ISO Standards Handbook 2, Units of Measurement, 2nd Ed., Geneva 1982.

Key to the Gmelin System of Elements and Compounds

HCl $ZnCl_2$

System Number	Symbol	Element
1		Noble Gases
2	H	Hydrogen
3	O	Oxygen
4	N	Nitrogen
5	F	Fluorine
6	**Cl**	**Chlorine**
7	Br	Bromine
8	I	Iodine
8a	At	Astatine
9	S	Sulfur
10	Se	Selenium
11	Te	Tellurium
12	Po	Polonium
13	B	Boron
14	C	Carbon
15	Si	Silicon
16	P	Phosphorus
17	As	Arsenic
18	Sb	Antimony
19	Bi	Bismuth
20	Li	Lithium
21	Na	Sodium
22	K	Potassium
23	NH_4	Ammonium
24	Rb	Rubidium
25	Cs	Caesium
25a	Fr	Francium
26	Be	Beryllium
27	Mg	Magnesium
28	Ca	Calcium
29	Sr	Strontium
30	Ba	Barium
31	Ra	Radium
32	**Zn**	**Zinc**
33	Cd	Cadmium
34	Hg	Mercury
35	Al	Aluminium
36	Ga	Gallium

$CrCl_2$ $ZnCrO_4$

System Number	Symbol	Element
37	In	Indium
38	Tl	Thallium
39	Sc, Y La–Lu	Rare Earth Elements
40	Ac	Actinium
41	Ti	Titanium
42	Zr	Zirconium
43	Hf	Hafnium
44	Th	Thorium
45	Ge	Germanium
46	Sn	Tin
47	Pb	Lead
48	V	Vanadium
49	Nb	Niobium
50	Ta	Tantalum
51	Pa	Protactinium
52	**Cr**	**Chromium**
53	Mo	Molybdenum
54	W	Tungsten
55	U	Uranium
56	Mn	Manganese
57	Ni	Nickel
58	Co	Cobalt
59	Fe	Iron
60	Cu	Copper
61	Ag	Silver
62	Au	Gold
63	Ru	Ruthenium
64	Rh	Rhodium
65	Pd	Palladium
66	Os	Osmium
67	Ir	Iridium
68	Pt	Platinum
69	Tc	Technetium[1]
70	Re	Rhenium
71	Np, Pu ...	Transuranium Elements

Material presented under each Gmelin System Number includes all information concerning the element(s) listed for that number plus the compounds with elements of lower System Number.

For example, zinc (System Number 32) as well as all zinc compounds with elements numbered from 1 to 31 are classified under number 32.

[1] A Gmelin volume titled "Masurium" was published with this System Number in 1941.

A Periodic Table of the Elements with the Gmelin System Numbers is given on the Inside Front Cover